Nuclear Technologies in a Sustainable Energy System

Selected Papers from an I IASA Workshop
Organized by W. Häfele and A. A. Harms

Editors:
G. S. Bauer and A. McDonald

With 127 Figures

Springer-Verlag Berlin Heidelberg New York 1983

G. S. BAUER
Projekt Spallations-Neutronenquelle
der Kernforschungsanlage Jülich GmbH
Postfach 1913
D-5173 Jülich 1
Federal Republic of Germany

A. McDONALD
American Academy of Arts and Sciences
Norton's Woods
136 Irving Street
Cambridge, MA 02138/U.S.A.

IIASA
International Institute for Applied
Systems Analysis
A-2361 Laxenburg
Austria

ISBN-13: 978-3-642-81990-2 e-ISBN-13: 978-3-642-81988-9
DOI: 10.1007/ 978-3-642-81988-9

Library of Congress Cataloging in Publication Data
Main entry under title:
Nuclear technologies in a sustainable energy system.
Based on presentations made at a workshop held May 25-27, 1981
at the International Institute for Applied Systems Analysis.
Includes bibliographical references and index.
1. Atomic power--Congresses. 2. Nuclear reactors--Congresses.
I. Bauer, G. S. (Günter Siegfried), 1941 -
II. McDonald, A. (Alan), 1951 -
TK9006.N83 1983 621.48 83.462

Softcover reprint of the hardcover 1st edition 1983

2061/3020 - 543210

FOREWORD

In March 1981 the International Institute for Applied Systems Analysis (IIASA) published the results of a global energy study looking fifty years into the future: *Energy in a Finite World: A Global Systems Analysis* (Cambridge, Massachusetts: Ballinger Publishing Co., 1981)*. Not surprisingly, this book raises almost as many questions as it answers; thus, it defines a broad range of research topics that might be taken up by IIASA or other research institutions around the world.

A 25–27 May 1981 workshop at IIASA entitled "A Perspective on Adaptive Nuclear Energy Evolutions: Towards a World of Neutron Abundance" was a beginning on one of these topics; it was organized by Wolf Häfele (Kernforschungsanlage Jülich, Jülich, Federal Republic of Germany, and IIASA) and Arkadius Archie Harms (McMaster University, Hamilton, Ontario, Canada). The origin of this workshop was the effort within the IIASA energy study to explore possible "sustainable" global energy systems that might eventually replace the current "consumptive" system. In investigating the possible contributions nuclear technologies might make to a sustainable energy system, it had become clear that it is not so much particular, distinct technologies within the nuclear family that should be examined as a question of particularly advantageous configurations of mutually complementary technologies. Only when one considers exploiting a whole spectrum of arrangements of fission breeders, fusion reactors, and accelerators does the true potential of nuclear power become apparent. The IIASA energy study had only been able to touch upon these ideas, and particularly in light of developments in both nuclear engineering and nuclear physics in the last several years, it was felt that the field deserved, if not demanded, a much more comprehensive and careful scrutiny. Much of that research would necessarily involve engineering work and energy systems analysis carried out at institutions other than IIASA, but it was felt that IIASA could make a substantial immediate contribution by convening a workshop to explore from a broad perspective long-range nuclear options going well beyond the time horizon usually considered and focussing on how the various strengths of fission breeders, fusion reactors, and accelerators can be combined advantageously.

This book includes twenty-one selected papers based on the presentations made at the workshop, as well as a report on the panel discussion that took place on the final day of the workshop. Our hope is that publishing this material will make available to a wide audience at least some of the insights and understandings that the workshop participants

*This is Volume 2 of the Report by the Energy Systems Program Group of IIASA, Wolf Häfele, Program Leader, *Energy in a Finite World*, written for technical readers; Volume 1, subtitled *Paths to a Sustainable Future*, is written for the more general reader and has the same publisher. A shorter version for the general reader is also available from IIASA: Alan McDonald, *Energy in a Finite World: Executive Summary*, Executive Report 4, May 1981.

shared with each other directly. We cannot hope to provide a complete record of all that was discussed, but we believe that we have captured in a reasonably concise way the most important features of the workshop.

The material is divided into six sections. Section One contains three papers that set the stage by addressing the broad aspects of the global energy problem and the potential of nuclear technologies to contribute to its resolution. The first paper, by Häfele, begins with a summary of the IIASA global energy study, which analyzed both resource potentials, including those of nuclear fuels, and several quantitative scenarios, each of which balanced world energy demand with supply over the next 50 years. This paper focusses in particular on the persistent, pervasive demand for liquid fuels that characterized all the scenarios and points to the need to adapt nuclear technologies appropriately, since they are currently conceived primarily as producers of heat or electricity, but not of liquid fuels. With this background, Harms' paper then establishes a more specific framework for addressing the driving question behind the workshop: How can various fission breeder, fusion reactor, and accelerator technologies be arranged to contribute to resolving the global energy problem described by Häfele? Harms' objective is to offer a framework that, rather than reinforcing current habits and focussing on the distinctions between the three categories, draws attention to promising possibilities for synergistic combinations. Marchetti's paper then proceeds to sketch an example of a future energy system, based on nuclear technologies, that responds to the liquid fuels problem by producing hydrogen to be used in synthesizing liquid fuels from solid fuels and is consistent with an independent analysis of sizing energy distribution grids.

The next three sections of the book then address a variety of ways in which specific nuclear technologies can be arranged to produce energy and fuels in useful forms. The range of perspectives taken by the different authors is, in keeping with the spirit of the workshop, quite broad, and we were faced with many possible alternatives for categorizing and ordering the papers. We have chosen to group them in three sections based on the nature of the predominant technology addressed by the paper. Thus Section Two includes papers dealing primarily with fission and fission breeding – i.e., with familiar, near-term technologies which are, however, relatively poor in neutrons. Section Three focuses on fusion technologies and dense plasma neutron sources, possibilities either less near-term or richer in neutrons. And Section Four addresses accelerator breeding, a technology that is perhaps more remote, but characterized by a greater abundance of neutrons.

In Section Five the focus shifts to the back end of the fuel cycle, and the fact that one of the key issues to be resolved when contemplating future nuclear strategies is how the waste should be handled. Ideally, it would be desirable to convert the "waste" back into "fuel", and it is precisely this topic that is a principle focus of the three papers in the section.

The final section, "Reflections and Summary", contains two items. First, there are the remarks presented on the second evening of the workshop by Freeman Dyson: in his marvelous way, he offers a perspective on the subject of the workshop from some important and compelling lessons of history. Second, there is the summary of the final day's panel discussion, in which Hans Bethe, Floyd Culler, Wolf Häfele, Archie Harms, Gerry Kulcinski, and Rudolf Schulten led a discussion that not only summarized much of what had been covered by the workshop, but also reanalyzed much of the technical material from broader perspectives covering national energy difficulties, the global energy problem as a whole, and the implications of more general social, political, and economic considerations.

We hope that in preparing this book we have been at least partially successful in meeting our objective – to provide a clear, concise, and timely set of material communicating to a wide audience most of what was explored at the workshop. We found the experience, both the participation in the workshop and the attempt to document it, enjoyable and exceptionally rewarding. Wolf Häfele and Archie Harms deserve special acknowledgment for their wisdom and foresight in organizing such a workshop and in collecting the outstanding group of participants that attended.

Finally we would like to thank all those who contributed to these proceedings for their cooperation in helping get the book to press as quickly as was possible.

GÜNTER BAUER
ALAN McDONALD

CONTENTS

SECTION FOUR: ACCELERATOR BREEDING

SECTION FIVE: HANDLING THE WASTE

SECTION SIX: REFLECTIONS AND SUMMARY

Section One

Global Strategies and Concepts

ENERGY STRATEGIES AND NUCLEAR POWER

Wolf Häfele
Energy Systems Program, International Institute for Applied Systems Analysis, Laxenburg (Austria)

ABSTRACT

The results of two quantitative scenarios balancing global energy supply with demand for the period 1980–2030 are reviewed briefly. The results suggest that during these 50 years there will be a persistent demand worldwide for liquid fuels, a continuing reliance on ever more expensive and "dirty" fossil fuels, and a limited penetration rate of nuclear generated electricity into the energy market. The paper therefore addresses a possible "second" grid driven by nuclear heat – a grid based not on electricity but on "clean" liquid fuels manufactured from gaseous and solid fossil fuels using nuclear power. Such a second grid would be an important complement to the electricity grid if the world is to progress towards a truly sustainable energy system after 2030.

1 INTRODUCTION

The Energy Systems Program Group at the International Institute for Applied Systems Analysis (IIASA) has investigated various aspects of energy demand, energy supplies, and their constraining parameters on a long-range and globally comprehensive basis. The concluding results have been published recently (Häfele, 1981). Among the various supply options considered were fossil fuels, solar power, the renewables, and nuclear power. Their respective supply potentials were considered by stretching present optimistic assumptions to their upper limits. In a second round it was then attempted to conceive two benchmark scenarios in which the world was subdivided into seven regions. In doing this, great emphasis was placed on internal self-consistency. Naturally, the two benchmark scenarios are not predictions but being self-consistent and detailed they can be used for general orientation and assessment of future options. An era of neutron abundance might eventually evolve in a somewhat distant future. It is therefore important to consider a context for such an era and future uses of nuclear power. It is the purpose of this paper to provide such a context through the use of the IIASA benchmark scenarios and on this basis to elaborate on some aspects of a combination of fossil and nuclear power.

2 THE TWO IIASA SCENARIOS

The principal elements of the IIASA energy study are its medium- to long-range perspective and its global comprehensiveness. In such an approach the growth of the world population in particular turns out to be a determining driving force. Figure 1

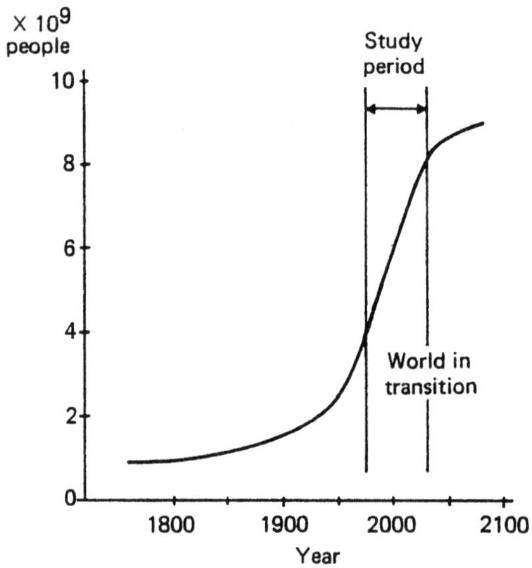

FIGURE 1 World population: projections to 2030 based on data from Keyfitz (1977).

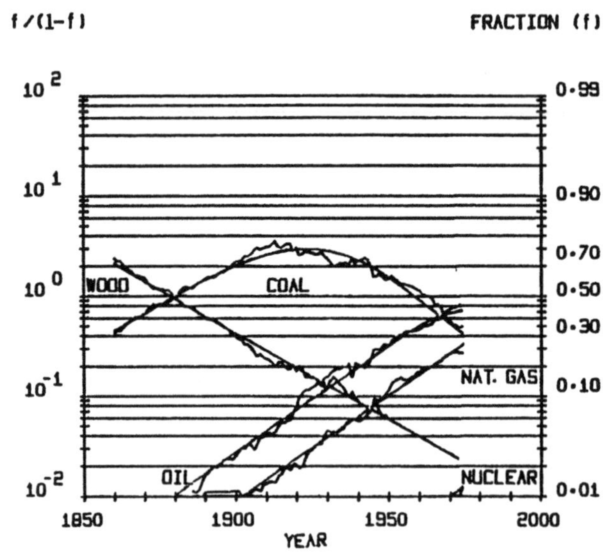

FIGURE 2 Global primary-energy substitution: a logarithmic plot of the transformation $f/(1-f)$ where f is the fractional market share. The smooth curves are model estimates of historical data; the zigzag lines are historical data; the straight lines show the logistic model substitution paths.

illustrates this population growth. It suggests a 50-year study period because the steepest population growth is expected in the next 50 years. The penetration rates of energy technologies into the market also suggest a 50-year period. Figure 2 illustrates the penetration of wood, coal, oil, and natural gas since 1860. This data plot reveals striking regularities, which prevail irrespective of world wars, revolutions, etc. The data are globally aggregated and point to the fact that within 50 years a market share of roughly 30% can be gained or lost. When changes in the infrastructure underlying systems of energy supply and consumption are considered a 30% change is clearly significant; therefore from this point of view also a 50-year study period appears to be reasonable. To go much further would be a questionable move since one would have to take into account possible technological breakthroughs, the unexpected. However, according to Figure 2,

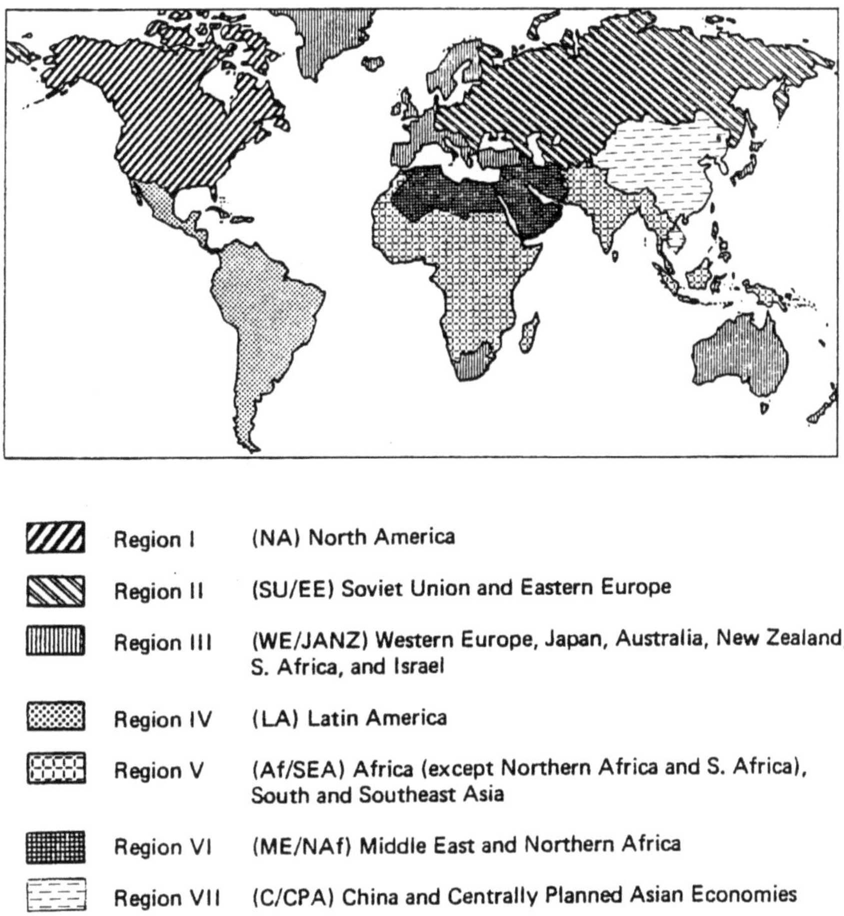

FIGURE 3 The IIASA world regions.

new technologies that are likely to contribute significantly in the next 50 years should already be visible now.

The IIASA study did not consider the world either as a unified civilization or as more than 150 individual nations. Instead seven IIASA world regions were considered (Figure 3). For each of these seven regions demand and supply were considered and certain patterns of, for example, oil trading were assumed.

Quantitative scenario writing requires mathematical models. Such use of mathematical models is different from the use of economic models for more near-term-oriented decision making (e.g. decisions on investment). While uncertainties in economic forecasts are generally recognized, it is, nevertheless, only one future development that is to be forecast as successfully as possible. Quantitative scenario writing instead addresses the topic of conceivable futures. It stresses the consideration of identifiable necessary conditions for such futures and, related to this, internal consistency. Figure 4 gives a brief overview of the set of individual models that was conceived, developed, and used in the IIASA energy study. However, it is not the purpose of this paper to elaborate further and the interested reader is referred to Häfele (1981) for a more extensive treatment.

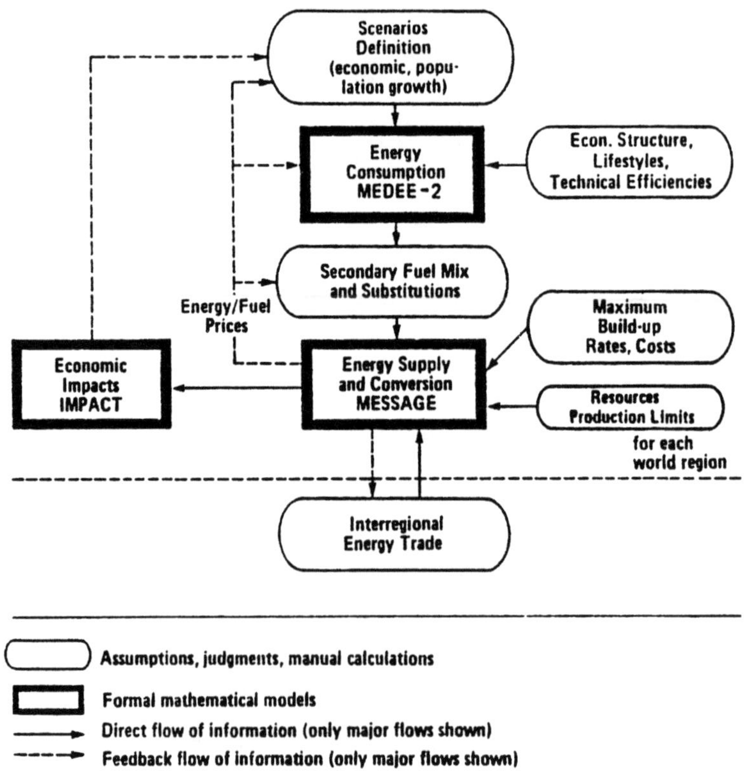

FIGURE 4 IIASA's set of energy models: a simplified representation.

The rather detailed considerations on energy demand in each of the seven world regions led to a few rather robust observations.

(1) In any event it will be necessary to institute fairly strong energy conservation measures. Space heating (and cooling) and transportation are the prime areas for this.

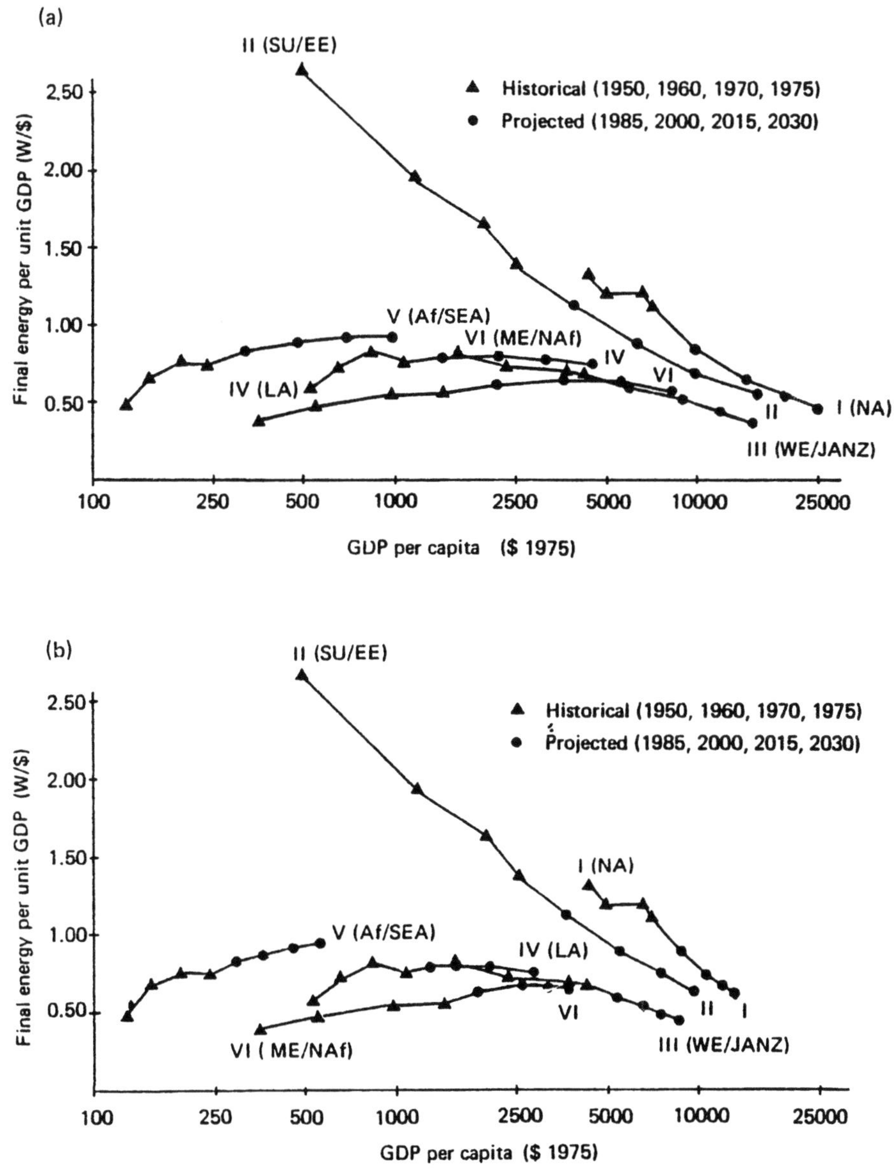

FIGURE 5 Energy intensiveness in six IIASA world regions: (a) high scenario; (b) low scenario.

Without considering a new degree of conservation effort it was found to be virtually impossible to draw a reasonable picture for the overall balancing of energy demand and supply. Figure 5 summarizes the features of anticipated energy conservation by considering the overall ratio of yearly energy inputs to Gross Domestic Product (GDP) outputs as a function of the GDP per capita (and thereby of time) for each of the world regions except China and the centrally planned Asian economies. It should be noted that a value of 0.5 W $\$^{-1}$ yr^{-1} is anticipated while today's figures are typically closer to 1 W $\$^{-1}$ yr^{-1}.

(2) There appears to be a problem within the energy problem: namely balancing the demand and supply of the liquid hydrocarbons. It becomes a clear necessity to increasingly restrict their uses to "noble" applications: i.e. motor fuels and feedstocks. This in turn means substitution by other forms of secondary energies including electricity. Accordingly the IIASA scenarios show an increasing share of electricity use (Table 1). For the world as a whole this share increases from 11% in 1975 to 17% in 2030. In the developed parts of the world it is larger, reaching as high as 24%. While 24% does not make an all-electric society, it does mean a full doubling of today's share of electricity in final energy demand. This is relevant later when we consider the case of nuclear power.

(3) The IIASA scenarios result in increases by a factor of 4.3 (high scenario) and 2.7 (low scenario) in the need for primary energy between the IIASA base year of 1975 and 2030. In absolute terms 35.69 TWyr and 22.30 TWyr, respectively, are required in 2030 compared to 8.2 TWyr in 1975; this is shown in Table 2 together with the disaggregated figures for the world regions.

TABLE 1 Shares of electricity and liquid fuels in final energy[a].

	Base year 1975	High scenario 2030	Low scenario 2030
Regions I and III			
Final energy (TWyr yr^{-1})	3.5	8.0	5.6
Electricity (%)	12	21	21
Liquids (%)	56	46	46
Region II			
Final energy (TWyr yr^{-1})	1.3	3.7	2.6
Electricity (%)	10	23	20
Liquids (%)	34	32	30
Regions IV, V, VI and VII			
Final energy (TWyr yr^{-1})	1.0	10.6	6.0
Electricity (%)	6	13	13
Liquids (%)	48	50	54
World			
Final energy (TWyr yr^{-1})	5.7	22.8	14.6
Electricity (%)	11	17	17
Liquids (%)	50	45	46
Motor fuel and feedstocks in liquid fuel			
Final demand (%)	64	92	88

[a] The regions are defined in Figure 3.

TABLE 2 Two supply scenarios: primary-energy demand by region, 1975–2030 (TWyr yr^{-1}).

Region[a]	Base year	High scenario		Low scenario	
	1975	2000	2030	2000	2030
I	2.65	3.89	6.02	3.31	4.37
II	1.84	3.69	7.33	3.31	5.00
III	2.26	4.29	7.14	3.39	4.54
IV	0.34	1.34	3.68	0.97	2.31
V	0.33	1.43	4.65	1.07	2.66
VI	0.13	0.77	2.38	0.56	1.23
VII	0.46	1.44	4.45	0.98	2.29
Total[b]	8.21[c]	16.84	35.65	13.59	22.39

[a] The regions are defined in Figure 3.
[b] Columns may not sum to totals because of rounding.
[c] Includes 0.21 TWyr yr^{-1} of bunkers.

Such energy-demand scenarios equip the analyst with a yardstick which is useful when considering supply potentials. Local uses of solar power and renewables (hydro, biogas, wood, wind, etc.) turn out to have a global supply potential of the order of 1–2 TWyr yr^{-1} and their total is not more than, say, 6–8 TWyr yr^{-1}. While this is significant it will not meet the potential demand of 20–35 TWyr yr^{-1}; we cannot therefore expect to live exclusively from the renewables in spite of their inexhaustability. By contrast, fossil fuels are in principle exhaustible. However, one of their most salient characteristics is their relative scarcity when clean and cheap resources are considered and their relative abundance when low-grade high-cost resources are considered. For the former a total share of 1000 TWyr yr^{-1} (coal, oil, and gas) is proposed in the IIASA study and for the latter a total share of 3000 TWyr yr^{-1}. Production rates are also limited. It is extremely unlikely that more than 9 TWyr yr^{-1} for oil, more than 7 TWyr yr^{-1} for gas, and, in particular, more than 12 TWyr yr^{-1} for coal will be achieved. Later we argue the case for nuclear power in greater detail but for reasons of comparison we mention here that it is hard to put forward a conceivable case of production of more than 17 TWyr yr^{-1} of (thermal) nuclear energy in the year 2030. However, when based on breeding, the resource potential will be very large, say, 300,000 TWyr.

Such is the "menu" from which energy supplies can be allocated to needs. In the IIASA model set this is done by a linear programming procedure, which in particular allows for the orderly accounting of constraints such as buildup rates for new technologies, costs of resources, etc. One of the more important results is shown in Table 3 where the shares of global primary energy by source are given in absolute terms. The relative shares are given in Figure 6. These results invite a number of robust observations.

(1) While the relative share of oil supply decreases if synfuel production comes into the picture around the year 2000, the absolute amount of oil increases significantly. We must go into oil, not out of oil, which can be accomplished only by moving to unconventional (i.e. low-grade, high-cost, and therefore dirty) fossil fuels. This appears to be a major trend of the future. Taken together with the absolute numbers, this does not imply merely "more of the same" when the related technologies are considered. Rather, it

TABLE 3 Global primary energy by source for two supply scenarios, 1975–2030 (TWyr yr^{-1}).

Primary source[a]	Base year	High scenario		Low scenario	
	1975	2000	2030	2000	2030
Oil	3.83	5.89	6.83	4.75	5.02
Gas	1.51	3.11	5.97	2.53	3.47
Coal	2.26	4.94	11.98	3.92	6.45
LWRs	0.12	1.70	3.21	1.27	1.89
FBRs	0	0.04	4.88	0.02	3.28
Hydroelectricity	0.50	0.83	1.46	0.83	1.46
Solar[b]	0	0.10	0.49	0.09	0.30
Other[c]	0	0.22	0.81	0.17	0.52
Total[d]	8.21	16.84	35.65	13.59	22.39

[a] Primary-fuel production or primary fuels as inputs to conversion or refining processes (e.g. coal used to make synthetic liquid fuel is counted in coal figures).
[b] Includes mostly "soft" solar energy (individual rooftop collectors) and also small amounts of centralized solar electricity.
[c] Includes biogas, geothermal, and commercial wood use.
[d] Columns may not sum to totals because of rounding.

means new technologies and new technological systems; nuclear engineers should bear this in mind. Tar sands, shale oil, and secondary and tertiary recovery schemes must be considered. Demand and supply can be matched only when these contributions are taken together with the restriction in the use of liquid hydrocarbons to "noble" uses and the continuation of exports from the Persian Gulf area.

(2) Coal use shows a tremendous increase, by as much as a factor of five (high scenario). Given toda;'s circumstances, this appears to be impossible. In view of the global comprehensiveness of the scenarios one must therefore search for a substitute.

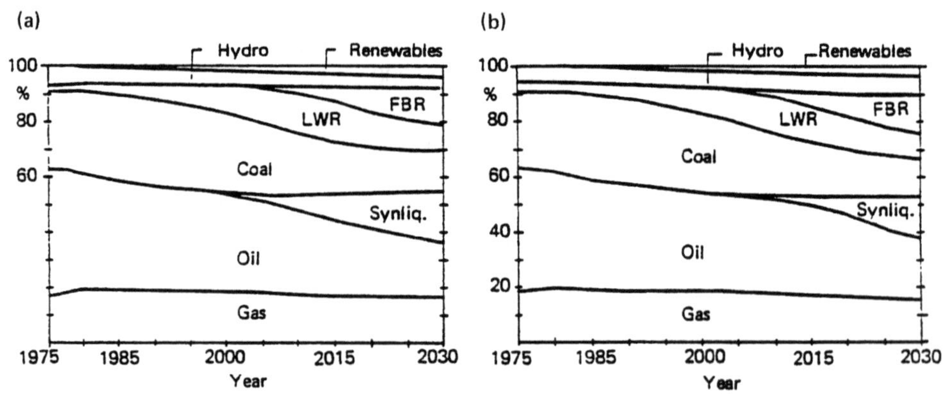

FIGURE 6 Global primary energy: shares by source for two supply scenarios, 1975–2030. (a) High scenario; (b) low scenario.

However, circumstances will change, and such large-scale use should not be discarded at the outset. Toward 2030 much of the additional coal use will go to the assumed autothermal coal-liquefaction schemes (Figure 7). In the IIASA study attention is given to the related CO_2 releases which constitute a potential danger. Again we refer to Häfele (1981) where this is explained in greater detail.

(3) Gas shows an increase similar to that for coal. Although this is not explicitly expressed in the IIASA scenarios, the future large-scale uses of gas will require long-distance intercontinental shipment and transportation.

(4) By 2030 nuclear power (Light-Water Reactors (LWRs) and Fast Breeder Reactors (FBRs)) has a total of 8.09 $TWyr\,yr^{-1}$ for the high scenario and 5.17 $TWyr\,yr^{-1}$ for the low scenario. Its relative share is close to 23% in either case. Earlier we mentioned briefly that its potential contribution could be as high as 17 $TWyr\,yr^{-1}$. This observation requires further elaboration in a paper such as this which tries to view nuclear power in a broader long-range perspective (see Section 3).

(5) The contributions of solar and other renewables are moderate. Actually, the contributions do not result from the linear programming procedure as the related costs of these technologies are too high for that. Their share is open to more general debate.

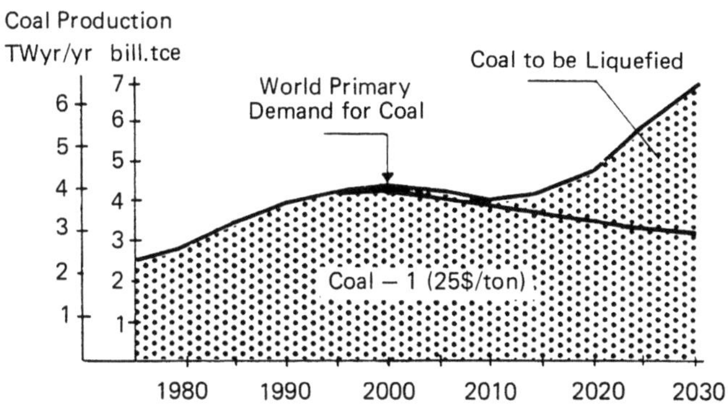

FIGURE 7 World coal demand and supply, 1975–2030: low scenario.

3 THE POSITION OF NUCLEAR POWER IN THE LONG-RANGE PERSPECTIVE OF IIASA

In the previous section we mentioned that 17 $TWyr\,yr^{-1}$ could be considered as the potential upper limit of nuclear power by 2030. At a load factor of 0.7 and a thermal-to-electrical conversion efficiency of 40%, this corresponds to an installed capacity of approximately 10 TWe. To reach 10 TWe would require an all-out industrial effort which today does not appear to be feasible, but for reasons outside the purely techno-economic domain. Such an all-out effort would imply the installation of more than 150 GW per year during the period 1995–2000. This would require a 50% expansion of supply capacity during the next 15 years. In order to put this number into perspective, it should be borne in mind that the global annual increment of power plants in general (nuclear or fossil) is presently about 100 GWe per year. The yearly installation of nuclear power would peak around 2020 at a value of 360 GW per year. Such additions result in the reference trajectory explained in Table 4.

TABLE 4 Trajectories for potential nuclear power installations worldwide (GWe)[a].

Trajectory	Year 1980	1990	2000	2010	2020	2030
IIASA high nuclear reference trajectory[b]	160	580	1,630	3,640	7,030	10,000
Annual addition[c]	24	64	154	305	359	252
Annual growth rate (%)[d]	15	11	9	8	4	1
Nuclear capacity, INFCE high	188	698	1,654			
Nuclear capacity, INFCE low	167	531	1,082			
Nuclear capacity, IAEA high	207	909	2,227			
Nuclear capacity, IAEA low	162	558	1,403			
Nuclear capacity, WEC	152[e]	521[e]	1,543		5,033	

[a] INFCE, International Nuclear Fuel Cycle Evaluation; IAEA, International Atomic Energy Agency; WEC, World Energy Conference. Data for the IAEA figures are from Lane et al. (1977), World Energy Conference (1978), and INFCE (1979).
[b] Equivalent electrical capacity, not necessarily for distribution on electrical grids.
[c] Includes replacement after 30 years of service.
[d] Net growth rate, after deduction of replacements.
[e] Interpolated by IIASA.

The question then arises of whether these plants can be fueled: how much uranium is available? The only honest answer is that no one knows. There is a far greater lack of basic information than in the case of fossil fuels. However, to arrive at a reference figure uranium finds and discovery rates were considered (Table 5). It was then asssumed that the rest of the world would have the same "discovery density" as the United States. This assumption would become the better the lower the expected grade of uranium in question. One thus arrives at Table 6, from which it seems reasonable to consider a figure of 15–20 million tons of uranium for our purposes here.

This could already have been consumed by 2020 if all the reactors making up our reference trajectory were LWRs. (Actually one might also consider heavy-water reactors or advanced gas-cooled reactors. For the present purposes this does not change the observations very much. Therefore the LWR, when mentioned here, stands generally for all burners.) It is therefore customary to consider the classical LWR–FBR strategy, in which any net plutonium coming out of an LWR or an FBR is used to form an initial core inventory for another FBR. This leads to a gradual takeover of the FBR (Figure 8), which permits the consumption of uranium to be limited to only 15 million tons, essentially because the stockpile of natural uranium will be transformed into a stockpile (an active inventory) of plutonium. However, such a strategy would require the early advent of breeders in a fashion that is presently discernible only in France and the Soviet Union. Nevertheless, there are ways and means of compensating for the delayed advent of the breeder. One scheme is the enhancement of the conversion ratio in the LWR by using

TABLE 5 Uranium finds and finding rate[a].

Country	Drill holes ($\times 10^3$ m)	Find[b] ($\times 10^3$ tons)	Finding rates (kg m^{-1})
Australia	1,100	345	314
Central African Empire	55	16	290
Argentina	200	42	212
India	306	54	178
Finland	25	3.2	128
Philippines	8	0.3	38
Italy	74	2.2	30
Turkey	196	4.1	28
FRG	205	5.6	28
Spain	638	16	25
France	5,151	123	24
USA	82,500	1918	23
Japan	393	7.7	23
Portugal	435	9.2	21
Mexico	723	7.1	9.8

[a] Based on data from OECD (NEA)–IAEA (1977).
[b] Finds include all resources at a forward cost of less than \$130 kg^{-1}, plus cumulative production.

TABLE 6 Adjusted uranium resource estimates.

IIASA world regions[a]	Area ($\times 10^6$ km^2)	OECD (NEA)–IAEA estimate ($\times 10^6$ tons)[b]	IIASA estimate ($\times 10^6$ tons)
I	21.5	2.53	3.87
II	23.5		4.23
III	15.5	1.26	2.79
IV	20.6	0.08	3.71
V	33.6	0.33	6.05
VI	9.8	0.08	1.76
VII	11.5		2.07
I–VII	136.0	4.29	24.48
Polar regions (including uninhabited islands)	12.5		2.25
World	148.5	4.29 (14.2–26.4)[c]	26.73

[a] The regions are defined in Figure 3.
[b] Excluding regions II and VII.
[c] Including the speculative resources given in OECD (NEA)–IAEA (1977).

^{233}U instead of ^{235}U. The ^{233}U would have to come from the radial blanket of an FBR (Figure 9), resulting in a conservation scheme as given in Figure 10.

It may be enlightening to consider a situation in which *no* breeders were employed. The ongoing consumption of natural uranium would make inevitable an ever-increasing trend towards the use of low-grade natural uranium. Bearing in mind that the ratio of the energy content of 1 g of fissile material to that of 1 g of carbon is 3×10^6, the

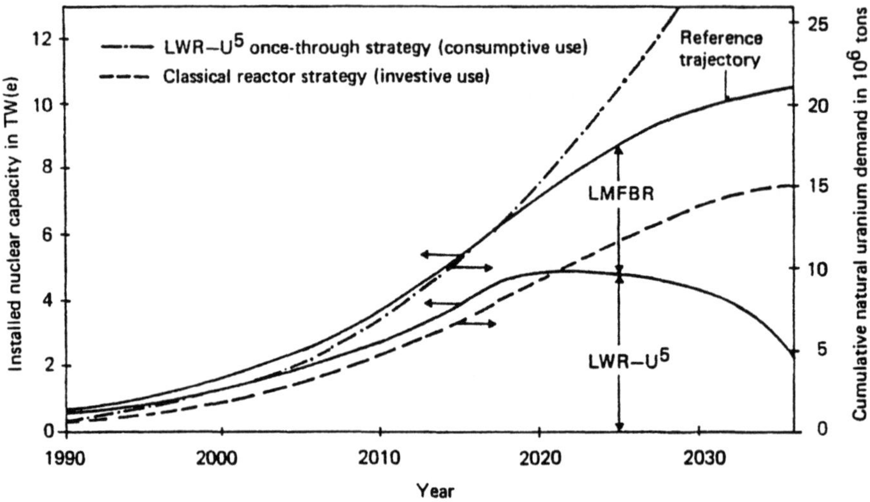

FIGURE 8 A classical reactor strategy.

necessity of using low-grade ores with concentrations of the order of 70 ppm on a large scale would result in the reduction of the ratio down to 1, as burners essentially use the natural uranium at only 0.5% efficiency. For a 70-ppm uranium shale this means a factor of $(3 \times 10^6) \times (5 \times 10^{-3}) \times (7 \times 10^{-5})$, in other words, approximately 1. Indeed, man would handle nuclear power in the way that he is used to handling coal, and since this involves the use of yellow cake one might well rename it "yellow coal". Table 7

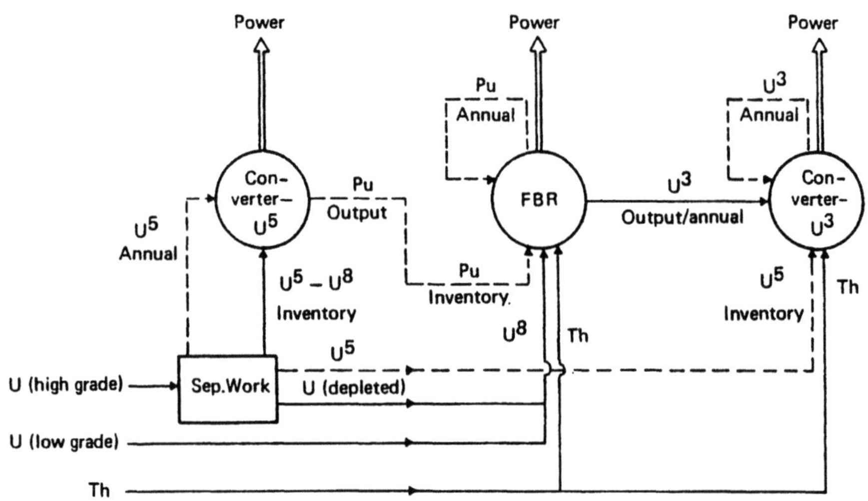

FIGURE 9 A converter–breeder system.

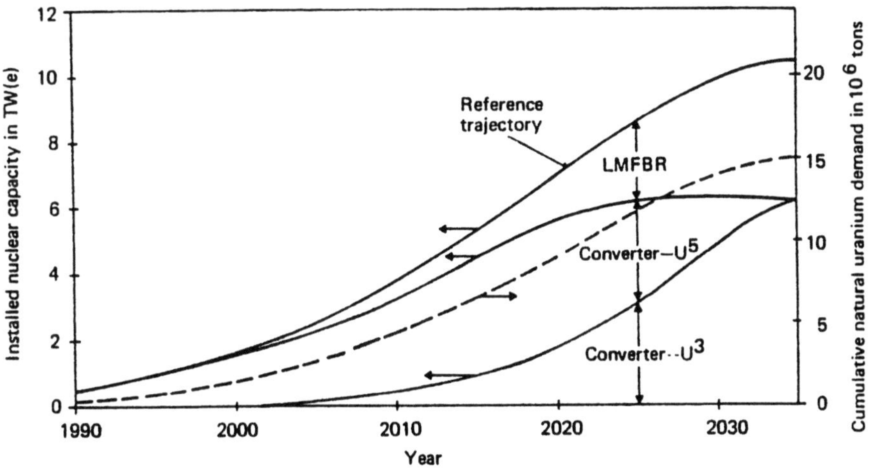

FIGURE 10 A converter–breeder strategy.

thus compares the requirements for the operation of a 1-GW power plant. The LWR, when based on 70-ppm ores, does indeed parallel coal-fired plants in terms of effort and environmental impact.

It has been an almost unchallenged position to consider nuclear power plants first and foremost, or even exclusively, as generators of electricity. By organizing and aggregating innumerable small and large end users, the electric grid permits the use of relatively large supply units which fit the inherent features of nuclear reactors. However, the 17% share of electricity in all final energy needs as explained earlier for the IIASA scenario permits the use of only 9.4 TWyr yr^{-1} (high scenario) or 5.8 TWyr yr^{-1} (low scenario) out of a conceivable total of 17 TWyr yr^{-1}, i.e. only 55% or 34%, respectively, of the potential.

These numbers are aggregated values for the world as a whole. In addition, one must consider the specific cases of the individual world regions. Of these, region III is

TABLE 7 Requirements for the operation of a 1-GWe power plant[a].

Power plant	Land, 30-year total (km^2)	Mining personnel (man-yr yr^{-1})[b]	Material handling involved, 30-year total (× 10^6 tons)
LWR (2000-ppm ore)	3	50	45[c]
Coal	10–20	500	321[d]
LWR (70-ppm ore)	33	300	360[d]

[a] Corresponds to an electricity chain producing 6.1 TWh with a 30-year life span.
[b] 1 man-yr = 2000 h.
[c] Overburden factor, 15 m^3 ton^{-1} (averaged).
[d] Overburden factor, 3 m^3 ton^{-1} (averaged).

in particularly bad shape and must import not only vast amounts of oil but also vast amounts of coal. In the high scenario this can be as much as 600 million tons, i.e. the present total annual production of coal in the United States! The need for additional energy is therefore extremely great, and the specific need is for liquid hydrocarbons. This must be borne in mind when considering nuclear power, even when for the case of, for example, region III the share in the final end use is not 17% but 24%. What is evidently needed is the application of nuclear power beyond the generation of electricity.

It is therefore natural to consider the generation of process heat, particularly high-temperature process heat. In doing so one is faced with a rather disappointing result: the individual applications are small scale when compared with nuclear power plants for electricity generation. What is lacking is the integration of the many small and large end users of energy into a grid, a second grid next to the original grid used for electricity. Such integration is at present achieved by the liquid nature of heating oil and gasoline. Not only do these provide the required energy as such (the calories) but they do it in a high-density form, be it by weight or by volume. This permits easy transportation, distribution, and in particular storage without much impeding and costly infrastructure. The integration brings in additional utilities beyond the utility of mere energy; in this it largely parallels electricity. Not only is the energy content utilized but there are the additional utilities of cleanliness and controllability, which can explain the more than proportional growth rates for liquid fuels. Thinking on additional applications of nuclear power beyond the generation of electricity must therefore center around liquids.

An obvious choice would be liquid hydrogen. While this could well be a solution eventually it poses formidable problems of adaptation of the existing infrastructure. For reasons of practicality one must look for an intermediate solution, and this intermediate solution must, in turn, incorporate the carbon atom. As shown earlier, one is driven into the extended use of the carbon atom in any event. The only question is how to do it most prudently. Such prudent approaches always appear to revolve around hydrogen and carbon.

One prudent approach centered around natural gas is described in the following. Natural gas may have a bright future, at least for some decades and particularly so if the promises of deep-lying unconventional gas should come true. However, it might not be a good idea just to burn it straightforwardly. Instead it might be more prudent to split it all, in the presence of water, into carbon monoxide and hydrogen using the high-temperature process heat of a nuclear reactor. It is then possible to convert these intermediates into methanol by the use of appropriate catalysts (Fischer–Tropsch, Mobil Oil). The reaction for the simultaneous liquefaction of methane and splitting of water is

$$\begin{array}{l} CH_4 + H_2O + (HT) \rightarrow CH_3OH + H_2 \\ 100 \quad : \quad 30 \; : \; 90 \quad : \quad 30 \end{array} \tag{1}$$

where HT is high-temperature process heat (note that zero emission and the transportability of liquids eliminate constraints on plant siting). As eqn. (1) indicates, a process of 90% efficiency takes place for the overall conversion of methane into methanol. At the same time, a water molecule is split and the energy balance indicates that this is done by nuclear power! Ideally all the nuclear high-temperature process heat finds itself in the liberated H_2. Technically this leads to a scheme as given in Figure 11. In fact 338 MW of

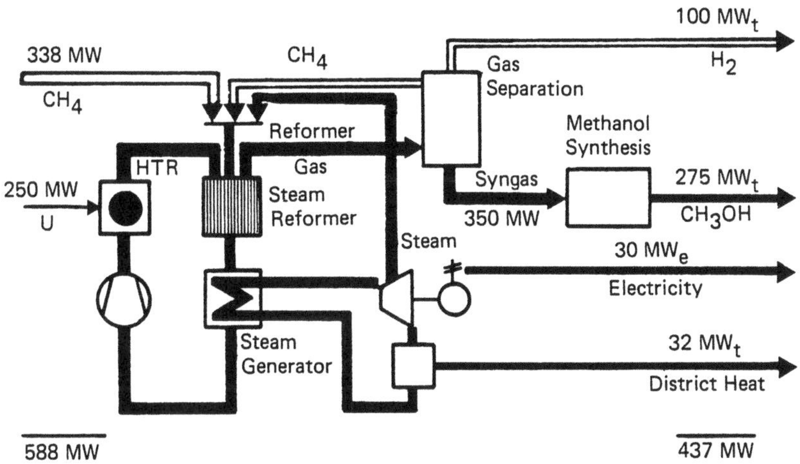

FIGURE 11 A schematic process for converting methane to methanol using an HTR.

methane are transformed into 275 MW of methanol while 100 MW of H_2 are produced. Under this scheme, the 30 MW of electricity and 32 MW of district heat also produced are truly by-products only. In Figure 11 we have assumed that the high-temperature reactor (HTR) will be fueled by an FBR; we have already seen that breeding is a necessity in any event.

There are many opportunities for using the liberated hydrogen. One is the hydrogasification of "dirty" fossil fuels whose use will be the predominant energy feature of the next few decades. It would permit a more environmentally conscious approach to that era. For the example of lignite use in the Federal Republic of Germany (FRG) this is explained in Figure 12. This results in the scheme of Figure 13. Here only 101 MW of natural gas are required, the balance of the 338 MW of eqn. (1) being made up by the use of lignite.

What happens when the gas runs out? It is then possible to shift this scheme to the splitting of water. Numerous stepwise chemical processes lend themselves to water

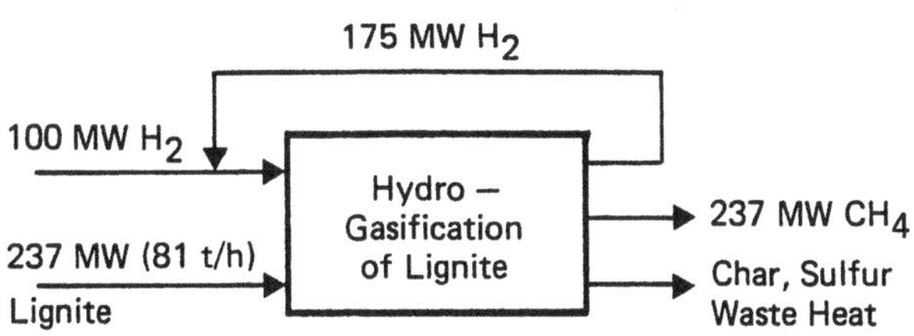

FIGURE 12 A schematic process for converting lignite to methane.

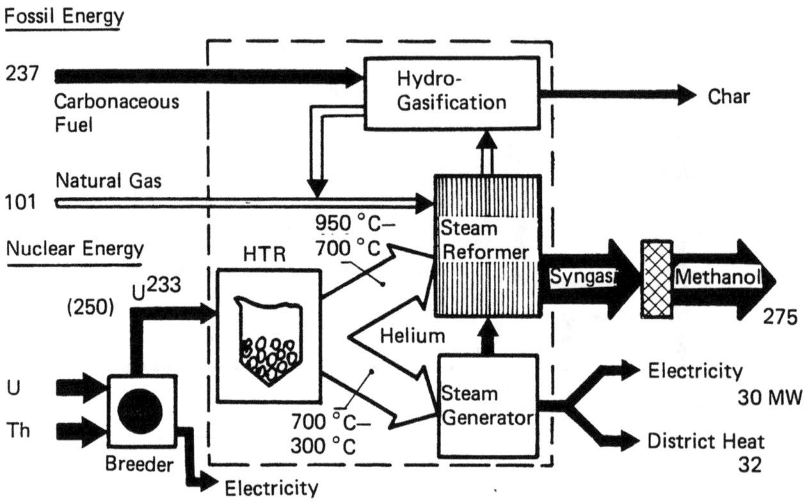

FIGURE 13 A zero-emission energy system for the conversion of dirt into secondary energy.

splitting. One closed-cycle water-splitting scheme which illustrates the extension of high conversion but still consumptive uses of methane into investive uses of methane is

$$
\begin{aligned}
&CH_4 + H_2O + (HT) &&\rightarrow CH_3OH + H_2\\
&CH_3OH + SO_2 &&\rightarrow CH_4 + SO_3\\
&SO_3 + (HT) &&\rightarrow SO_2 + \tfrac{1}{2}O_2
\end{aligned}
\qquad (2)
$$

$$
\begin{aligned}
&CH_4 \quad : {}^{235}U \wedge 300\ \mathrm{TWyr}\\
&CH_3OH : {}^{239}Pu; H_2O : {}^{238}U, Th
\end{aligned}
$$

where *HT* is high-temperature process heat. Methanol would be recycled and one would have to use only hydrogen (and oxygen). This would be possible only a few decades from now but more importantly it will also be necessary then. Natural gas would bridge the gap. Although process (2) is already possible in the laboratory, it is not yet technically possible because of severe materials problems. Also, there are still materials problems with methane splitting (1). However, the point is that these problems are less severe than those associated with the SO_3 splitting in scheme (2); by using methane splitting first, we can buy time to attack the more difficult SO_3-splitting problems. In such a fashion, prudent energy strategies are translated into strategies for the development of appropriate materials!

4 CONCLUDING REMARKS

We have explained that the uses of nuclear power for the generation of electricity seem to have a bright but limited future. Nuclear power stations that feed the first (i.e. the electrical) grid have started their buildup as a group in the 1970s, as shown in

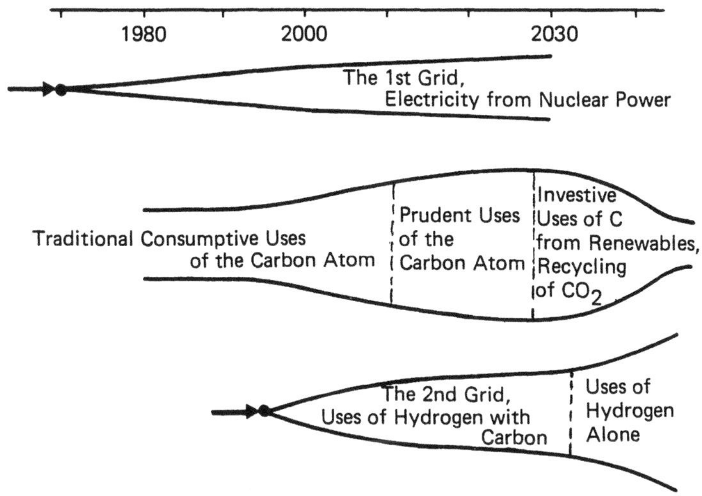

FIGURE 14 The prudent uses of the carbon atom, hydrogen, and nuclear power.

Figure 14. What must soon be considered is the extension of the traditional uses of the carbon atom to prudent uses of the carbon atom where not autothermal but rather allothermal schemes gradually prevail, with the use of high-temperature heat not based on fossil fuels and the efficient use of the hydrogen thus generated. Practically, the use of methanol or liquid hydrocarbons of a higher order (e.g. gasoline) would facilitate the use of a second grid for the application of nuclear power. However, this should be done with an overall strategy in mind. Eventually, beyond 2030, we must arrive at the use of hydrogen for end-use purposes. Hydrocarbons should stay in the picture only to the extent that biomass is available, as this implies essentially a recycling of the related CO_2. At this workshop on neutron abundance which is considering various long-range perspectives for nuclear power, we feel that such strategic considerations must be borne in mind.

REFERENCES

Häfele, W. (Program leader) (1981). Energy in a Finite World. Ballinger, Cambridge, Massachusetts.

INFCE (1979). Final Report. Draft. INFCE/W6.1/17. Working Group 1 on Availability of Nuclear Fuel and Heavy Water. International Nuclear Fuel Cycle Evaluation, Vienna.

Keyfitz, N. (1977). Population of the world and its regions 1975–2050. WP-77-7. International Institute for Applied Systems Analysis, Laxenburg, Austria.

Lane, J.A., et al. (1977). Nuclear power in developing countries. In Proc. Int. Conf. on Nuclear Power and its Fuel Cycle. IAEA-CN-36/500. International Atomic Energy Agency, Vienna.

OECD (NEA)–IAEA (1977). Uranium resources production and demand. A Joint Report by the Organization for Economic Cooperation and Development and the International Atomic Energy Agency. Organization for Economic Cooperation and Development, Paris.

World Energy Conference (1978). World Energy Resources, 1985–2020. IPC Press, Guildford.

THE NUCLEAR ENERGY CONTINUUM AND ITS SPECTRUM OF CHOICES

A.A. Harms
McMaster University, Hamilton, Ontario (Canada)

ABSTRACT

Nuclear energy options are generally perceived as consisting of a choice between two systems: fusion or fission. Recent developments, however, suggest that an expanded perspective of nuclear reaction systematics can lead to a spectrum of nuclear energy options in which stand-alone fission or fusion reactors are but two of many possibilities. The underlying basis for this synergistic perspective of alternatives is the potential for an enhanced utilization of complementary nuclear reactions and their integration into a coherent systems framework. Specific examples are hybrid reactors, spallation breeders, and various kinds of symbionts. These alternative nuclear energy systems hold the promise of more appealing and efficient methods of extracting energy from the nucleus.

1 INTRODUCTION

Nuclear energy, like other developing technologies, bears the imprints of the institutions which nurtured it and the marks of its own intrinsic logic. Superimposed on these evolutionary features is the characteristic of the scientific method which seeks to group processes and phenomena into distinguishable sets. The combination of these influences has led to a broadly based consensus of thought on the technology which tends to mask a subtle though fundamental continuity of nuclear energetics and a consequent expanded range of available systems choices.

The underlying physical basis for the extraction of energy from the nucleus is associated with the rearrangement of nuclear structure based on processes which may be represented by the following:

$$N_1 + N_2 \rightarrow \sum_i N_i' + Q \tag{1}$$

Here N_1 and N_2 are nuclear fuel nuclei or nucleons and N_i' is one of several reaction products; the symbol Q stands for the energy released in the process and has its origin in the mass decrement of this reaction. Evidently, both fission and fusion are special cases of this representation.

The ultimate utility of a reaction of this type must be gauged by issues such as the following: (i) Is the reaction density sufficiently high for economy and functionality? (ii) Is the recirculation energy component sufficiently low to minimize energy losses? (iii) Is the nuclear fuel available in sufficient quantities? (iv) Is the program for managing the reaction products adequate?

As a minimum requirement, energetics considerations impose the imperative that the total energy necessary to sustain the reaction must be less than the net energy eventually used for societal purposes. While this energetics consideration provides an obvious bound for the useful implementation of a nuclear energy reaction process, no similar bound relating to nuclear mass considerations – excepting the actual availability of a nuclear fuel – is generally imposed, and least of all on the associated reaction products. The radiological hazards of reaction products or the strategic importance of nuclear fuels are normally regarded as debatable questions in the implementation of a particular set of nuclear reactions as a source of energy; such front-end and back-end aspects of the nuclear fuel cycle are assumed to be manageable by institutional control and regulation.

Dissociating ourselves for a moment from specific exoergic nuclear reactions and specific types of nuclear reactors, we may assert that the ideal nuclear energy system might be characterized as follows:

$$\begin{bmatrix}\text{Supply of}\\ \text{readily}\\ \text{available}\\ \text{isotopes}\end{bmatrix} \longrightarrow \begin{bmatrix}\text{Set of}\\ \text{integrated}\\ \text{complementary}\\ \text{nuclear reactions}\end{bmatrix} \begin{matrix}\longrightarrow [\text{Energy}]\\ \\ \longrightarrow \begin{bmatrix}\text{Stable}\\ \text{isotopes}\end{bmatrix}\end{matrix} \tag{2}$$

This ultimate ideal constitutes an objective to be sought even though the very nature of radioactivity means that the goal will be only partially reached.

To complement this point of emphasis, the view is advanced here that from the numerous nuclear reactions known it may be possible to identify a reaction set which by selected intrinsic compensatory processes tends in the direction suggested in eqn. (2). The physical basis for this is that, rather than isolating exoergic reactions on the curve of binding energy (Figure 1), one seeks to couple them so as to arrive at a more holistic schema of nuclear energy production. Interestingly, both historical records and recent developments suggest the existence of "currents of thought" in this direction. The underlying fabric seems to be the recognition of a continuum of energy systems options.

2 THE NUCLEAR TRIAD

Contemporary nuclear energy enterprises may conveniently be classified as follows. (1) *Nuclear fission technology*: this technology is defined by the present generation of fast and thermal fission reactors and their associated support industries. (2) *Nuclear fusion technology*: this technology is associated with the long-term realization of obtaining usable energy from nuclear fusion, with present emphasis on magnetic or inertial confinement. (3) *Beam–accelerator technology:* by this we mean electromagnetic wave generators (e.g. lasers) and particle accelerators for light and heavy ions. These three technologies are generally viewed as essentially independent and may be represented as in Figure 2.

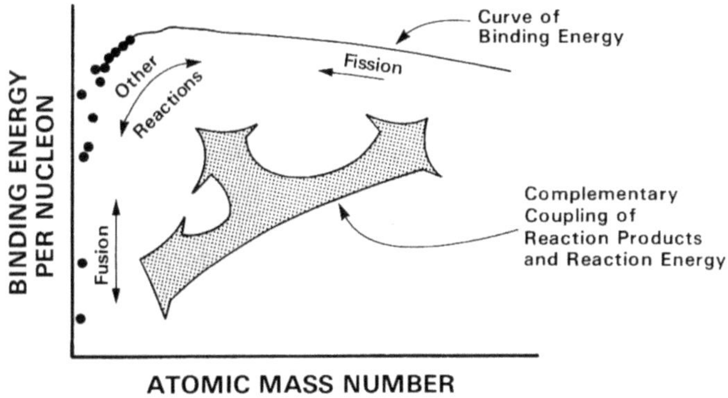

FIGURE 1 The curve of binding energy displaying the relative location of atomic nuclei; fission and fusion processes as well as potentially complementary processes are indicated.

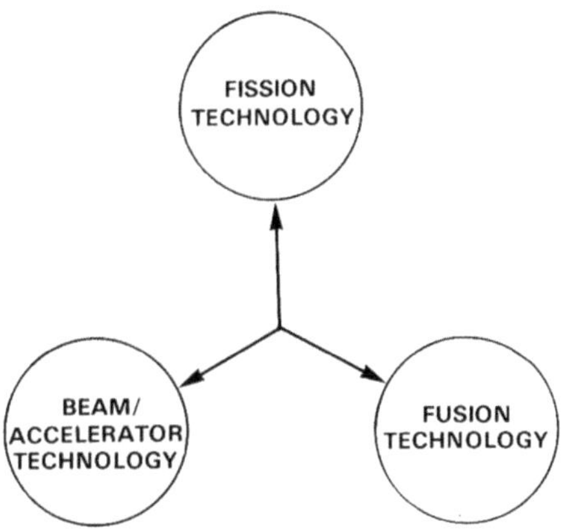

FIGURE 2 A schematic diagram of the triad of existing nuclear technologies which are generally viewed as independent.

It is revealing to recognize how firmly these three distinct technologies are ingrained into our consciousness. Research laboratories are structured according to these divisions; government policy and funding agencies have administrative organizations which separate these functions; scientific–professional conferences and academic programs also view these technologies as separate and distinct specializations.

Throughout the development of the nuclear energy program during the last 30 years, however, a strand of scientific thought can be identified which has sought to merge this triad of technologies; this has been based on the recognition that certain benefits, which would not be possible otherwise, could accrue from the joint consideration of the technologies. We now review these historical trends and relate them to our present objective.

3 SOME HISTORICAL MARKERS

The earliest document which suggests a specific merging of two of the nuclear technologies is a report by Lewis (1952). On the basis of experimental results and theoretical considerations associated with neutron emission from heavy nuclei excited by means of a high-energy ion accelerator, Lewis concluded that if the neutron yield per incident proton is sufficiently high then these neutrons could well be used to breed fissile fuel for a fission-reactor economy. The energy for the accelerator would be supplied by the reactor. This schema is illustrated in Figure 3(a) and, in terms of our triad of technologies, provides for the existence of that portion of a "rim" which connects fission-reactor technology with beam–accelerator technology in Figure 2.

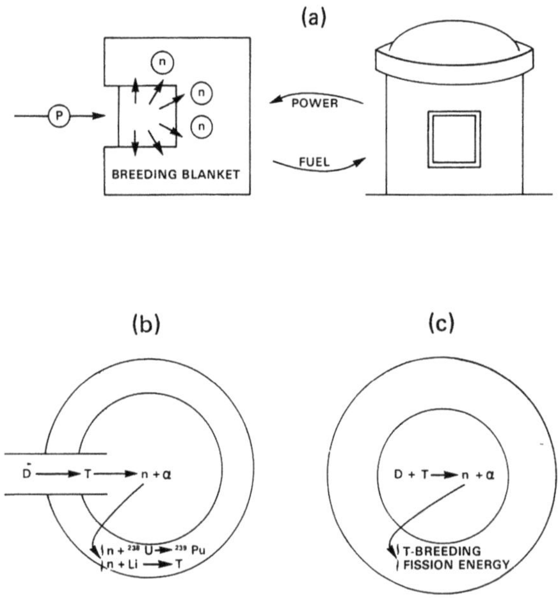

FIGURE 3 A schematic depiction of nuclear energy systems concepts combining (a) accelerators and fission processes, (b) accelerators and fusion processes, and (c) fusion and fission processes.

This spallation–fission symbiosis evidently involves the coupling of spallation reactions with fission and breeding processes. Listing only the members of the nuclear reaction chains of interest, and using the thorium fuel cycle, as an example, we can represent the reaction systematics as follows:

$$\begin{array}{ll} {}^{235}\mathrm{U} + \mathrm{n} \rightarrow \nu_f \mathrm{n} & \text{Fission core} \\ {}^{232}\mathrm{Th} + \mathrm{n} \rightarrow {}^{233}\mathrm{Th} \rightarrow {}^{233}\mathrm{Pa} \rightarrow {}^{233}\mathrm{U} & \text{Fission reactor and} \\ \vec{p} + Z \rightarrow \nu_s \mathrm{n} & \text{accelerator blanket} \end{array} \qquad (3)$$

This fuel cycle for a spallation-supported fission economy may in principle be fully closed for either a steady or an expanding state and depends only on the relative rates of fission, breeding, and spallation. The feature to note is that, because of the high spallation-neutron yield, the fission technology which has always been handicapped by being neutron poor has now become neutron rich; the concepts of these synergetic systems are further elaborated by Vasil'kov et al. (1970), Harms and Gordon (1977), Kouts and Steinberg (1977), and Grand (1979).

Continuing our historical review, in the year after Lewis's paper the California Research Company submitted a proposal (Powell, 1953) to develop an accelerator equipped with a preionized plasma target. The expectation was that if the plasma target contained tritium then the accelerated deuterium ions might lead to a sufficient level of fusion reactions in the target to yield a net energy gain. Thus, using $\vec{D}$ as an accelerated deuteron, the principle fusion reaction involving an accelerator and a plasma could be written in the usual form involving fusion of deuterium and tritium. Additionally, the plasma target was to be surrounded by a blanket containing uranium and lithium; the former was expected to provide a degree of neutron multiplication and ^{239}Pu breeding while the latter would replenish the burned tritium by neutron absorption. We provide an illustration of this schema in Figure 3(b) and suggest the following coupled and integrated nuclear reaction pattern:

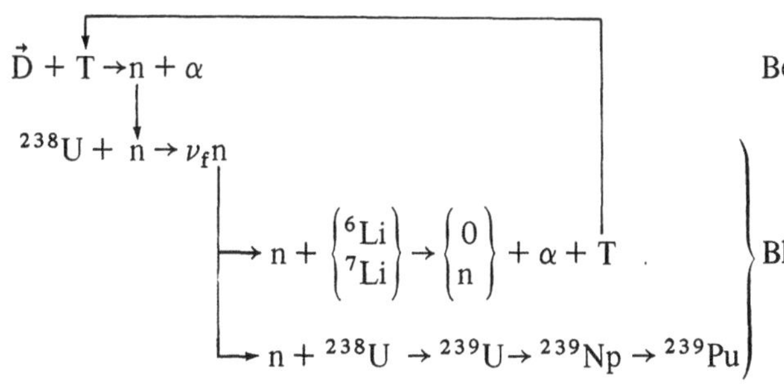

(4)

Returning to our intent of completing a "rim" in Figure 2, we now add this scheme which provides for a connection between beam–accelerator technology and fusion technology.

The next relevant suggestion in support of the continuum of synergetic nuclear energy systems can be attributed to Lawson (1955). In his paper Lawson discussed various means of accomplishing controlled fusion and suggested that, however the D–T reaction is brought about, the fusion neutron should enter a surrounding uranium blanket. Since the energy of this 14-MeV fusion neutron is above the fission threshold for ^{238}U it should then be possible not only to meet the requirement of tritium replenishment and to breed fissile fuel but also to generate fission energy in the blanket. In effect the blanket could have the characteristics of a fission reactor. Lawson calculated, in particular, that for every unit of fusion energy an average of seven units of fission energy could be generated in the blanket. Thus the important idea of energy multiplication in the blanket and the concept of a hybrid reactor were born.

We illustrate the Lawson fusion–fission concept in Figure 3(c) and suggest a reaction schema (with *FP* standing for fission products) as follows:

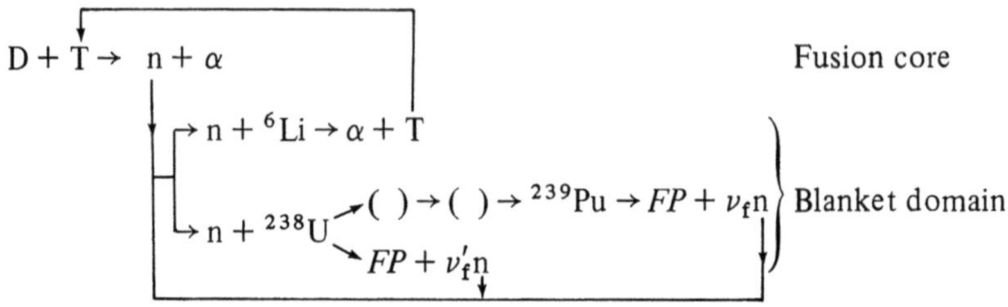

(5)

The Lawson scheme may be viewed in two interesting ways: either a fission blanket surrounding a fusion core or as the placement of a fusion reactor inside a fission core. In either case it illustrates the establishment of a connection between the fusion and fission technologies of Figure 2 based on the coupling of accessible reactions as suggested by Figure 1.

The general concept of combining the "neutron richness" of D–T fusion with the intrinsic "neutron poverty" of fission reactors as suggested in these early proposals has in recent years provided a rich field of research inquiry. As mentioned earlier, the works of Lewis (1952), Vasil'kov et al. (1970), Harms and Gordon (1977), Kouts and Steinberg (1977), and Grand (1979) are indicative of the scope for the spallation–fission interface; starting with two papers by Jung (1969) and Lidsky (1969), considerable interest has been shown in the fusion–fission combination (Leonard, 1973; Kolesnichenko and Reznik, 1974; Harms, 1975, 1976; Horoshko et al., 1974; Maniscalco, 1976; Bethe, 1978; Blinkin and Novikov, 1978; Velikhov et al., 1978). The label "hybrid reactor" is frequently associated with the fusion–fission concept for both magnetic and inertial confinement schemes.

4 THE EMERGING CONTINUUM

Although the historical record implies considerable distinctiveness about these various synergetic nuclear energy systems a closer examination suggests the basis for an energy systems continuum.

Consider first the Lawson fusion–fission hybrid (Figure 3(c)). Though the same device contains both fusion and fission processes, no restrictions on their relative magnitudes need be imposed. By suitable design and operation the blanket could serve mainly the function of producing energy or that of producing fuel; i.e. it could be an "energy factory" or a "fuel factory" or some combination of both. Functionally, a fusion–fission reactor could in principle be placed anywhere on the rim between the fission and fusion technology spokes depending only on the relative contributions of the fission-like or fusion-like processes (Figure 4).

A similar continuum between the beam–accelerator and fission technologies can be formulated; the only difference involves the mechanism for the production of the non-fission neutrons. The continuum characterization between beam–acceleration technology and fusion technology could similarly be represented by the relative overlap of these two technologies (Figure 4).

To introduce sufficient generality we revise Figure 4 into the form represented by Figure 5 where we also list frequently used names and generic labels associated with current systems concepts.

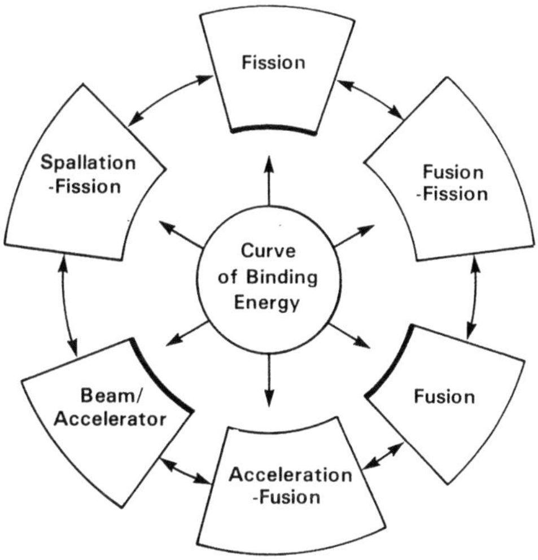

FIGURE 4 An expanded representation of the independent triad of nuclear technologies (Figure 2) which includes the integrated synergistic nuclear energy systems concepts. In effect a continuum of systems choices is thus generated.

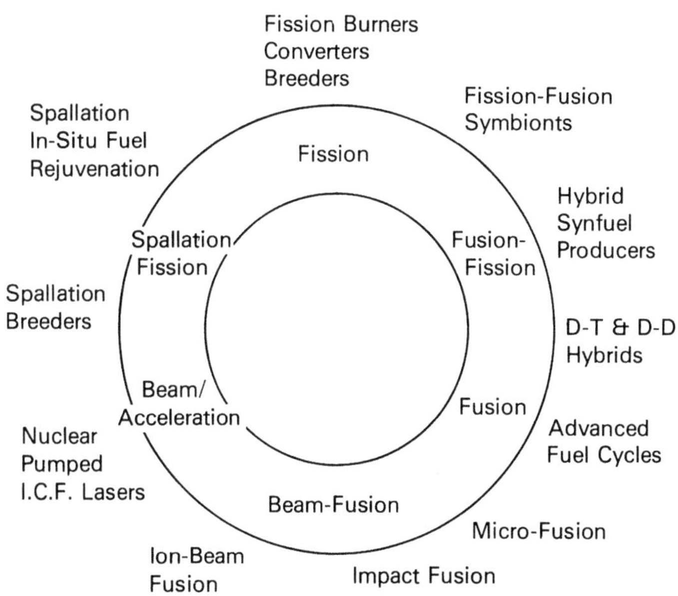

FIGURE 5 A detailed itemization of the emerging nuclear energy continuum. The systems included are a partial listing only and are meant to illustrate the diversity of emerging concepts.

At this juncture, one may recognize another circumferential level of continuity of the synergetic nuclear energy formulation. While the foregoing discussion identified systems options which exist specifically between the three reference technologies of Figure 2, here we point to some emerging nuclear perspectives which are best described by spanning a larger part of the rim. For example, both spallation neutrons and fusion neutrons were identified as – in a sense – being external neutrons which may support a steady-state reaction process in a subcritical fission assembly; this defines in effect the domain of so-called driven fission reactors. In terms of our continuum (Figure 5) we may – since the origin of these external neutrons is in a sense of secondary interest – place these driven reactors between the spallation and fusion technologies with fission technology at their middle (Figure 6).

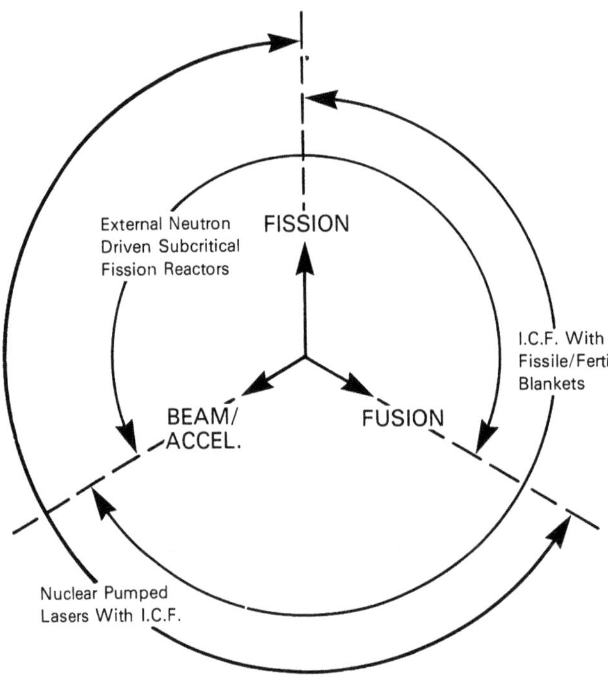

FIGURE 6 A suggested representation of the breadth of technologies encompassed by particular systems concepts. The abbreviation I.C.F. stands for inertial-confinement fusion and could imply the use of lasers or ion accelerators.

We may identify additional, more broadly based synergetic concepts. One particularly evident case involves inertial-confinement fusion for which the target pellet chamber is surrounded by a fissile–fertile–fusile blanket, i.e. the blanket maintains tritium breeding as well as fission reactions and fissile-fuel breeding. Clearly this spans the domain from beam–accelerator technology to fission technology with fusion at the center (Figure 6).

Finally, for reasons of completeness, we cite nuclear pumped lasers (in which fission fragments are used to pump the laser) used to induce fusion by inertial confinement

(Kruger and Miley, 1978). Evidently, this now involves fission and fusion technologies with the beam–accelerator technology at their center (Figure 6).

5 THE BROADER PERSPECTIVE

The illustration used constitutes but a low-dimensional representation of what is essentially a multidimensional structure. For example, in addition to the degree of inclusion of any of the three reference technologies in a synergetic system, choices specifically associated with subtechnologies of these reference technologies and the overall system function or functions can be identified as follows.

(1) The fission part of a synergetic system may possess a resemblance to existing reactor concepts; the fuel cycles could cover the range from low to high enrichment and may involve the uranium cycle, the thorium cycle, or possibly a combination of both.

(2) For the fusion part, the systems choices may involve such broad technologies as magnetic or inertial confinement as well as less emphasized subtechnologies such as colliding beams and impact fusion. The specific system choices may be equally diverse (tokamaks, mirrors, surmacs, and numerous less conventional types) with fuels initially being D–T and subsequently D–D with its variants.

(3) The dominant choice in the beam–accelerator part is the driver. Though the two classes are either particle or electromagnetic wave drivers, further choices are extensive; the particles could be electrons, protons, or light or heavy ions with the devices being possibly linacs, cyclotrons, etc.

(4) The resultant synergetic nuclear energy system may have one or several functions: e.g. electricity production, district heating, synthetic-fuel production, spent-fuel rejuvenation, fission-product and actinide destruction, and industrial steam supply.

6 NUCLEAR CATALYSIS

Basic to the incorporation of an optimally beneficial fusion–fission–spallation symbiosis is the fullest possible utilization of the blanket which surrounds a fusion core or spallation target. Both experimental and analytical considerations suggest that there are indeed many and distinct blanket design options serving a variety of purposes. Fundamental for these various purposes are the following neutron–nucleus characteristics associated with these neutron-driven blankets: (i) high initial neutron source energies allow further extensive enhancement of neutron multiplication by fast fission and (n, xn) reactions; (ii) high initial neutron source energies allow more latitude in shaping the neutron energy spectrum for selected purposes.

Clearly the blankets should operate as subcritical assemblies, thus eliminating many problems of reactor safety and control. Then, in a broader context, one can identify a number of specific as well as general characteristics.

(1) A fission reactor which is suitably coupled by energy and isotope linkages to either a spallation accelerator or a fusion reactor may be viewed as a synergetic breeder. However, in contrast to a conventional fast breeder, the effective breeding capacity of a synergetic breeder can be exceedingly large and requires no initial fissile inventory, provided that electricity is initially available from other sources.

(2) Operating in a symbiotic support capacity to a fission reactor, neither the fusion companion nor the spallation companion needs to be a net energy producer. These support systems may be viewed as representing a fractional energy burden on the fission reactor in much the same way as a uranium mine or fuel-enrichment plant represents an energy load on the fission reactor. The key issue here is not to allow the energy cost of the externally supplied neutrons to become too large.

(3) The spallation target and blanket offer some new and promising possibilities for nuclear fuel management. Two new options can be identified.

First, since neutron yield in a spallation process increases with atomic mass number, the actinides accumulating in the fission-reactor wastes could be effectively destroyed in the target while concurrently increasing the neutron yield. Little experimental information exists on the isotopic composition of spallation products from actinides though theoretical considerations suggest a more constant mass distribution which contrasts with the bimodal fission-product distribution.

Second, it is known that the capture-to-absorption cross-section ratios of fertile–fissile isotope pairs (i.e. σ_c(Th-232)/σ_a(U-233) and σ_c(U-238)/σ_a(Pu-239)) exhibit a significant high plateau in the neutron-energy domain 100–500 keV. With a sufficiently higher energy source-neutron spectrum available it should be possible to tailor a neutron spectrum in the blanket so as to allow effective spent-fuel rejuvenation; in other words, it should be possible to place the entire spent fuel element (which contains a depleted fissile concentration while still retaining a high fertile composition) in a blanket position with a suitably tailored neutron spectrum to enrich the fissile concentration by neutron capture in the fertile nuclei (Harms and Hartmann, 1978; Takahashi et al., 1978). Subsequently, assuming no significant adverse structural effects, the fuel could be

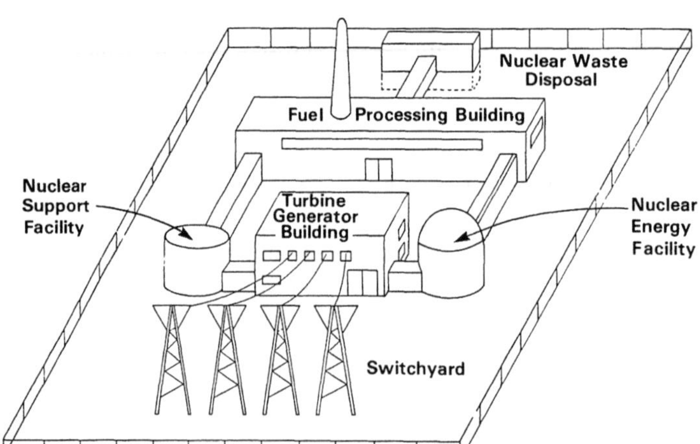

FIGURE 7 A schematic representation of a synergetic nuclear energy system. As indicated, one facility is optimized for energy production while another provides a support capacity. As suggested here, intrinsically safe fertile materials only would be transported into this energy park and energy is supplied to a power grid.

returned to the reactor and could eventually proceed through a repeat, or repeatable, rejuvenation cycle. This function is also possible in a fusion blanket (Conn et al., 1980).

(4) The existence of an ample supply of neutrons would suggest that many structural components for reactor cores and blankets could be designed with durability and reliability – rather than neutron economy – as the dominant criteria.

7 CONCLUDING COMMENT

One particularly appealing feature of these synergetic nuclear systems is that they provide the essential basis for autarkic and syntonic nuclear energy parks. These energy parks would have the feature of being totally self-sufficient in fuel and would contain an internal nuclear waste transmutation capacity. Figure 7 provides a schematic representation for any one nuclear energy system chosen from the continuum suggested here.

REFERENCES

Bethe, H.A. (1978). The fusion hybrid. Nuclear News, 27(4):41.

Blinkin, V.L. and Novikov, V.M. (1978). Symbiotic system of a fusion and fission reactor with very simple fuel reprocessing. Nuclear Fusion, 18:7.

Conn, R.W., Kantowitz, F., and Vogelsang, W.F. (1980). Hybrid for direct enrichment and self-protected fissile fuel production. Nuclear Technology, 49:458.

Grand, P. (1979). The use of high energy accelerators in the nuclear fuel cycle. Nature, 278:693.

Harms, A.A. (1975). Hierarchial systematics of fusion–fission energy systems. Nuclear Fusion, 15:939.

Harms, A.A. (1976). Upper bounds of fissile fuel yield with fusion breeders. Canadian Journal of Physics, 54:1673.

Harms, A.A. and Gordon, C.W. (1977). A parametric analysis of the spallation breeder. Nuclear Science and Engineering, 63:336.

Harms, A.A. and Hartmann, W.J. (1978). Spent nuclear fuel re-enrichment without reprocessing. Annals of Nuclear Energy, 5:213.

Horoshko, R.N., Hurwitz, H., and Zmora, H. (1974). Application of laser fusion to the production of fissile materials. Annals of Nuclear Science and Engineering, I:233.

Jung, H. (1969). Two suggestions regarding controlled fusion: approximation equations for fusion–fission reactors and fusion–reactor–tube reactors. Nuclear Fusion, 9:169.

Kolesnichenko, Y.I. and Reznik, S.N. (1974). D–T plasma as a source of neutrons for the combustion of uranium-238. Nuclear Fusion, 14:114.

Kouts, H.J.C. and Steinberg, M. (Editors) (1977). Proceedings of an Information Meeting on Accelerator Breeding. CONF-770107. Brookhaven National Laboratory, Upton, New York.

Kruger, A.W. and Miley, G.H. (1978). Nuclear pumped laser coupling to inertial confinement fusion. Transactions of the American Nuclear Society, 30:25.

Lawson, J.D. (1955). A survey of some suggested methods of realizing fusion reactors. AERE-CP/M 185. Atomic Energy Research Establishment, Harwell, UK.

Leonard, B.R., Jr. (1973). A review of fusion–fission (hybrid) concepts. Nuclear Technology, 20:161.

Lewis, W.B. (1952). The significance of the yield of neutrons from heavy elements excited to high energies. DR-24. Chalk River Nuclear Laboratories, Chalk River, Canada

Lidsky, L.M. (1969). Fission–fusion symbiosis: general considerations and a specific example. In Proceedings of a Conference on Nuclear Fusion Reactors, Culham Laboratory, 17–19 September 1969. UKAEA Culham Laboratory, Culham, UK, for the British Nuclear Energy Society.

Maniscalco, J. (1976). Fusion–fission hybrid concepts for laser induced fusion. Nuclear Technology, 28:98.

Powell, F. (1953). Proposal for a driven thermonuclear reactor. Report LWS-24920. California Research and Development Company, Livermore, California.

Takahashi, H., Steinberg, M., Grand, P., Powell, J., and Kouts, H. (1978). Analysis of the nonprocessing two-cycle scenario for LWR fuel regenerated by linear accelerator reactor. Transactions of the American Nuclear Society, 30:345.

Vasil'kov, V.G., Gol'danskii, V.I., Dzhelepov, V.P., and Dmitrievskii, V.P. (1970). The electronuclear method of generating neutrons and producing fissionable materials. Soviet Atomic Energy, 29:858.

Velikhov, E.P., Glukhikh, V.A., Gur'ev, V.V., Kadomtsev, B.B., Kolbasov, B.N., Kotov, V.V., Monoszon, N.A., Netecha, M.E., Orlov, V.V., Pistunovich, V.I., Ulasevich, V.K., Churakov, G.F., and Shatalov, G.E. (1978). Tokamak-type thermonuclear hybrid reactor for the production of fission fuel and electric energy. Soviet Atomic Energy, 45(1):653.

ON THE INTERNAL LOGIC OF ENERGY ISLANDS

C. Marchetti
Energy Systems Program, International Institute for Applied Systems Analysis, Laxenburg (Austria)

ABSTRACT

The structure of an energy system, from production through to consumption, is fundamentally influenced by the physical properties of the energy medium, and in particular its transportability. For oil, which is highly transportable, the system optimizes on a world scale. Electrical systems, in contrast, optimize over areas hundreds of kilometers in diameter. In future systems nuclear reactors may be interfaced with hydrogen as the energy medium. If the hydrogen is in gaseous form, the optimal system configuration will be at the level of a continent; if the hydrogen is liquid, then, as in the case of oil, the system optimizes at the world level. Generating centers will then be optimally very large: the paper gives a rapid overview of the main problems associated with such "energy islands" together with some possible solutions.

1 INTRODUCTION

The production, transportation, and distribution of something faces boundary conditions and optimization constraints which are formally independent of the nature of the thing produced, be it ammonia or electricity. Consequently it is possible to analyze the process in a general form and to check the quality of the formulation using examples taken from very diverse areas.

In order to minimize the cost of the unit of final product delivered one has to compromise between the economies of scale that come from concentrating production at a few points and the diseconomies of transportation that come from the consequent lengthening of the feed lines. Because economies of scale usually fall in a relatively restricted range, what really determines the spatial structure of a production system is the transportability of the product, which can in fact vary by orders of magnitude.

These concepts have for a long time been organized by geographical economists under the name of "central-place theory". I will use them to describe the energy system and to predict some of its long-term features if certain switches are turned on. I will make more than a prediction in the strict sense; I will in fact describe an attractor configuration. (Here I use "attractor configuration" to represent a finally-evolved configuration toward which all intermediate configurations tend.)

In order to simplify the argument I have taken an actual system, the electrical system, as a reference case and have described the other systems in relative terms. Such a parametric analysis avoids actual optimizations, which are nonessential at this level of analysis.

A fundamental characteristic of nuclear (or fusion) energy sources is that they provide energy in the form of (low entropy) heat, unlike the fossil sources that preceded them, where high-free-energy chemicals were at play. Heat is not very convenient from the point of view of transportation, storage, and distribution; consequently the problem of an optimal energy system based on nuclear energy sources needs to be thought over.

In Table 1 the basic characteristics of energy systems have been tabulated in a semi-quantitative form, basically to establish some rank between them and to give an aid to intuition. The table provides an extremely simplified picture; for example, the spatial density of energy consumption is assumed to be the same for all cases and equal to that of electricity. Transportability is measured in terms of the distance where transportation costs equal 20% of production costs but does not take account of economies of scale in transportation. However, the scattering of the data does not permit a better numerical approximation. In spite of these drawbacks I think that the fundamental information about rank is preserved.

TABLE 1 Energy transportability and generation-plant size[a].

Energy type	Transportability (km)	Technical maximum (km)	Size of generation plant (GW)
Hot water	≈ 2	50	0.2
Electricity	100	1000	1
H_2	1000	3000	100
Compressed air	2–3	10	10^{-3}
Adam and Eva[b]	20	200	0.04
Natural gas	1000	3000	100
Oil	10^4	10^4	2000[c]

[a] The table is an attempt to rank energy vectors by their ability to spread. The transportability indicator is estimated as the distance at which transportation costs are about 20% of value. The size of plant is estimated with reference to an electric plant, taken as unity, and assuming the same spatial density of consumption as for electricity.

[b] Adam and Eva is based on the process

$$CH_4 + H_2O \xrightleftharpoons{40\ \text{kcal}} CO + 3H_2$$

with the reaction running (endothermally) to the right at reactor level and (exothermally) to the left at consumer level.

[c] Possible production from a field.

As can be seen from the table, the poor transportability of hot water calls for small central generators, contrary to the tendency of reactors to be big due to the very high technical and systemic economies of scale. It is in fact only the existence of an electric network of sufficient density and transportability to accommodate 1-GW generators that has made nuclear energy a practical proposition.

The mean distance traveled by electricity in developed nations is around 100 km; this gives a reference dimension for the "unit cell", i.e. the area where, conceptually, electrical energy is produced and consumed.

The electrical system cannot improve substantially without a breakthrough in transportability, e.g. via a room-temperature superconductor. Its configuration and performances are therefore basically frozen except for the normal slow evolution linked to spatial intensification of use and the consequent possibility of going one step higher in transmission voltages.

Although transportability of the energy vector is the most important characteristic in determining the overall configuration of the system, other factors may contribute essentially to its efficiency and competitiveness. Storability of the energy vector comes first, and this is a weak point of electrical systems where storage per se is impossible. Consequently the whole system has to be geared to satisfy maximum demand, on line. The utilization factor is therefore low (around 50%) because of the large daily and seasonal variations in demand. A storable energy vector, like hydrogen or the Adam and Eva mixture of H_2 and CO, is obviously an advantage because it can raise the utilization factor of very capital-intensive structures to the technical limits.

My thesis is that system properties make hydrogen plus electricity the most probable intermediates or energy vectors for a system based on nuclear or fusion reactors as sources of primary energy. I will try to delineate the final configuration of such a system toward which the intermediate configuration will be more or less inevitably attracted.

Before proceeding and because my considerations are of a very general character, I will give some examples taken from life in order to illustrate how precise is the effect of system micro-optimization on the system's overall configuration. In Figure 1 the size of ethylene-production units is shown versus the size of the market served. Since ethylene is usually carried in railway tanks there is no economy of scale in transportation and consequently the configuration should not change with volume. In fact the size of the units grows in excellent syntony with market size. Figure 2 shows the same analysis for an electrical system, with similar results. In this case, however, there is an economy of scale in transportation if the time span examined allows for a step in high-voltage transmission. In fact, examining the development of the US electric grid since 1900, one sees electricity consumption doubling every 8–9 years or so and the size of generators doubling every 6–7 years. Because of continuous reoptimization, therefore, the number of generators actually *decreases* with increasing electrical consumption over a large area (e.g. a nation). Many other similar examples can be found. The mechanism operates with astonishing precision.

Let us come back to an energy system based on nuclear fission or fusion reactors and assume that hydrogen plus electricity are the vectors to distribute energy. What will be the configuration of the system? Hydrogen will most probably be distributed as a gas penetrating the infrastructure that methane is now creating (Figure 3). Also, trunk transportation will most probably be by pipeline. In this case the data of Table 1 will be valid with all their limitations, which do not, however, modify the outlines of the picture.

Table 1 shows a reference transportability for hydrogen of 1000 km, which is the same as that of natural gas. As a consequence of this, the "unit cell" served by a generating station, or a cluster of stations, is a hundred times larger than the cell of an electrical

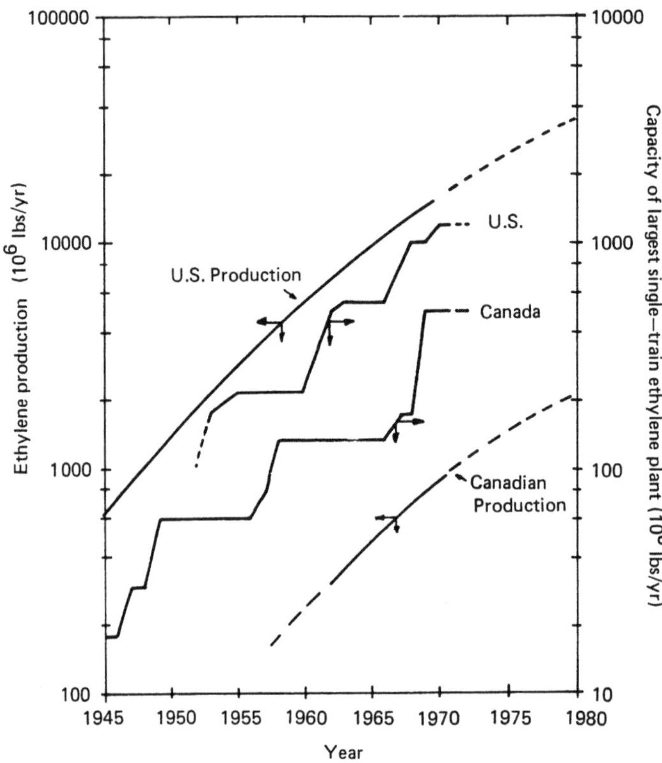

FIGURE 1 The size of the largest production units (stepped curves) and the size of the market (smoothed curves) for ethylene in the United States and in Canada. Technology is obviously "hunting" the market.

system which has a reference transportability of 100 km. The size of the generating stations will then optimize in the 100-GW output range.

To nuclear engineers this may appear staggering, but the history of industrial development shows that size (or better, capacity) has never really been a deterrent if we look at things in the proper time scale. After all Edison did christen his 100-lamp (10-kW) generator "Jumbo". With a pitiful lack of foresight he ran out of words to christen properly the 10^6-kW generators of today. A more modest example shows that oil tankers have increased in tonnage by a factor of 50 (Figure 4) in response to a proportional increase in oil traffic. Incidentally this is another case of system internal optimization at work.

With cells of 1000 km $\times$ 1000 km, continents will be served by a handful of generating sites, presumably located on the seashore to provide for the huge amounts of cooling water necessary. These stations could work round the year irrespective of demand, appropriate storage taking care of the matching of production and demand. Appropriate storage will be in geological structures, as is now largely the practice for natural gas. To give a reference example, the Gröningen structure, where most of the

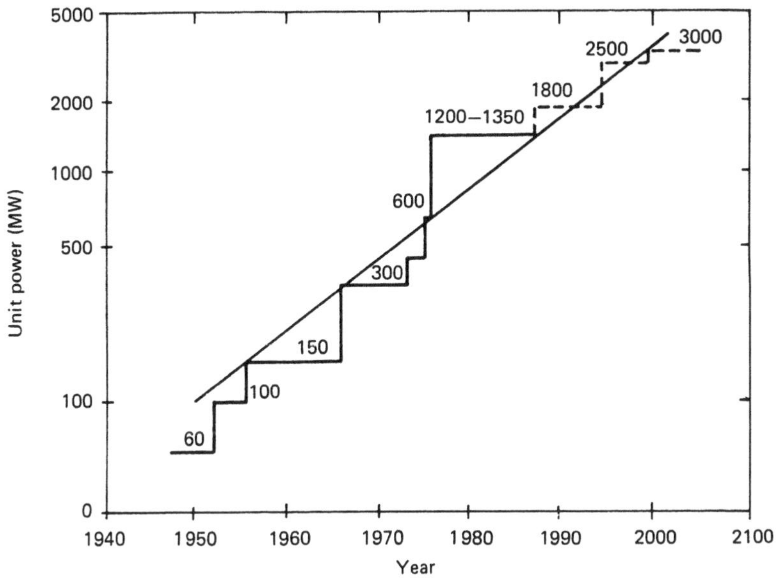

FIGURE 2 The relation between the size of electric generators and electrical energy consumption in the Federal Republic of Germany. For relatively short intervals, the size of generators is always proportional to consumption.

Dutch gas lies, if filled with hydrogen, would provide three years' energy for the whole of Western Europe, at the present level of energy consumption.

Such a cheap and practically unlimited storage capacity is of primary importance in determining the evolutionary advantage of a system, particularly when the capital investment in the generation area is very high, as in the case of fission (or fusion) reactors.

The selective forces of future energy systems, where manpower use will be extremely low are basically related to capital minimization; storage at various hierarchical levels will minimize investment not only in generation but also in transportation.

So far so good, if the energy vector is a gas with a transportability of 1000 km. However, if investigators find a way to produce a synthetic liquid fuel with a transportability similar to that of oil then the unit cell becomes the world. Let us suppose, to keep things conceptually simple, that this liquid is just liquid hydrogen carried in cryotankers, the way that Liquefied Natural Gas (LNG) is carried nowadays.

Since the transportability is an order of magnitude larger than that for gas, the system will coalesce into an order-of-magnitude larger generating sites. We call these generating sites *energy islands* and have tried to guess what they might look like. The most important result of this exercise is probably the fact that the machinery may well be constructable using basically present-day technologies.

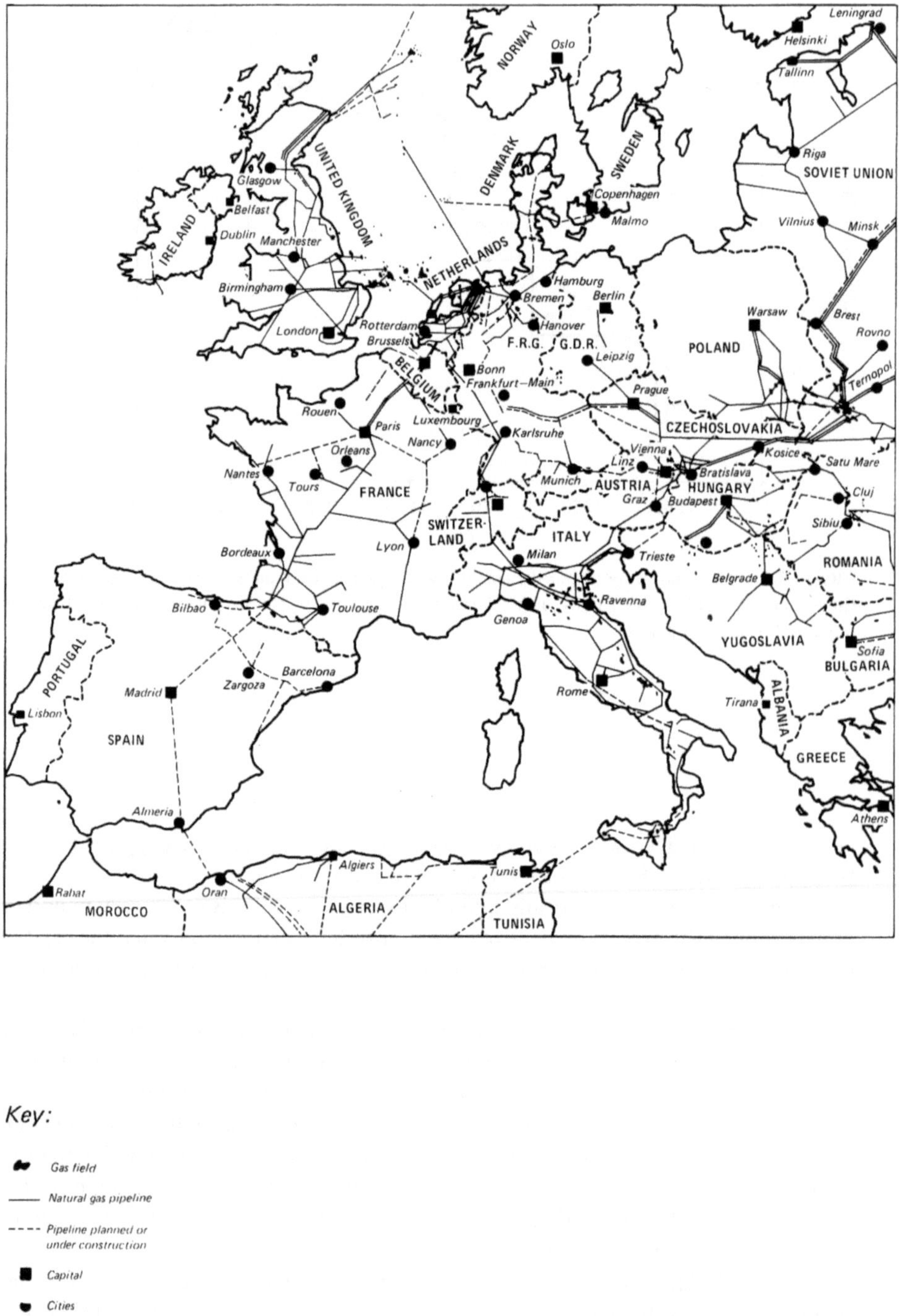

FIGURE 3 The European natural-gas grid. The present-day structure of the trunk-line and first-level gas-transportation system gives an idea of the potential of this grid, which is still in its infancy, to transport a synthetic gaseous fuel.

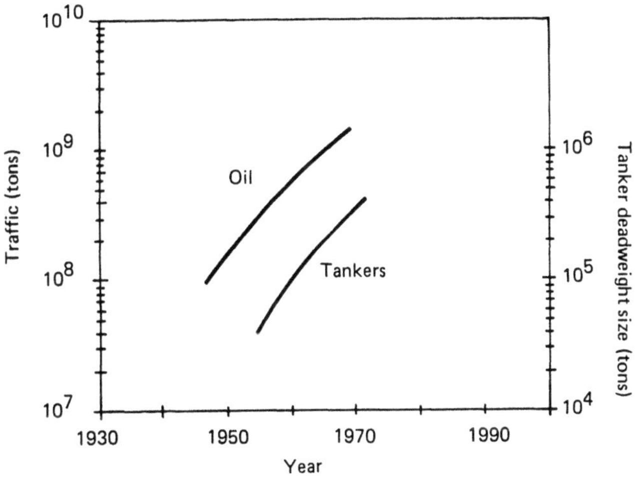

FIGURE 4 The evolutions of "largest-size" tankers and of the overseas oil market, showing a remarkable tuning. The number of tankers is actually to a large degree independent of the level of traffic.

2 BASIC OPTIONS

I will list here the basic options for our energy island together with some of the arguments supporting the choice. Since the time horizon is beyond the end of the century, technologies in full development now, such as that of LNG tankers, are assumed to be mature.

Installed Power: Approximately 1 TWth

The level of installed power comes from the present world energy consumption of approximately 8 TW and the systemic requirement that no more than 10% of the eggs be in the same basket. In case one island is shut off for some reason, the other nine should shoulder the extra load. This can be done by some overstretching of the plants and by the excess capacity that is always present in systems that grow by large increments. Since the cost of the hydrogen is due almost entirely to capital charges it will be a good tactic always to use all the capacity available and to store hydrogen in exhausted gas fields in the continents.

Reactor Type: High Temperature Gas-Cooled Reactor (HTGR), Pebble-Bed Type

Because of their high thermodynamic potential, heat from these reactors is the most suitable to operate chemical systems and in any case permits the highest efficiencies. The choice of the pebble-bed version stems from the opinion that it is more suitable for scale-up than the prismatic version.

Reactor Mix: HTGR Burners plus Breeders

The system can operate with HTGRs only. Breeders can be added to obtain a mean conversion ratio of unity. In this case the system does not grow but burns all the uranium and thorium introduced. The great systemic advantage would be that the necessary uranium could be extracted from the cooling water (i.e. seawater).

Reactor Size: 200-GWth Toroidal Vessel

With current diameters for the vessel, a diameter of 100 m for the torus, and the usual core power densities one can obtain the power indicated (Figure 5). The size of such a reactor is certainly mind-boggling to nuclear engineers, as would be the sight of a 1000-MW generator to Thomas Edison.

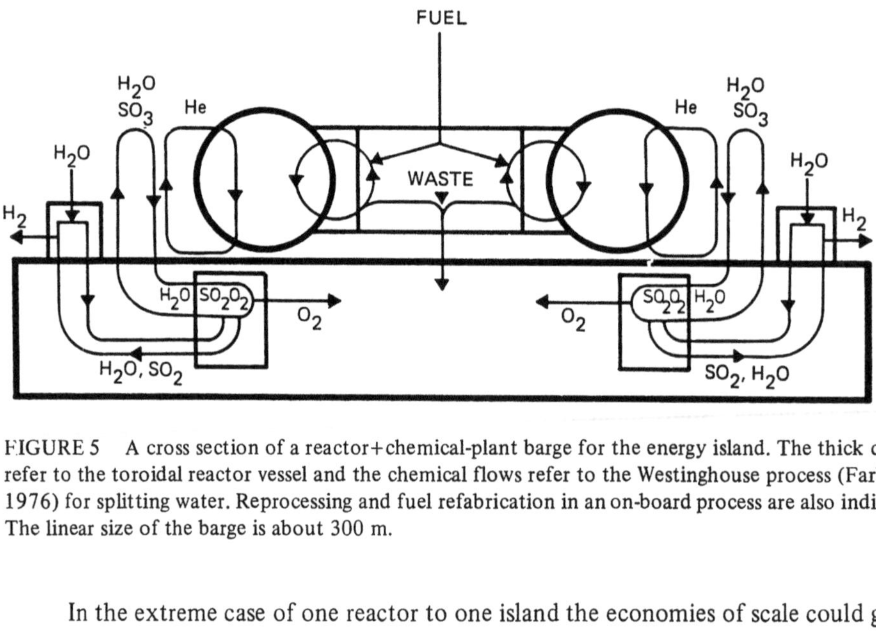

FIGURE 5 A cross section of a reactor+chemical-plant barge for the energy island. The thick circles refer to the toroidal reactor vessel and the chemical flows refer to the Westinghouse process (Farbman, 1976) for splitting water. Reprocessing and fuel refabrication in an on-board process are also indicated. The linear size of the barge is about 300 m.

In the extreme case of one reactor to one island the economies of scale could give a further reduction by a factor of 2 in the specific investments and consequently in the cost of the heat generated, but obviously the system would be much less flexible. Another reason for the preference for a reactor of this size is that it matches the economic scale of a reprocessing plant. The fuel cycle can then be completely contained in the reactor building.

Reactor Mounting: A Prestressed Concrete Barge

The concept of barge mounting has many advantages. The reactors can be built in a shipyard by a stable organization in a rich technological environment. The site can be changed following redeployment and reoptimization of the system. Final disposal of plants after the end of their useful life can be rationally organized at proper sites. The technology of large (prestressed) concrete barges is under rapid development, especially given the demand for offshore drilling for oil. They have expected lives of more than 50 years. The barges that I am talking about are in the displacement range of millions of tons.

Reference Site: An Equatorial Atoll (Canton Island)

The requirements for the site appear to be satisfactorily fulfilled by Canton Island (Figure 6): (a) the atoll provides shelter for the barges and sufficient draft in the lagoon; (b) deep cold waters for cooling purposes are easily accessible; (c) the lagoon is large enough (~ 10 km × 15 km) to accommodate complex infrastructure, possibly including the uranium-extraction plant, be it chemical or biological; (d) the area is outside the hurricane belt and atmospheric perturbations in general tend to be mild (the island, being located in a region of constant air subsidence, should have a desert climate); (e) the basalt core of the island is considered a suitable place for the final disposal of fission products and radioactive material (Figure 7).

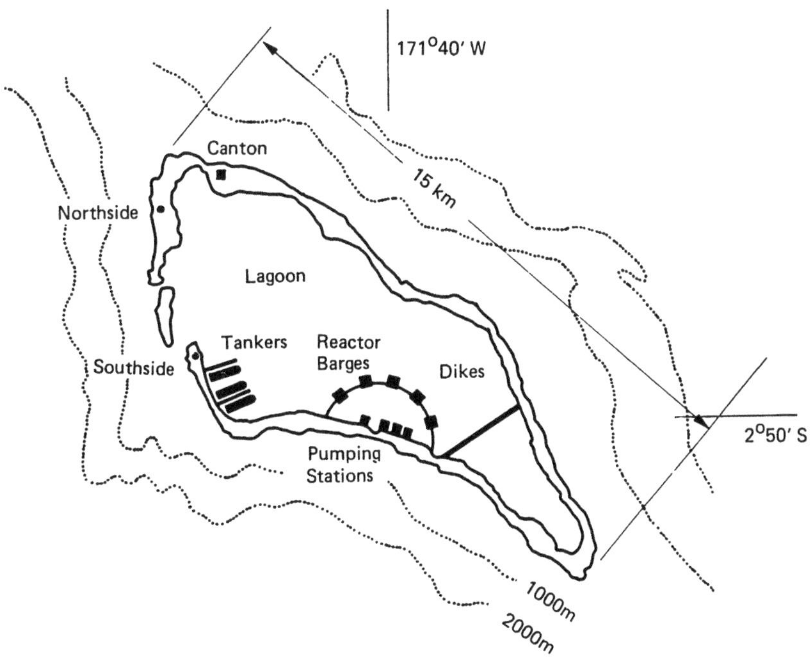

FIGURE 6 A plan view of Canton Island, with the installation drawn to scale. Each barge represents 200 GWth of primary energy. The island is located in the central equatorial Pacific (171° 40′ W, 2° 50′ S).

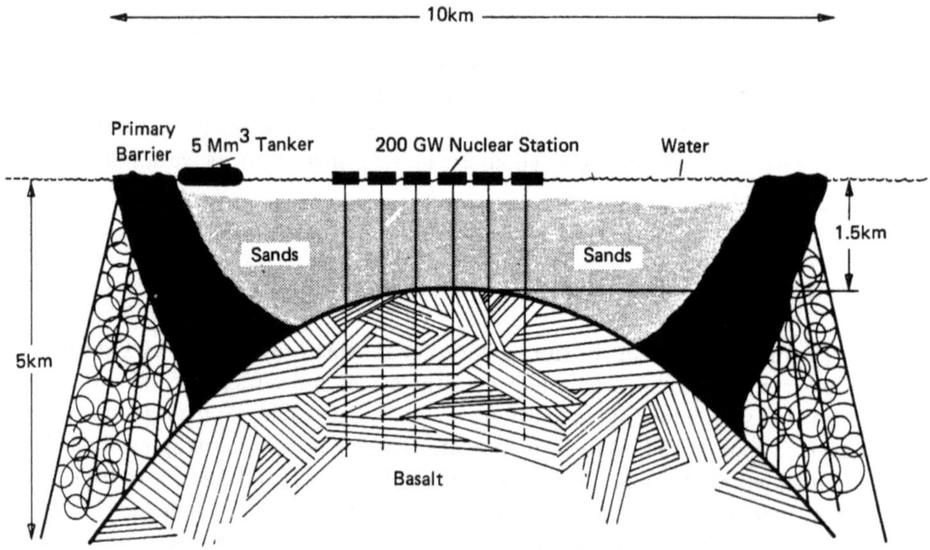

FIGURE 7 A scale cross section of Canton "Energy Island" with its presumed geological structure. The full lines under the power station barges represent permanent radioactive-waste sink holes and the broken extensions represent the free trajectories of the sinking capsules.

Waste Heat and the Cooling System: Using Water from the Deep

The system takes full advantage of the thermal gradient of the ocean, making a fine art of what is usually the brutal rejection of waste heat. By pumping water from the proper depth, temperature differentials with surface waters can reach 20° C in most equatorial regions and at Canton Island in particular. This means one can expel cooling water at surface temperatures or lower. If cooling water is expelled at surface temperature, no thermal plume – with its consequent potential damage of the biosphere – will appear. These waters will, however, be nutrient rich; this will generate intensive algae and fish growth, which are characteristic of upwelling. In the second case, where cooling water is expelled at temperatures lower than surface temperature, the water will sink to its equilibrium buoyancy level in the thermocline. The equatorial ocean currents will carry this water over long distances and the heat will be finally released to the atmosphere where strong winds and low temperatures thin out the thermocline.

This possibility of modulating time, geographical position, and rate of heat transfer to the atmosphere provides a rich interface with the work of meteorologists and climatologists, and many subtle ways for getting rid of the heat with a minimum of disturbance of the geosystem become possible.

The Fuel Cycle: Fully Contained in Reactor Building

The most important characteristic of the fuel cycle will be its complete containment in the reactor building. This is a natural consequence of the size of the reactor and has no economic penalty. The reprocessing and fuel fabrication should be mostly automatic, based on sol–gel and coated particles. We assume that carbon is also recycled. In this scheme fissionable material, including plutonium, is always mixed with radioactive products and never appears in pure form. Since the reactors are of the continuous-charge type, no large stocks of fuel will be present at any time. Theft of plutonium appears to be intrinsically very difficult as it means stealing radioactive fuels.

The Hydrogen Plant: Thermochemical Westinghouse Sulfur Cycle (Farbman, 1976)

The Westinghouse sulfur cycle, based on the thermal decomposition of sulfuric acid followed by the anodic oxidation of sulfurous acid, has been taken as the reference because of its great simplicity and promise (Figure 8).

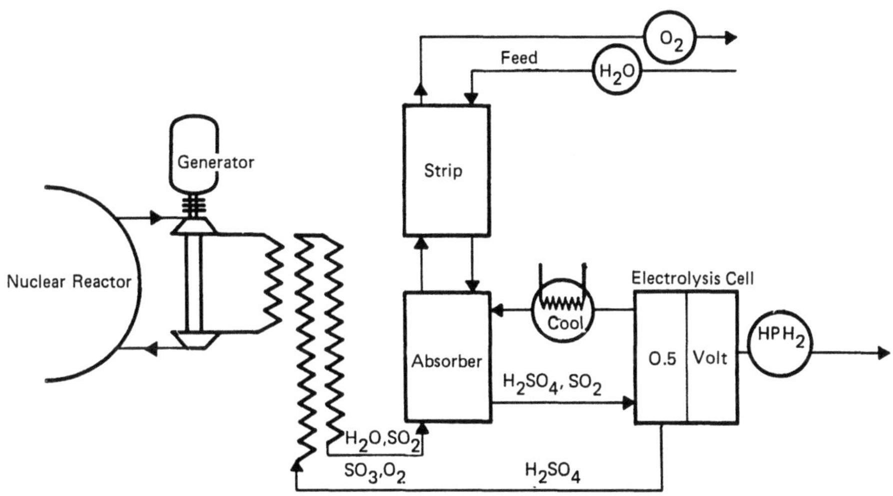

FIGURE 8 A possible block flow sheet for the realization of the Westinghouse sulfur cycle (Farbman, 1976). Sulfuric acid is decomposed thermally and SO_2 + SO_3 are stripped in an absorption column where the electrolyte of a cell is circulated. SO_2 is transformed into SO_3 electrolytically.

The ideal configuration would be that of a power cycle (using SO_3 and its decomposition products as working fluids) interfacing with the reactor. SO_2 could then be fed umbilically to the chemical plant (located on another barge) to be used in conjunction with electricity to produce hydrogen. In this way the chemical plant would be in a sense disconnected from the nuclear system, the interaction being through storable chemicals. It is probably appropriate to have the H_2-liquefaction plant associated with the chemical plant, with liquid hydrogen (LH_2) being transported to the ocean terminal by a pipeline.

Hydrogen Transportation: in LH_2 Tankers

The use of cryogenic tankers appears to be the best option to date. Cryogenic tankers with capacities of up to 150,000 m^3 have been built to transport LNG. If demand exists, rates of growth like that for oil tankers, which had a doubling time of six years for tonnage, may be technically feasible. This would produce a tanker of millions of cubic meters capacity in the late 1990s.

The volumes of LH_2 to be transported are very high, owing in part to the low density of LH_2 (≈ 0.07). A 1-TW island would produce approximately 10^9 m^3 of LH_2 per year. Assuming three tankers under charge all the time and a charging time of 24 h, this would call for 2-Mm^3 ships. This large size is not particularly distressing. The low density makes for low draft. The design which I prefer is that of a double-cylinder prestressed-concrete catamaran with the cargo partly stored in the hulls and partly on the deck. The low draft should permit access to many sites on all the continents (Figure 9).

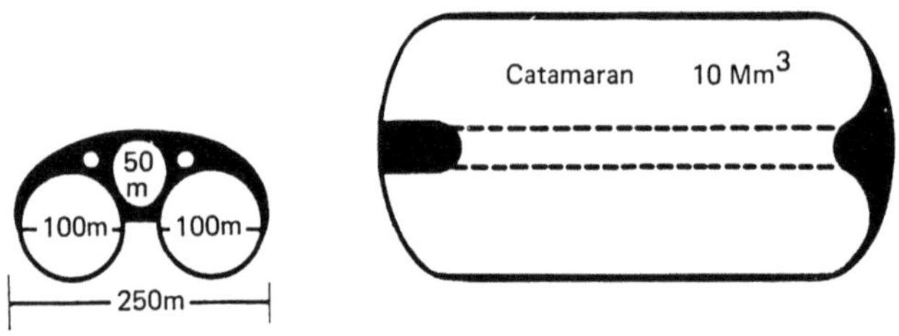

FIGURE 9 Vertical and horizontal cross sections of a possible LH_2 tanker.

Some consideration has been given to the case of an accident leading to spillage and hydrogen fire. Hydrogen fires are far less dangerous than oil fires because they last for very short times and because their flame is almost radiationless, radiation being the most important vehicle for flame propagation in oil and hydrocarbon fires.

Cursory attention has been given to the alternative transport of hydrogen by a special airship which carries it partly as gas but mostly as LH_2. At first sight the system does not appear particularly attractive.

Fuel Procurement: Extraction from Seawater

If the breeder–burner combination and the uranium extraction can be economically realized then the island would become a sort of bootstrap operation eliminating the ecological and political problems of uranium procurement. The uranium needed (500 t yr^{-1}) is about 20% of the uranium carried in by the cooling water.

Waste Disposal: Self-Sinking Capsules

After actinide separation and recycling, waste will be concentrated at levels of 10–20 vol.% of fission products. With a container of linear dimension of the order of a meter, or a bunch of smaller capsules thermally mimicking it (Figure 10), heat generated by the fission products is then capable of melting the ground in which the container is buried; if its density is higher, the container will sink for tens of years at initial rates of the orders of meters per day (Figure 11). The system automatically provides the necessary cooling by melting of the rock and the necessary sealing by solidification of the rock above the capsule. Each reactor on the island should have a disposal shaft drilled through the coral overburden and reaching the basaltic core of the island. The shaft will be instrumented and filled with salt so that descent can be monitored for the first year or two and the capsules are in principle recoverable during that time.

The system is intrinsically tamper-proof. Final nonretrievability is probably the best insurance against criminal use of fission products or inadvertant removal by future generations.

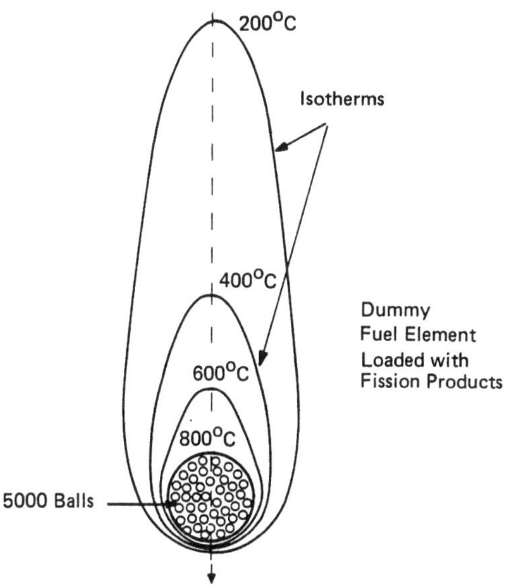

FIGURE 10 A "cooperative" sinking capsule, proposed because of the difficulty of building a large sphere capable of mechanically and chemically resisting the hard environment of molten rock under pressure: a large sphere is replaced by a large bunch of small spheres made, with the same technology, of the fuel elements of a pebble-bed nuclear reactor.

The Operation: Ad Hoc Multinational

The scale and complexity of the operation on the technical, commercial, financial, and political sides could well absorb the potential and the ambitions of an ad hoc

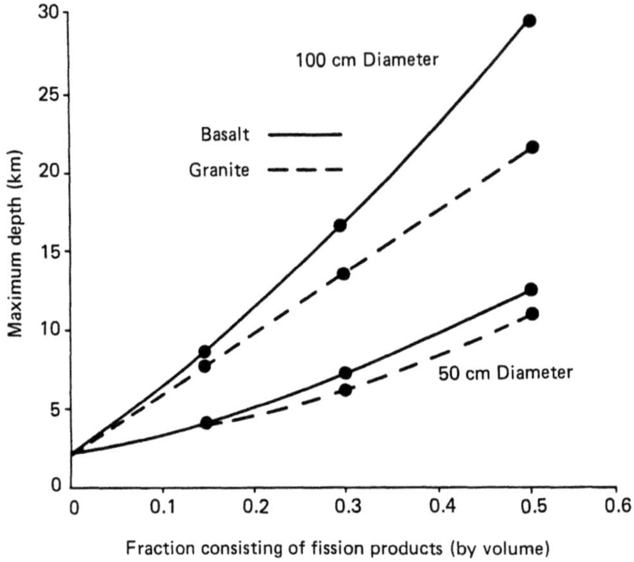

FIGURE 11 Capsules of various diameters, containing fission products of initial age six months, melt the surrounding rock by nuclear decay heat and thus sink. The final depth of penetration is shown as a function of the fraction by volume of fission products, for each combination of capsule diameter and surrounding rock.

multinational. Multinationality would come from the market served, from the source of finance and control, and from the staffing. It is an essential requirement for the final deployment of the scheme.

The model could be that of an oil company, fading somewhat into the image of a public utility. Investments of the order of \$100 billion are well within the capability of such organizations. The system will presumably be manned by engineers only, in the range of 1000 people. With an investment of \$100 million per man, salaries could be high, selection rigorous, and turnover brisk. With so restricted a staff, however, most of the repairs and maintenance would have to be made by substitution of plug-in components.

3 CONCLUSIONS

The options listed make a self-consistent set and provide the basis for solving to a large degree the many problems related to an ever-increasing use of energy. What I was aiming at is only a topological description of the system.

An obvious question is: What will the energy finally cost at the consumer level? For that I made a parametric analysis of the system and reached the very interesting result that the economies of scale will compensate for all the complexities of the transformations and transportation involved. In other words, a landed kilocalorie of LH_2 will cost the same as the *thermal* kilocalorie out of a land reactor of the gigawatt size. Thus the mobility, storability, and flexibility of hydrogen fuel is formally acquired at zero incremental cost with respect to the base heat by properly redeploying the structure of the system.

REFERENCE

Farbman, G.H. (1976). Studies on the use of heat from high temperature nuclear sources for hydrogen production processes. NASA: CR-134918. NASA Lewis Research Center.

Section Two

Fission and Fission Breeding

SELF-SUSTAINING SYSTEMS OF REACTORS

Bernard I. Spinrad
Department of Nuclear Engineering, Oregon State University, Corvallis, Oregon (USA)

ABSTRACT

Nuclear fission is capable of making a very large contribution to world energy needs. This capability rests on the assumption that the nuclear fuel supply can be sustained by using a mixture of reactor types (mainly advanced converters and breeders). In this way an inventory of available fissile material can be maintained with almost trivial inputs of freshly-mined uranium and thorium. To use the nuclear energy to its best advantage, a diversity of end uses should also be considered, including the following: central-station electricity and process-heat parks; dispersed electrical stations to serve dispersed major electricity consumers such as urban centers and large, fixed-site industries; process-heat sources for secondary fuel production from coal or biomass; total energy systems for remote settlements; and marine, space, and other special uses. A mixture of high-temperature gas-cooled reactors, liquid-metal-cooled fast breeders, light-water reactors, and various "pure burner" special reactors is explored for achieving the twin goals of fuel self-sufficiency and diversity of end use, and it appears to be capable of doing the job.

1 INTRODUCTION

Nuclear power will have a long run of utility if and only if it can function asymptotically using economically superabundant resources. The heavy elements uranium and thorium represent such superabundant resources. Effectively infinite quantities of these elements are contained in low-grade deposits such as shales and granites.

The key to using these materials is to achieve their fissioning almost quantitatively. Only then would the energy obtained from them cover the costs (including energy inputs) of recovering very dilute materials from nature at a generous enough multiplication of value to make the scheme economically feasible. One method for achieving this quantitative fissioning is by the use of breeder reactors. Such reactors first transmute the elements into fissile nuclides (i.e. nuclides that easily undergo fission) and then use the new fissile material as fuel. Enough extra fissile material is made during this process to permit the fueling of other reactors as well.

There are other ways to transmute the originally "fertile" nuclides of the heavy elements into fissile nuclides. ("Fertile" is the adjective used to describe nuclides which become, or quickly decay into, fissile nuclides after they have captured a neutron.) One could use the blankets of fusion reactors that produce copious supplies of neutrons from the D–T fusion reaction or one could manufacture neutrons for the transmutation in an accelerator. At present, though, the breeder reactor is by far the cheapest way of doing the job, and it will probably remain so.

This paper explores the possibility that breeder reactors win the competition that is implied in the foregoing. What, then, would a system of fission reactors producing a large part of the world's commercial heat and electricity supplies be like?

2 SYSTEM DEMANDS – TYPES OF POWER PLANT

Today's nuclear reactors serve two principal purposes: generation of electricity and propulsion of naval vessels. If a large fraction of the world's energy were to come from fission, however, it would be necessary to match more carefully the nuclear system to end uses of energy. Therefore the fission system should provide propulsion capability for cargo shipping, high-temperature heat for large chemical and metallurgical plants, lower-temperature heat for miscellaneous industrial and domestic uses, electricity, and even power and propulsion capability for spacecraft.

Such a broad spectrum of demands can only be satisfied by a variety of nuclear power plants. Such dual-purpose systems can be contemplated within the package, of course; thus one can conceive of central electricity stations as furnishing relatively large amounts of low-quality heat as well. However, only a fraction of the potential heat supply could be accepted in the vicinity of the power plant, and it is not economic to transmit low-quality heat over long distances. An ideal match might be between nuclear power plants and seawater-desalination units or between nuclear power plants and large chemical plants. We must, however, plan for a considerable demand for dispersed heating supplies.

Then we must consider satisfying the electricity *and* heating demands of isolated communities and industries. It turns out that the type of reactor that one might envisage for this purpose is not too dissimilar from the type that might be attractive for shipping.

Finally we have to leave room in our thinking for the very special-purpose types of nuclear system, which spaceship reactors might symbolize. In summary, our system must consist of large and small reactors, centralized and dispersed reactors, hot and merely warm reactors.

For these purposes a variety of reactor system types have been proposed. Most of them are feasible, but to keep our considerations finite I will limit further discussion to just three reactor types: light water reactors (LWRs), liquid-metal (sodium) cooled fast breeder reactors (LMFBRs) and high-temperature gas-cooled graphite-moderated reactors (HTGRs). Additionally I will add a category of "pure burners" of unknown design – thermal reactors that use fuel whose fissile isotope content is so concentrated that regeneration of fissile material is inconsequential.

Lest my purposes be misunderstood, it must be stressed that other reactor systems such as the already commercial heavy-water reactors (HWRs) and the intriguing molten-salt reactors (MSRs) will undoubtedly be a part of a future mix of reactor types according to their technical and economic merits and their adaptability to types of energy demands.

3 ROLES OF REACTOR TYPES

Based on their particular properties, one can estimate the probable uses of the different reactor types.

3.1 LMFBRs

FBRs will be used primarily for the generation of large amounts of electricity in central stations. This generalization arises from several properties of LMFBRs. (a) They are complicated many-component systems that stand to gain more than simpler reactors from economies of scale. (b) As they have thin-shelled primary-system containers – a characteristic which arises from the absence of high pressure in the primary system – scaling up is technically more feasible for these reactors than for other systems. (c) As reactors which make large demands on associated fuel-cycle facilities, particularly on fuel fabrication and reprocessing plants, they are the first choice for location within nuclear fuel-cycle "parks". (d) As reactor systems which, along with their associated nuclear fuel-cycle plants, turn over a large throughput of quite concentrated "weapons-usable" material, the safeguarding of this material is of prime importance; physical security and ease of inspection measures favor, again, the location of breeders in industrial parks.

Breeder reactors will probably raise steam at temperatures in the range 500–600° C. This is a sufficiently high temperature range to permit efficient generation of electricity; conversion efficiencies might be 37–42%. The temperature is somewhat below that which is of interest to high-temperature processing industries. Moreover, the centralized location would restrict the access of industries to direct heat. Small amounts of steam might be furnished as a by-product to the fuel-cycle facilities on the site but the basic product of these reactors and reactor parks would be electricity. Four types of "customers" could be served.

(1) General electricity grids. Although we envisage nuclear parks as being in "remote" sites, remoteness is a relative matter. A distance of 100 km from large population centers is remote both from the point of view of population safety against reactor accidents and from the point of view that a guarded site ought not to intrude on people's everyday activities. Yet, such a distance is small enough for it to be eminently feasible to transmit electricity into urban grids with only small losses.

(2) Existing electricity-intensive industries. The manufacture of reactive metals such as aluminum, magnesium, calcium, and sodium is done electrolytically, and these products all have a good growth potential. As natural gas becomes scarcer, the incentive to produce hydrogen electrolytically for chemical processes such as the Haber process for ammonia becomes stronger.

(3) Production of pipeline hydrogen by electrolysis. Hydrogen is a viable substitute for natural gas in the gas grid.

(4) Where location is conveniently close to sources of coal, electrolytic hydrogen might be specifically used for converting coal to usable methane (gas fuel) or methanol (liquid fuel).

3.2 LWRs

LWRs would be used mainly in mobile or dispersed settings. Asymptotically, they are likely to be typical "small" reactors if we interpret the word "small" generously to cover thermal powers between about 10 and 2000 MW. Properties of LWRs that lead to this extrapolation are as follows. (a) They tend to be simpler systems than other reactors, which is probably why they were the first type to be commercialized; simpler systems tend to be less penalized by diseconomies of small scale. (b) Their fuel, be it low-enriched uranium or even ^{233}U-enriched or plutonium-enriched uranium, is much less attractive to those interested in diverting material for use in weapons than is LMFBR fuel; this means that fuel shipment is much more feasible for these reactors. (c) They can be made relatively compact, which is one reason why they have been accepted for ship propulsion; extra engineered safety, including if necessary such special features as underground siting, can most easily be added to these reactors, and this would permit their location close to centers of demand.

Uses of LWRs could be of several types.

(1) As basic energy systems for islands and other locations that cannot be easily connected to large central grids. They could be used both for electrical generation and for the provision of district heat.

(2) As integrated energy systems for individual industries that require large quantities of low-grade heat. Food processing is one such industry, and there are a variety of chemical products that are made using steam at temperatures below 400°C. Electricity could be cogenerated in these systems.

(3) There are a number of potential uses of locally generated steam in extractive industries. The Frasch process for sulfur is a well-known prototype. Interesting possibilities exist in heavy-oil extraction, in secondary and tertiary oil recovery from oil fields, and elsewhere in the liquid-fuel industry. It is not beyond the bounds of possibility that steam produced by nuclear reactors could be used to generate town gas in situ from underground deposits of coal.

(4) The oil "crisis" spawned the supertanker and with it, the commercial feasibility of nuclear propulsion of ships. The world of the future is likely to see super-cargo-ships of other types, carrying iron ore, bauxite, and similar resources. An affluent world might see the revival of large passenger liners for travelers who have more leisure time. All these uses are well satisfied by LWRs designed as propulsion units.

(5) The industrial development of small countries creates a demand for medium- and low-power reactors, which industrialized countries have not needed. Only after large regional power grids have been established does the setting up of energy parks become feasible. During the energy-buildup period, LWRs are a suitable electricity source.

3.3 HTGRs

HTGRs represent both a preferred system for process chemistry and a nuclear compromise between LMFBRs and LWRs. The properties that support these conclusions are as follows. (a) HTGRs can generate heat at temperatures well above 600° C; moreover, the carrier of this heat is an inert gas, helium, which does not introduce new corrosion problems in processes using this heat. (b) HTGRs seem to have many safety advantages and in particular they are relatively tolerant to certain loss-of-coolant accidents. (c) While it does not seem to be commercially feasible to design HTGRs as breeders, they can be designed as relatively efficient converter reactors; i.e. it appears to be feasible to design HTGRs to manufacture 0.8–0.9 atoms of new fuel per atom of old fuel destroyed, whereas comparable numbers for LWRs are 0.6–0.8. Based on these characteristics, HTGRs have the following apparent uses.

(1) As process-heat sources for many chemical, refining, and even metallurgical industries, with electrical cogeneration as a secondary product.

(2) As a potential source of a new type of energy grid system based on the "Adam and Eva" process. In this process methane and steam are endothermally reacted at about 650° C to produce carbon monoxide and hydrogen. These products are piped into a grid. Users of the grid can catalytically combine the CO and H_2 in an exothermic reaction to reform methane and steam at about 450° C. The methane is piped back to the heat source after its heat content has been extracted.

(3) A particular use of great importance would be to achieve the production of hydrogen from water by a process that is, at least predominantly, thermochemical. Several such processes are mentioned in the literature. This hydrogen could be used for the "hydrogen economy" in the same way as electrolytic hydrogen from LMFBRs. What makes HTGR hydrogen important is that these reactors could be sited at the mine wherever a remote coal deposit exists. Local hydrogen would permit the on-site production of liquid and gaseous fuels, which would make such coal "transportable" and would simultaneously avoid the export of pollutants.

(4) To the extent that we must be concerned with the fissile-material economy of the whole nuclear system, HTGRs will tend to displace LWRs as general energy systems for the larger remote centers of demand.

3.4 Pure Burner Reactors

Pure burner reactors are reactors whose fissile material is *not* replaced by conversion of fertile nuclides to fissile nuclides during operation. Their fuel would be primarily fissile nuclides (^{233}U, ^{235}U, ^{239}Pu) diluted with whatever other materials best serve the reactors' applications. For example, for many research reactors a higher availability of neutrons for experimental purposes can be achieved by using highly enriched ^{235}U only, in conjunction with aluminum, zirconium, carbon, heavy water, or some other low-neutron-absorbing material. Space-propulsion reactors made primarily of graphite or space-electric systems using refractory metals are other examples. Use of pure burners is indicated whenever some extreme of performance is to be achieved and when that performance would be compromised by inclusion of fertile materials. Possible applications include those already mentioned and other special-purpose uses that may be developed. Any revival of the "plowshare" type of application, i.e. the use of nuclear explosives for civil operations of earth moving and rock fracturing, would fall into this category of operation, as would specialized power sources for seafloor exploration and mining, space bases, and other remote installations posing unusual environmental conditions.

4 A REACTOR SCENARIO

As was done in "Energy in a Finite World – A Global Systems Analysis" (Häfele et al., 1981), in a "multipurpose strategy" we look for a nuclear system to meet the desired capability using a mix of reactor types. The selected system was there described as equivalent to 10 TWe of electrical capacity, featuring 5000 GWe of HTGRs, 3500 GWe of (LM)FBRs, 1000 GWe of LWRs, and 500 GWe of pure burner reactors.

However, this description is inadequate. We are describing a multipurpose energy system in today's terms, and today nuclear energy is thought of only as a source of electricity. We have to decompose these numbers back into annual energy outputs, keeping in mind the different thermal qualities and capacity factors of the various types. When this is done the balance becomes, in *thermal* output, 8750 GWyr yr^{-1} from HTGRs, 6125 GWyr yr^{-1} from FBRs, 1450 GWyr yr^{-1} from LWRs, and 515 GWyr yr^{-1} from pure burners. The sum is 16,900 GWyr yr^{-1} (rounded off), which may be further rounded to yield the 17 TWyr yr^{-1} mentioned in "Energy in a Finite World" (Häfele et al., 1981), representing the actual thermal output of the nominal 10 TWe installed capacity.

In accordance with our description of the role of these reactors we can now describe their settings and useful outputs.

4.1 FBRs

The FBRs would be located in approximately 200 energy and nuclear fuel-cycle parks deployed throughout the world – approximately one such park per 40 million (40,000,000) people. Recall that we are talking of an asymptotic world population of 8 billion people. The FBRs would generate about 2400 GWyr yr^{-1} of electricity, i.e. about 0.6 kWyr yr^{-1} per person served. This would be the baseload of the world electrical system. Associated with each park would be the nuclear fuel processing facilities for *all* the nuclear activity demanded by the energy needs of the associated population.

There might be some dedication of electrical energy to hydrogen production from FBRs but it would be likely to be only a small fraction of the energy generated. A clustering of electricity-intensive industries, particularly those dealing with electrometallic reduction, can be conceived as associated with these energy parks.

4.2 HTGRs

The basic role of HTGRs would seem to be in the generation of hydrogen for chemical-conversion industries, including the manufacture of fluid fuels from coal and/or biomass, the production of ammonia, and petrochemical organic syntheses. Accordingly, one would expect HTGRs to be deployed in energy parks close to sources of raw materials, particularly coal deposits and cultivated forests. About 120 of these parks producing about 30 GWyr yr^{-1} each of high-temperature heat could conceivably be located in such areas. It is difficult to project the split between thermolytic processes and electrolytic processes for hydrogen production, so the fraction of the energy converted to electricity must be left open. What is significant about this segment of a world nuclear enterprise is that it is dedicated energy production; the entire output is to be used in the process industries that are served.

This leaves about 2500 GWyr yr^{-1} of HTGR installation for distributed use. A mix of applications would be called for: as localized sources of grid energy, particularly district heat via the "Adam and Eva" process or some equivalent method; as energy sources for smaller industrial complexes that utilize high-temperature heat; as more dispersed sources of electricity for urban conglomerations (in this case power stations would supply both grid energy and satellite industrial heat, as in the previous applications mentioned; as electrical stations, they would play an important role in balancing the electrical grid).

4.3 LWRs

Although we are used to thinking of LWRs as being primarily sources of electricity for highly industrialized regions of the world and of their being deployed in large sizes in energy parks, I am convinced that their long-term role will be as dispersed reactors in smaller sizes that are suitable for ship propulsion, total energy supply of areas that cannot be serviced by large grids, and related uses. Some fraction of LWRs, particularly those that remain serviceable after the present phase of nuclear reactor deployment is completed, will continue to be used for the balancing of electrical loads in regions of high demand. For the long term, however, the future of LWRs must reside in the smaller sizes that are suitable for ship propulsion (where, to paraphrase Hyman Rickover, a water reactor is in its element) or for the provision of electricity and district heat to island cities, jungle cities, arctic cities, or floating cities.

4.4 Pure Burner Reactors

I am a space enthusiast, and I am myself quite willing to envisage the entire 515 GWyr yr^{-1} allocated to burner reactors as used in space – for thermal propulsion (ROVER), for electrical propulsion, and, particularly, for planetary base operations. This last application is particularly amusing to me, for I see good possibilities of using nuclear energy as a bootstrapping source of energy for the ultimate conversion of planetary bases to solar energy. In brief, I consider that the development of planetary bases will have two phases: first, using nuclear energy which is compact and easily supplied to build a manufacturing capability; second, using a part of that capability to construct a solar energy supply system.

Of course, as we proceed outward through the solar system during the approaching third millenium of the Common Era, sunlight will become less intense and harder to tap for energy; so just about the time we are decommissioning our nuclear plants on Luna and Mars we may be commissioning new, and permanent, ones on Ganymede or Titan!

5 THE FISSILE-MATERIAL INVENTORY – GROSS ESTIMATE

A large part of Chapter 4 of "Energy in a Finite World" was devoted to exploring the terms under which a nuclear system of the type just described could be deployed without requiring continued large demands of nuclear fuel from nature. The matter at issue is not the amount of nuclear fuel being burned: 17 TW of power are produced by the fission of less than 20 t d^{-1} of heavy elements. The problem is that we must consider asymptotically obtaining this material from rocks where its concentration may be, say, 50 ppm. Even this is not much: we are talking of mining and processing 400,000 t d^{-1} of rock to provide a large part of an expanded fuel supply for the whole world. For comparison, we are mining over 2,000,000 t d^{-1} of coal in the United States now. What is to be avoided is inefficient use of the heavy elements mined. If, as with LWRs today, we fission less than 1% of the uranium mined then we would need to mine about 50,000,000 t d^{-1} of rock to support the "asymptotic" nuclear industry, and this is unreasonable, environmentally and economically.

The difference between LWRs and the asymptotic nuclear industry is that today's LWRs burn the equivalent of the fissile ^{235}U content only of the natural uranium. Even if we recycled the spent fuel, which contains usable fissile plutonium isotopes, over and over again, we could achieve only about 2% fissioning of the uranium mined. Then fresh natural uranium must be brought in to supply more ^{235}U to start the process over.

The key factor is to use the rest of the uranium – the other 98%, which is entirely composed of the fertile isotope ^{238}U. The way to achieve this is to use ^{238}U as a source material to make fissile plutonium. This is not so hard to do. A neutron chain reaction produces many more neutrons than are needed to keep the chain going. The extra neutrons can be reacted with ^{238}U to form plutonium. An alternative system would be to react the neutrons with natural thorium to produce the easily fissioned (i.e. fissile) ^{233}U. Thorium is an even more abundant element in the earth's crust than uranium, and ^{233}U is a better fuel than plutonium for thermal reactors.

Once ^{238}U or thorium has been converted into a fissile atom it can of course be fissioned as a nuclear fuel. If the conversion of fertile atoms to fissile atoms occurs as rapidly as the fissile atoms are destroyed, the system will never need to be resupplied with fissile atoms. The fertile materials then become the resource that must be supplied. However, these are the abundant atoms: over 99% of all uranium atoms and 100% of all thorium atoms are fertile.

Some reactors, e.g. LWRs of present design, have only about one-half neutron to spare per fission and their fissile-material content runs down quite rapidly. Advanced converter reactors run down slowly, and breeder reactors actually convert more fertile atoms than they burn fissile atoms. This excess from breeders can then be fed back into the converter reactors to retain a critical mass for the whole system. It is the retention of this critical mass that permits us to consider, as fuel, the fertile material that we mine.

The principles by which this balance may be calculated are well known. The crude description that follows is only approximate but serves for qualitative understanding. We characterize each reactor type by its fissile-material content and its conversion ratio CR. The rate at which fissile material is lost is roughly proportional to its power multiplied by $(1 - CR)$. Thus for a system of reactors to retain a constant fissile-material content it is necessary to have a zero sum:

$$\sum_i P_i(1 - CR_i) = 0 \tag{1}$$

where the subscript i refers to reactor type i and P is power. On an annual basis we replace power P_i, by annual energy production E_i.

The summation of eqn. (1) can only be zero if at least one of the CR_i values is greater than 1. For such a reactor type we have breeding, and a conversion ratio greater than 1 is known as a breeding ratio (BR). If we have just one component, say component 1, as a breeder, we can rearrange eqn. (1) to obtain

$$E_1(BR_1 - 1) = \sum_{i>1} E_i(1 - CR_i) \tag{2}$$

Equation (2) expresses the point we made previously. The left-hand side, $E_1(BR_1 - 1)$, represents the rate at which excess fissile material is produced by breeders and the right-hand side represents the net consumption rate of fissile material in converters.

Equation (2) represents an absolute condition on a self-sustaining system of reactors – absolute, that is, when system- and fuel-specific multipliers of order of magnitude unity are associated with the E_i. We can therefore obtain a set of E_i that are feasible for the CR_i with which we started. The system of FBRs, HTGRs, LWRs, and pure burners that has been postulated meets this characteristic. The pertinent numbers that were assumed are listed in Table 1.

TABLE 1 Reactor parameters of a steady-state scenario.

Reactor type	Annual energy rate (GWth)	Conversion ratio
FBR	6125	1.25
HTGR	8750	0.90
LWR	1450	0.8
Burner	515	0

Of the parameters in Table 1, the only number that is exceptional in the light of what has been already discussed is the conversion ratio of LWRs. As compared with the performance of today's LWRs, with conversion ratios of 0.5–0.6, the LWRs in asymptotic use would be expected to be designed with less parasitic neutron absorption than is the case now. This would be a natural design direction as we move from today's conditions, under which fissile atoms are cheap, to asymptotic conditions under which fissile atoms are expensive. Also, asymptotically the fissile atoms in LWRs would be ^{233}U whereas today we use the less efficient ^{235}U. Incidentally, it is the annual energy rate rather than the total installed electrical capacity that should be used in working out the fissile-material balance.

Once we have a balanced system such as that exhibited in Table 1 we can work out its "critical mass" in the following way.

First we associate a specific inventory (in kilograms of fissile material per KWth) with each reactor type. Next we determine how long the fuel stays in the reactor under assumed limits of burnup (a measure of how much energy is extracted from the fuel before isotopic changes and physical deterioration require it to be replaced) and of capacity factor. Then we consider how long it takes to "turn around" fuel between unloading the reactor and putting refurbished fuel back, as compared to how long the fuel stays in the reactor. This cycle time for cooling, reprocessing, and refabricating is called T_c, whereas the time it spends being irradiated in the reactor is called T_{ir}. The factor $(1 + T_c/T_{ir})$ tells us how much fuel is committed to the system per unit of fuel actually located in reactors.

TABLE 2 Factors affecting the fissile inventory.

Reactor type	Burnup (FIFA)[a]	Capacity factor	Specific inventory ($g\ kWth^{-1}$)[b]	T_{ir} (d)	T_c (d)[c]
FBR	0.40	0.70	1.91	550	1500
HTGR	1.00	0.70	2.04	1950	1500
LWR	1.00	0.45	2.32	2180	1000
Burner	0.40	0.33	4.88	1950	1000

[a] The units of FIFA (Fissions per Initial Fissile Atom) are dimensionless: atoms fissioned per atom of fissile material originally loaded.
[b] The units are grams of fissile material loaded per kilowatt of thermal power.
[c] The out-of-reactor times are considered to be rather long (see text).

Pertinent numbers for this calculation are listed in Table 2 for our suggested asymptotic system. The numbers listed in the table are not those used (Spinrad, 1979) in Chapter 4 of "Energy in a Finite World". In particular, the out-of-reactor times are considerably greater, reflecting my own later judgement that reactor fuels will be left to cool for an appreciable length of time before reprocessing.

These numbers can then be used as just described to yield the "endowments" of fissile material that could be expected to yield 17 $TWyr\ yr^{-1}$ of nuclear power. These endowments are of fissile plutonium for the FBR and of ^{233}U for the other three reactor types. They are listed in Table 3.

TABLE 3 Fissile-material endowments for an asymptotic capability of 17 $TWyr\ yr^{-1}$.

Reactor type	Fuel isotope	Capacity ($TWyr\ yr^{-1}$)	Fissile endowment (t)
FBR	^{239}Pu	6.125	43,500
HTGR	^{233}U	8.750	31,600
LWR	^{233}U	1.450	4,900
Burner	^{233}U	0.515	3,800
Total			83,800

To complete the picture presented here we can attempt to envisage the amount of ^{235}U and of natural uranium that is represented by this endowment. To do this we assume that (1) it takes 1.2 kg of ^{235}U in an FBR, recycled indefinitely, to achieve an asymptotic 1.0 kg of ^{239}Pu, (2) it takes, on the average, insertion of 1.3 kg of ^{235}U into the thermal reactor system to achieve an asymptotic 1.0 kg of ^{233}U, and (3) all the ^{235}U that is mined is ultimately used in the system (this requires recycling of any enrichment-plant "tails" as blanket material for FBRs or diluent for ^{233}U in thermal reactors).

This then makes a total endowment of about 105,000 tonnes of ^{235}U, contained in about 14.5 million tonnes of natural uranium.

6 THE FISSILE-MATERIAL INVENTORY – ADJUSTMENTS AND CONSIDERATIONS

Our previously published results (Häfele et al., 1981; Spinrad, 1979) indicate that the scenario that we envisage here could be achieved in an orderly schedule with ultimate consumption of 12.5 million tonnes of natural uranium. There are differences between the methods used in arriving at these two numbers so that the relatively good agreement is partly an accident, although the two results tend to confirm the order of magnitude of the resource to be supplied. In particular, the present estimate envisages leaping directly into the asymptotic system whereas the previously published results move gradually toward it. There are two consequences of the simplified way of estimating requirements which seem roughly to cancel. (1) From a uranium-utilization viewpoint it is more efficient to invest highly enriched uranium in reactors of good neutron efficiency in order to achieve the ultimate inventory than to use uranium in precursor reactors (e.g. thermal reactors fueled with slightly enriched uranium) that are less efficient. (2) A scenario which permits a lengthy buildup time before it reaches its asymptote can be scheduled for relatively early buildup of breeder reactors. This scheduling permits breeders to increase the fissile inventory of the world before the asymptote is reached, thus reducing the initial investment.

As noted in the previous section, I have considerably increased the out-of-reactor inventories that are to be carried, particularly for the breeder-reactor fuel cycle. This would be considered pessimistic by today's standards although it must be noted that these standards are of expectation rather than practice. Clearly, if we can and do attain reactor systems of low specific inventory and fuel cycles of short turnaround time then the required fissile inventory will be well below 100,000 t.

7 SOME IMPORTANT CONSIDERATIONS

The picture just painted is quite conservative with respect to fuel inventories. However, in one sense today's reactor-fuel-cycle technologists would consider it to be extremely optimistic. This is because losses in the fuel cycle have been ignored.

We have a set of existing practices that I do not understand. We routinely recover uranium from ores at around 95% yield when the ore contains less than 1/10 of 1% (one part in a thousand) of uranium. Yet it is present practice to expect losses of the order of one or a few percent of the heavy material in such processes as fuel fabrication and fuel reprocessing. This results in our rejecting, as "waste", material that is enriched, material that contains plutonium, and so on. This waste has a far higher uranium concentration than the original ore! To illustrate how wasteful even a little such waste is, it is normally estimated that, since we recycle fuel many times in a breeder system before it is consumed, the cumulative waste from processes that are about 98% efficient amounts to half of the heavy metal originally extracted from nature.

After this explanation, it is evident that an extremely important condition for the asymptotic nuclear fuel cycle is that it shall be very efficient. It is necessary that 99.9% of the heavy metal unloaded from reactors should be returned to reactors under the asymptotic system. This would permit a very respectable 95% utilization of the uranium and thorium that is mined.

An efficient fuel cycle would also greatly improve the assurance that we will be able to undertake to keep long-lived radioactive wastes buried, to prevent them from returning to our environment in radiologically significant amounts. The reduction in waste that is contemplated would reduce the amounts buried by a factor of 20 since the heavy elements are the only important long-lived wastes. And we must efficiently recycle not only uranium, plutonium, and thorium, but also the other actinides: neptunium, americium, and curium. Fuel cycles that fail to do this are simply not acceptable for an asymptotic nuclear industry.

Of course, though, the overriding condition for an asymptotic nuclear industry is a world that is truly at peace. Right now, the balance is favorable toward continuing to exploit nuclear energy since its contributions to world energy supply have an important role in reducing the tensions that lead to war. However, so long as war is a thinkable option, the world is at some risk of having its reactor fuel diverted to bombs. Sooner or later we must make the choice: war *or* nuclear energy; we cannot have both.

8 REDUCING THE SCENARIO

It seems that the world will need to maintain an inventory of about 100,000 tonnes of fissile material if a goal of 17 TWyr yr^{-1} is to be reached; a number of different techniques for building up to that inventory all lead to an ultimate demand for 10–15 million tonnes of reasonably high-concentration uranium to be extracted from nature. What would happen if the world were to yield only about 5 million tonnes of uranium – a quantity that corresponds to our present reasonably assured resource estimates?

There are two ways of proceeding. One is to reduce the scale of our target scenario; the other is to increase the length of the buildup period and to deploy breeders and advanced converters earlier. These two measures actually work together.

To explore the scenario of reduced expectations we use as a guide the demand scenario of "Energy in a Finite World". Table 17.11 of that source (Basile et al., 1981) lists 9.9 TW of electrical capacity as required in a high energy scenario by the year 2030, but only 4.4 TW of this are to be nuclear. Of the remainder, roughly 0.95 TW are assumed to be hydroelectric (Figure 17.16), about 0.1 TW solar, and the remainder (4.5 TW) comes from various forms of fossil fuel. The percentage of electricity generated is more strongly nuclear and hydro, as the actual generation corresponds to 2.9 TWyr yr^{-1} of nuclear and only 1.9 TWyr yr^{-1} of other sources; clearly, nuclear power is considered primarily for baseload, and the rest for interruptable, intermediate, and peak loads, and reserve.

It is therefore logical to scale down the target figures for nuclear supply by a factor of 2 to about 5 TWe of capacity and 8.5 TWyr yr^{-1} of energy. The question then arises:

which parts of the comprehensive system of reactors should be scaled down for this target? For this we must look at trends. The trend by 2030 is in the direction of phasing out LWRs and replacing them with FBRs (to permit growth by breeding of fissile atoms) and HTGRs (for use in thermochemical processes). The nuclear industry would still be in a total growth phase.

The logic that I use is that a high target for FBR installation should be retained, but the LWRs and HTGRs could be reduced and the pure burners replaced by LWRs. This choice is predicated on the assumption that any further increase in nuclear energy past the year 2030 should not make resource demands on nature and that growth should be in the HTGR sector after 2030 so as to contribute to the production of metals and synfuels while reducing fossil-fuel consumption for these purposes. The target for the year 2030 then becomes about 3 TWe from FBRs, about 1 TWe from LWRs, and about 1 TWe from HTGRs. This is not very different from the simplified breakdown in Chapter 17 of "Energy in a Finite World", which is 2.6 TWe from FBRs and 1.8 TWe from LWRs. The principal difference is the assumption that it would be unwise to build, and thus to institutionalize, more LWR capacity than might be needed asymptotically; instead, an HTGR (or an equivalent advanced converter) industry would be worth developing with a view to future expansion.

The endowment of fissile material required for this scenario can be calculated from the data of Table 3. It amounts to 37,000 t of fissile plutonium and 11,000 t of ^{233}U, i.e. 48,000 t in all. This endowment could be supplied from conversion of about 60,000 t of ^{235}U, procurable from about 8.5 million tonnes of natural uranium. This comes closer to the sort of uranium resource whose recovery we can expect to achieve with high confidence.

This reduced-scale target could grow to a larger scale with very limited inputs from nature. The breeding ratio of the system is about 1.10, which means that each year about 350 t of additional fissile isotopes are created. If no more fresh uranium were available from nature, a slow but orderly growth of the nuclear enterprise could achieve the level of 10 TWe of capacity in another century. Acceleration of this growth would still be possible if the first additional components of the system were breeders or if more cheap uranium from nature became available.

To recapitulate, realistic nuclear energy demands by the year 2030 can be accommodated with uranium investments that are in the range of resources that we are most likely to find available. If the nuclear investment during the next half-century emphasizes breeder reactors, the system in place by the year 2030 could increase the nuclear contribution without further demands on natural uranium. This increase would, however, be slow.

A corollary of this reduced scenario would be that some of the more futuristic applications (of HTGRs for advanced chemistry and of pure burners for high-adventure enterprises) would be delayed.

9 REACTOR SYSTEM INTERACTIONS

Up to now FBRs have seemed to be supreme as the fissile-material factories of an advanced nuclear system. We also see nothing on the horizon to displace them in that role, barring the possibility of an economic breakthrough for accelerator breeders or an economic–technical breakthrough on fusion–fission hybrids. However, the type of FBR might be appreciably different from what is contemplated as a standard FBR today.

The reason is that, as pointed out by Fortescue (1978) and Spinrad (1978a), a principal function of FBRs is to achieve high-efficiency production of ^{233}U with which to fuel thermal reactors such as LWRs and HTGRs. The HTGR is likely to feature a pure ^{233}U–Th fuel cycle, requiring only an input of ^{233}U from FBRs, after the system of reactor types has reached steady state. However, the dispersed use of LWRs argues for consideration of a denatured fuel cycle. In this cycle ^{238}U is used as a diluent for ^{233}U and the result is that the reactor is a producer of plutonium.

In consequence, while the role of the FBR is to breed, it is not necessarily proper for it to breed plutonium. It is only necessary for it to keep its own inventory of plutonium constant or even, if there is a significant amount of plutonium available from LWRs, slowly decreasing. To the maximum feasible extent, all of the breeding gain would be realized as ^{233}U for the thermal reactors. The nuclear design of reactors for this purpose could be quite different from that of most of today's concepts (Spinrad, 1978b). The reactors would feature high-concentration ($>$ 25 wt.%) plutonium in their cores and considerable quantities of thorium and depleted uranium in internal and external blankets, and they could conceivably run to quite high breeding gains. The relatively low specific power and large fissile inventory that I have shown in Table 2 reflect trends in that direction while the breeding gain that I have used in Table 1 might, under these circumstances, be quite conservative.

10 ALTERNATIVE REACTORS

There are at least two other reactor types that deserve mention as competitors for future large-scale deployment. These are HWRs and MSRs. I conceive of the HWR as an alternative to the LWR and the MSR as an alternative to the HTGR with regard to their roles in asymptotic systems of reactors. Both the HWR and the MSR have some potential for becoming thermal breeders but it is likely that both will operate most economically as "near-breeders" (Spinrad, 1978b).

HWRs are already in commercial use in some countries. Adaptation of their design to be near-breeders, particularly on a ^{233}U–^{238}U fueling cycle, seems to be straightforward, although specific reactor development as well as fuel-cycle development would be needed. Their potential for success arises from two considerations of their capability relative to LWRs: their fueling costs are lower; their demands for enriched fuel are less, and their conversion ratios are higher. Relative to LWRs, they require a higher initial investment, and this has inhibited their application in countries where high investment-charge rates are the rule. However, this may be due to faulty accounting (Spinrad, 1980). In any case, they seem attractive as uranium becomes scarcer and more expensive (CONAES, 1979). Because their thermodynamic performance is essentially identical to that of LWRs – the properties of light and heavy water at high temperature are very similar – it is not impossible that the group of reactors labeled LWRs in the main discussion of this paper could be HWRs instead. This substitution would ease the strain on uranium resources during a buildup to 5-TWe capabilities and would increase the rate at which nuclear power could grow from that point, for it seems reasonable that efficient HWRs could run at conversion ratios close to 0.9, ultimately requiring less ^{233}U from breeders.

MSRs are in a very early stage of development and we cannot yet predict their success with any confidence. If they are successfully developed they would be somewhat competitive with HTGRs because (1) it is projected that they would deliver heat to a secondary salt at over 1200° F (around 700° C), which begins to be an interesting temperature for process chemistry, and (2) since they were originally designed to breed, a conversion ratio of 0.95 might be quite readily achieved.

The temperature capability of the MSR is not quite as good as that of the HTGR so it would not be a complete substitute for the HTGR; however, significant penetration of what has been postulated to be a large market could be achieved.

11 SUMMARY

A reactor system that is significant on a world scale would have to be self-sustaining with regard to its fissile inventory. Only then could it be fueled indefinitely without making inordinate demands on nature for uranium. The system should also in the long run be a multipurpose energy source, supplying heat both for conversion to electricity and directly for large-scale chemical processes. It should, finally, be capable of fueling a variety of special-purpose demands for space exploration and other exotic settings.

Such a system is feasible. It would be based on breeder reactors whose function would be to generate electricity for continental grids and to supply the fissile material for other reactors. By concentrating on neutron efficiency in these other reactors the system could actually generate more energy from nonbreeders than from breeders. Advanced HTGRs and LWRs can be projected to meet requirements, respectively, for chemical process heat and electricity and for small-network electricity and general energy while still permitting the system to allocate fuel for pure burning in specialty uses. All three of the reactor types just named would be significantly evolved from present thinking: breeders would put out most of their fissile surplus in the form of ^{233}U and would be designed with more attention to breeding gain and less to doubling time; HTGRs would be expected to exhibit high conversion ($CR \approx 0.9$) on a ^{233}U–Th fuel cycle; LWRs would also be expected to exhibit high conversion ($CR \approx 0.8$) on a ^{233}U–^{238}U cycle.

An examination of demand requirements as defined by the IIASA high scenario (Häfele et al., 1981) for the year 2030 suggests that a realistic target for the next 50 years would be to have in place by 2030 about 3 TWe of breeders and 1 TWe each of HTGRs and LWRs, as nominal capacity. Such a system would require an input of much less than 10 million tonnes of natural uranium as an endowment and could grow slowly without further resource demands, with a doubling time of about 80 years. Doubling could be faster if additional conventional (i.e. inexpensive) natural uranium were to become available.

If systems competitive with the LWR and HTGR are developed, resource constraints on the rate of nuclear system buildup become less severe. HWRs are very credible competitors for the LWRs while MSRs are more speculative competitors for part of the HTGR market.

REFERENCES

Basile, P., Agnew, M., Eddington, J., Papin, A., and Schrattenholzer, L. (1981). Energy supply and conversion. In W. Häfele (Program Leader), Energy in a Finite World – A Global Systems Analysis. Ballinger, Cambridge, Massachusetts, Chap. 17, pp. 576–582.

CONAES (Committee on Nuclear and Alternative Energy Systems, National Research Council) (1980). Energy in Transition 1985–2010. Freeman, San Francisco, Chap. 5, Table 5–6.

Fortescue, P. (1978). The role of thermal and fast reactors in sustaining an adequately safeguarded energy supply. In O.K. Kadiroglu, A. Perlmutter, and L. Scott (Editors), Nuclear Energy and Alternatives. Ballinger, Cambridge, Massachusetts, Chap. 19, pp. 191–200.

Häfele, W., Spinrad, B., Cohen, K., Perry, A., and Nakicenovic, N. (1981). The nuclear option. In W. Häfele (Program Leader), Energy in a Finite World – A Global Systems Analysis. Ballinger, Cambridge, Massachusetts, Chap. 4, pp. 129–132.

Spinrad, B.I. (1978a). Plutonium-fueled fast reactors as secure fuel cycles. In O.K. Kadiroglu, A. Perlmutter, and L. Scott (Editors), Nuclear Energy and Alternatives. Ballinger, Cambridge, Massachusetts, Chap. 13, pp. 127–142.

Spinrad, B.I. (1978b). Alternative breeder reactor technologies. In J.M. Hollander, M.K. Simmons, and D.O.Wood (Editors), Annual Review of Energy, Vol. 3. Annual Reviews Inc., Palo Alto, California, pp. 147–179.

Spinrad, B.I. (1979). Resource-efficient nuclear scenarios. WP-79-62. International Institute for Applied Systems Analysis, Laxenburg, Austria.

Spinrad, B.I. (1980). Effects of accounting rules on utility choices of energy technology in the United States. RR-80-27. International Institute for Applied Systems Analysis, Laxenburg, Austria.

SUSTAINABLE MINIREACTORS: A FRAMEWORK FOR DECENTRALIZED NUCLEAR ENERGY SYSTEMS

A.A. Harms
McMaster University, Hamilton, Ontario (Canada)

W.W. Sassin
Energy Systems Program, International Institute for Applied Systems Analysis, Laxenburg (Austria)

ABSTRACT

The concept of a nuclear energy system consisting of numerous small, specialized nuclear reactors providing heat or electricity for localized/regional purposes is considered. It is envisaged that a "parent" nuclear facility would sustain the fuel needs of many small nuclear energy "satellites" and possibly provide other fuel-management services. The choice of fuel cycle and the operational features of these satellites may be determined by the form of energy required, public and social preferences, and institutional factors. Three distinct classes of distributed systems, each based on extensions of existing nuclear technology, are identified and discussed.

1 INTRODUCTION

The current civilian nuclear power program has developed with a dominant emphasis on the development of large central-station electricity-producing power plants in the 2000–5000 MWth range. While the need to serve correspondingly large and interconnected grids is generally evident, it is also recognized that there exists a need for small power sources serving localized and special energy requirements.

We consider here a nuclear energy system consisting of numerous small minireactors in the 1–10 MW capacity range. Such small nuclear reactors could supply energy for small isolated communities, localized industrial activity, apartment buildings, and institutional centers such as schools and small hospitals. These minireactors would therefore need to operate safely with high reliability and, as we emphasize here, be assured of a sustainable supply of easily transportable fuel. A central nuclear "parent" facility is envisaged for this latter purpose.

By separating "fuel-supply" facilities from "fuel-burning" facilities, optimization with regard to safety and reliability may be imposed as well as adaptability to siting and technological considerations.

2 NUCLEAR SYNERGISM

In order to develop the minireactor systems concept we refer briefly to the underlying concept of nuclear energy synergism (Harms, 1980; Harms and Häfele, 1981). By this term we imply the integrated operation of distinct nuclear processes and separate nuclear energy components so that the total system displays advantages that are not possible otherwise. Nature provides many examples of synergism whereby one component in an ecological cycle provides a feedstock for another. Man-made systems, if suitably designed and efficiently integrated, may similarly display synergetic characteristics.

The concept of synergism may be illustrated by reference to a general exoergic nuclear reaction

$$\mathrm{A} + \mathrm{B} \rightarrow \sum_i R_i + Q \tag{1}$$

Here A and B are selected nuclei or nucleons and R_i is one of several reaction products; Q is the energy released in the process to be recovered. Both fission and fusion reactions are special cases of such reactions.

Nuclear synergism can be realized if the reaction products serve an added and dominant energy-producing process. By general extension of eqn. (1), we conceive two distinct possibilities: one is

$$\begin{array}{l} \mathrm{A} + \mathrm{B} \rightarrow \mathrm{C} + \mathrm{D} \\ \qquad\quad \downarrow \\ \qquad\quad \mathrm{C} + \mathrm{A} \rightarrow Q + \ldots \end{array} \tag{2}$$

in which one of the reaction products combines with one of the initial reactants to generate energy. The second case

$$\begin{array}{l} \mathrm{A} + \mathrm{B} \rightarrow \mathrm{C} + \mathrm{D} \\ \qquad\qquad\quad \downarrow \\ \qquad\qquad\quad \mathrm{D} + \mathrm{E} \rightarrow Q + \ldots \end{array} \tag{3}$$

is one in which a reaction product is able to combine with another readily available nucleus or nucleon to release significant amounts of energy.

These two examples will be used in the following discussion in which a "parent" process maintains the first parts of eqns. (2) and (3) while the second parts of these equations occur in "satellite" processes.

3 AN ACCELERATOR–FISSION SYSTEM

When a sufficiently energetic proton strikes a heavy nucleus a significant number of nucleons may be released; these emitted nucleons may excite other nuclei in the target, thus contributing to secondary neutron production. This so-called spallation process may be represented by

$$\vec{p} + Z \rightarrow \nu_{sp}\mathrm{n} + \sum_i R_i \tag{4}$$

where $\vec{p}$ is the accelerated proton, Z is some medium-to-heavy target nucleus, and ν_{sp} is the total number of neutrons released per incident proton. Depending on proton energy, target size, and target composition, ν_{sp} may be as high as 70.

The neutrons thus released may migrate out of the target domain and enter a surrounding blanket. If this blanket has a sufficient density of fertile nuclei (i.e. ^{232}Th, ^{238}U) then neutron capture may take place to initiate a transmutation chain leading to the production of fissile fuel (i.e. ^{233}U, ^{239}Pu). The bred fissile fuel is thus available to supply suitable fission reactors. The entire sequence (nuclear synergism) may be compactly represented by

$$\begin{array}{lll} \vec{p} + Z \rightarrow \nu_{sp}\mathrm{n} & \text{Accelerator target} & \\ \qquad\downarrow & & \text{Parent} \\ \qquad\mathrm{n} + Fb \rightarrow (\ ) \rightarrow (\ ) \rightarrow Fi & \text{Accelerator blanket} & \text{facility} \\ \qquad\qquad\qquad\qquad\quad\downarrow & & \\ \qquad\qquad\qquad\qquad\quad Fi + \mathrm{n} \rightarrow \nu_{fi}\mathrm{n} + Q & \text{Fission minireactors} & (5) \end{array}$$

In this representation Fb refers to fertile nuclei and Fi to fissile nuclei.

The potential role of high-energy proton accelerators as producers of fissile fuel in support of a fission-reactor economy was suggested as early as 1952 (Lewis, 1952) and has in recent years generated considerable interest (Vasil'kov et al., 1970; Kouts and Steinberg, 1977; Harms and Gordon, 1977; Steinberg et al., 1980).

Our emphasis here, however, is on the deployment of minireactors to receive the accelerator-bred fuel. We envisage that these mini-fission-reactors would be in the power range of up to 2 MWth and would consist of a small compact core in a pool maintained at a temperature below 100° C. These small heat sources could be patterned after the several operating SLOWPOKE reactors (Kay et al., 1973); the significant feature of these small reactors is that at these temperature conditions the power coefficient of reactivity is sufficient to provide total self-regulation so as to permit licensing for unattended operation. These minireactors are evidently restricted to low-temperature heat supply and would therefore need to be used only for space-heating purposes.

Figure 1 gives a schematic representation of this accelerator-supported fission minireactor system. Recent assessments (Hilborn and Glen, 1981; Harms et al., 1981) suggest that an efficient accelerator system could indeed provide sufficient fuel on a continuing basis to support several hundred of these small satellites. The power consumption of the accelerator is likely to be 20–30% of the total thermal production capacity, thus providing an acceptable level of system efficiency.

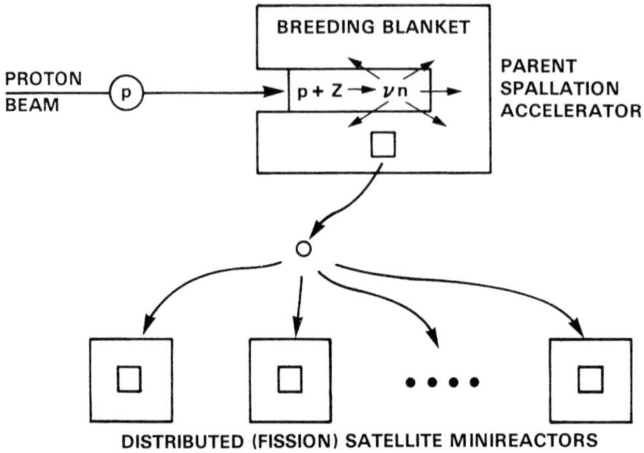

FIGURE 1 A schematic representation of an accelerator-supported system of fission minireactors.

4 A FUSION–FUSION SYSTEM

An alternative system could be based on the supply of fusion fuel to decentralized minireactors by a parent deuterium-fueled fusion reactor. It is known that D + D fusion occurs via two routes of almost identical probability:

$$D + D \begin{cases} \nearrow T + p \\ \searrow {}^3He + n \end{cases} \tag{6}$$

The cross section for in situ fusion of the bred tritium is sufficiently high to ensure its burning. Effective in situ fusion of ^3He, however, requires different plasma conditions (McNally, 1978), which suggests that a substantial ^3He inventory would build up in a D + D reactor. It has therefore been suggested (Miley and Gilligan, 1979) that small-size special-purpose reactors should be designed to be fueled by ^3He and deuterium.

By analogy to eqn. (5) we represent the isotope fuel balance and reaction processes as follows:

$$\left.\begin{array}{l} D + D \rightarrow T + p \\ \quad\quad\downarrow \\ \quad D + T \rightarrow n + \alpha \\ \\ D + D \rightarrow {}^3He + n \end{array}\right\} \text{Parent facility} \tag{7}$$

$$\begin{array}{ll} \quad\quad\downarrow & \\ \quad D + {}^3He \rightarrow p + \alpha & \quad\quad \text{Fusion minireactors} \end{array}$$

Figure 2 provides a schematic representation of this fusion–fusion parent–satellite energy system.

The dominant and necessary function of the parent reactor is evidently ^3He production for the satellites; any extractable energy should first be assigned for

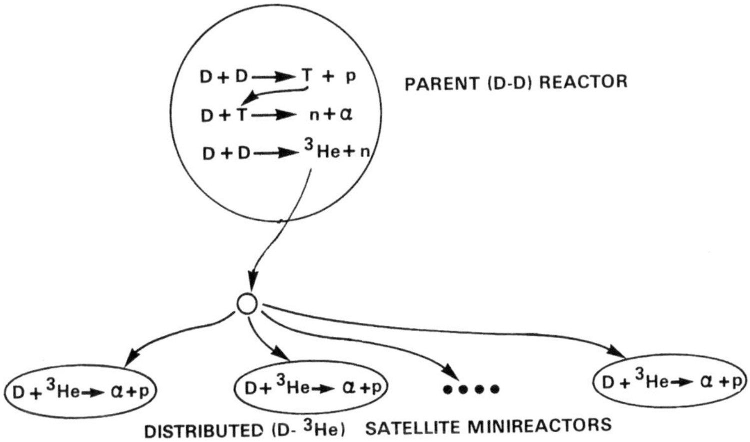

FIGURE 2 A depiction of a set of D–^{3}He fusion minireactors supported by a ^{3}He-breeding fusion reactor.

recirculation purposes to support the D + D fusion process. If its resulting Q value is greater than unity then it could similarly supply energy to an external market, but, by analogy with the preceding accelerator facility, it might even be a net energy consumer provided that its net power requirement is a sufficiently small fraction of the satellite power.

It appears that the field-reversed mirror reactor constitutes an excellent candidate for the D–^{3}He-burning satellites (Miley and Gilligan, 1979); these reactors are small, compact, and suitable for sizes less than 10 MW.

One of the most appealing features of these D–^{3}He-burning reactors is that they are close to being radiologically clean: protons and α particles are the only normal reaction products; neutron production could be kept small, attributable only to the residual D + D reactions taking place. Since the reaction products are charged particles only, efficient direct conversion of the charged particles into electricity might be considered.

5 A FUSION–FISSION SYSTEM

The third system that we introduce here combines both fusion and fission processes; it follows by extension of the preceding systems concept.

We propose that a D + D-burning reactor not only should produce ^{3}He for satellite fusion reactors but should also be equipped with a suitable blanket (Crowley et al., 1980) so as to utilize the fusion neutrons to supply fissile fuel for small fission minireactors (Schöpf and Harms, 1981). The isotope and fuel balance formulation is obtained by rewriting the first three lines of eqn. (7) and extracting the two neutrons and the ^{3}He as follows:

$$
\begin{array}{lll}
5\mathrm{D} + \mathrm{p} + \alpha + {}^{3}\mathrm{He} + 2\mathrm{n} + Q & \text{Parent core} & \\
\quad \downarrow & & \text{Parent facility} \\
\quad \mathrm{n} + Fb + Fi & \text{Parent blanket} & \\
\qquad \downarrow & & \\
\qquad Fi + \mathrm{n} \rightarrow \nu_{\mathrm{fi}}\mathrm{n} + Q & \text{Fission satellite} & \\
{}^{3}\mathrm{He} + \mathrm{D} \rightarrow \alpha + \mathrm{p} + Q & \text{Fusion satellite} &
\end{array}
\tag{8}
$$

We illustrate this system in Figure 3. It seems that its reactor-physics characteristics are consistent with the requirements of a sustainable system in the sense of the present analysis.

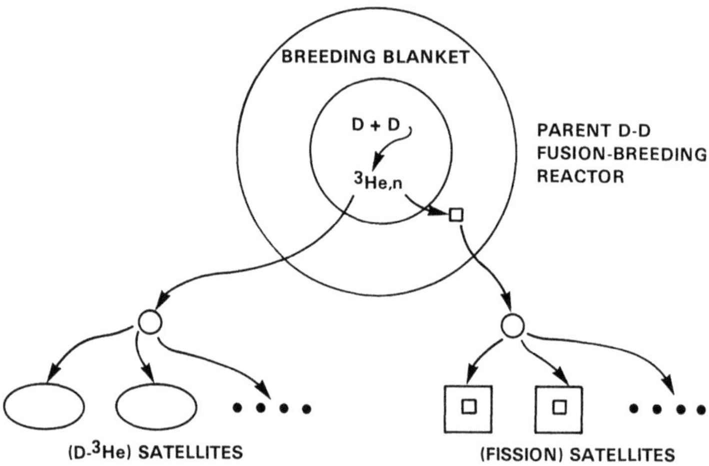

FIGURE 3 An illustration of a D–D-fueled hybrid reactor supporting both fusion and fission mini-reactors.

6 SYNERGISM EXPANDED

In the preceding discussion the focus has been on the front end of the fuel cycle. Though a sustainable fuel supply could evidently be assured, no reference has been made to any synergistic applications to the back end of the fuel cycle. We now address this issue and point to some interesting and potentially significant implications.

Some new back-end fuel-management practices appear feasible because the spallation neutrons and fusion neutrons possess a source spectrum that is harder than a fission-neutron spectrum and it might be possible to achieve a higher neutron flux than that possible in a fission reactor. The latter feature suggests that fission-product destruction in the spallation reactor or the fusion reactor may become feasible. The availability of the harder initial neutron spectrum would allow effective in situ fuel enrichment and fuel rejuvenation, as we discuss next.

It is known that the capture-to-absorption cross-section ratio of fertile–fissile isotope pairs (i.e. σ_c(Th-232)/σ_a(U-233) and σ_c(U-238)/σ_a(Pu-239)) possesses a significant high plateau in the 100–500 keV neutron-energy domain. With the energy of the primary source neutrons sufficiently higher than this value, it should be possible to tailor the spectrum hitting the blanket such as to suppress fission to a large extent and hence make effective spent-fuel rejuvenation possible. A spent fuel element – which contains a depleted fissile concentration while still retaining a high fertile composition – can thus be placed in a blanket position with such a suitably tailored neutron spectrum as to achieve enrichment of the fissile concentration by neutron capture in the fertile nuclei (Steinberg et al., 1980; Harms and Hartmann, 1978; Dorning and Gunnison, 1978; Conn et al., 1980). Subsequently, assuming no significant adverse structural effects, the fuel could be returned to the reactor and could eventually proceed through a repeated, or repeatable, rejuvenation cycle.

The possibility of in situ rejuvenation and enriching has two major operational and policy consequences. On the operational level it will allow access to a broader range of additional fuel cycles while on the policy level in situ enriching could reduce the hazards that are currently perceived in the production, storage, and transportation of nonradioactive enriched fuels (i.e. such fuels being diverted for non-peaceful purposes).

If there exists an ample supply of neutrons at these parent facilities, many structural components for reactor cores and blankets could be designed with durability and reliability – rather than neutron economy – as the dominant criterion.

7 CONCLUDING COMMENT

In addition to the points emphasized concerning the types of minireactors and the fuel cycles chosen, it is important to recognize the potential for mass-production of these smaller facilities. Also, if the fuel-consuming part of the system is widely distributed geographically and if the fuel can be stored, the simultaneous failure of substantial parts of the energy supply system seems unlikely. Finally, if there were a local need for medium-power facilities, provision for the stacking of minireactors to attain a specified power level could be introduced.

REFERENCES

Conn, R.W., Kantrowitz, F., and Vogelsang, W.F. (1980). Hybrids for direct enrichment and self-protected fissile fuel production. Nuclear Technology, 49:458.

Crowley, J.H., Pavleno, G.F., and Kaminski, R.S. (Editors) (1980). 1980 Source Book for Fusion–Fission Hybrid Systems. UEC-DOE/EPRI 783112. United Engineers and Contractors, USA.

Dorning, J. and Gunnison, F. (1978). On the feasibility of an accelerator driven rejuvenator for light-water reactor fuel. Transactions of the American Nuclear Society, 32:799.

Harms, A.A. (1980). Synergetic nuclear energy systems concepts. Advances in Nuclear Science and Technology, 12:211.

Harms, A.A. and Gordon, C.W. (1977). A parametric analysis of the spallation breeder. Nuclear Science and Engineering, 63:336.

Harms, A.A. and Häfele, W. (1981). Nuclear synergism. American Scientist, 69(3):310.

Harms, A.A. and Hartmann, W.J. (1978). Spent nuclear fuel re-enrichment without reprocessing. Annals of Nuclear Energy, 5:213.

Harms, A.A., Krenciglowa, E.M., Sassin, W.W., and Hilborn, J.W. (1981). Accelerator-supported regionally distributed nuclear heat system with nuclear waste abatement capacity. Annals of Nuclear Energy, 8:431.

Hilborn, J.W. and Glen, J.S. (1981). Small reactors for low temperature heating. In the Proceedings of the Canadian Nuclear Society Conference, 1981. Canadian Nuclear Society, Toronto, Ontario.

Kay, R.E., Stevens-Guille, P.D., Hilborn, J.W., and Jervis, R.E. (1973). SLOWPOKE: A new low-cost laboratory reactor. International Journal of Applied Radiation and Isotopes, 24:509.

Kouts, H.J.C. and Steinberg, M. (Editors) (1977). Proceedings of the Information Meeting on Accelerator Breeding, Brookhaven National Laboratory, Upton, New York, January 1977. CONF-770107. Brookhaven National Laboratory, Upton, New York.

Lewis, W.B. (1952). The significance of the yield of neutrons from elements excited to high energies. DR-24. Chalk River Nuclear Laboratories, Chalk River, Ontario, Canada.

McNally, J.R., Jr. (1978). D–^{3}He as a "clean" nuclear fusion reactor. Nuclear Fusion, 18(1):133.

Miley, G.H. and Gilligan, J.G. (1979). A possible route to small flexible fusion units. Energy, 4:163.

Schöpf, K.F. and Harms, A.A. (1981). Fusion-hybrid supported satellite reactors. Transactions of the American Nuclear Society, 38:537.

Steinberg, M., Powell, J.R., Takahashi, H., Grand, P., and Kouts, H.J.C. (1980). The linear accelerator fuel enricher regenerator (LAFER) and fission product transmuter (APEX). Atomkernenergie/Kerntechnik, 36:42.

Vasil'kov, V.G., Gol'danskii, V.I., Dzhelepov, V.P., and Dmitrievskii, V.P. (1970). The electronuclear method of generating neutrons and producing fissionable materials. Soviet Atomic Energy, 29:858.

SMALL REACTORS IN A NEUTRON-ABUNDANT WORLD

J.W. Hilborn
Reactor Physics Branch, Atomic Energy of Canada Limited, Chalk River, Ontario KOJ 1JO (Canada)

ABSTRACT

In a future world of neutron abundance, fissile fuel for small reactors could be produced from uranium and thorium using breeder reactors, fusion reactors, and spallation accelerators. Although the most successful application of small reactors to date has been in nuclear submarines, there appears to be a growing need for small (< 250 MWe) nuclear energy sources in developing countries and remote locations. However, in recent years only the Soviet Union and India have actually built multiple commercial units. Several dual-purpose reactors and heat-only reactors have been proposed, but future commitments are uncertain. Assuming the long-term availability of fissile fuels at a present-day cost of $200/g in a fabricated core, low-temperature heat from small reactors could probably compete with oil heat, if the price of oil were to double.

1 NEUTRONS FOR THE PEOPLE

Decentralized nuclear energy systems have been proposed by Jassby (1981) and Harms et al. (1981) whereby small mass-produced reactors would be used for heat and electricity production in applications where oil and natural gas are presently used. In a future world of neutron abundance, fissile fuel for the small reactors could be produced from uranium and thorium using breeder reactors, fusion reactors, and spallation accelerators. A more likely possibility is that hybrid combinations of fission, fusion, and spallation devices will prove to be more effective than any of the three separately (Harms and Häfele, 1981).

An interesting feature of the proposal by Harms et al. is in situ fuel rejuvenation, which means increasing the fissile concentration in spent fuel while it is still in the original cladding. For reactors in unit sizes up to approximately 30 MWth it may be possible to rejuvenate the entire core as a unit. It is envisaged (Figure 1) that the spent fuel from a number of reactors would be transported to a regional fuel center where it would be exposed to neutrons of a specified energy in a spallation accelerator. Eventually the fuel would have to be chemically processed to remove fission products, but the goal is to reduce greatly the volume and complexity of chemical reprocessing.

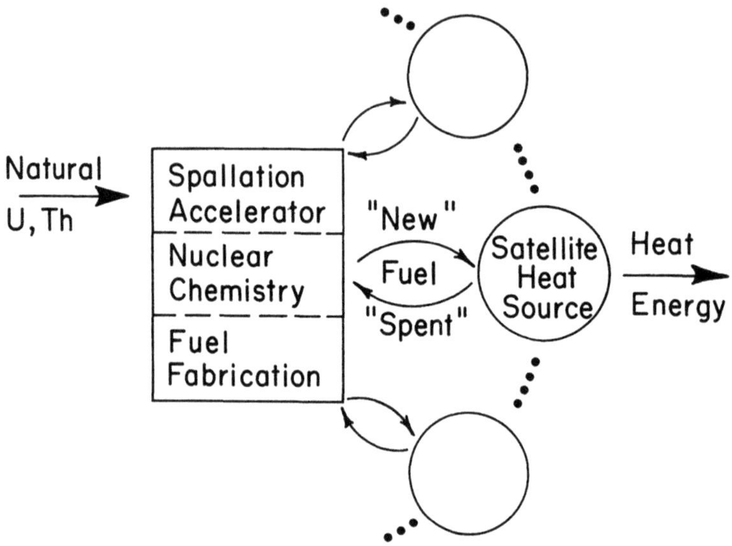

FIGURE 1 A graphical depiction of a central facility supplying a number of small heat sources with rejuvenated fuel. The power requirements for the central facility – which form part of the overall power balance – are not shown.

At first sight these bold conjectures would appear to be completely out of touch with the nuclear realities of the present-day world. It is as if our pioneering ancestors were planning to cross the ocean and had just learned how to build a raft.

However, assuming a future abundance of fissile material, how credible is a decentralized nuclear energy system? This paper will attempt to answer the following five questions. (1) If large reactors provide the cheapest electricity, why do we need small reactors? (2) What is the world experience in building and operating small reactors? (3) Can we provide the necessary technological and industrial infrastructure to support the operation of hundreds of small reactors in towns, cities, and remote locations? (4) What recent advances have taken place in small-reactor design that would give us confidence in their safety and reliability? (5) Can local nuclear heating compete with oil heating?

2 SMALL IS BEAUTIFUL – SOMETIMES

The most obvious role for small reactors is in developing countries and remote regions where the total electricity demand is too small to accommodate a full-scale nuclear power plant. Mehta et al. (1977) pointed out that in situations where the unit size of a nuclear plant is a significant fraction of the grid capacity (more than 15% say) outages have a crippling impact. They also concluded that if the load factor is less than 60% the cost of energy generation in certain cases will be less for a double-unit nuclear plant than for a single-unit plant. India, for example, is planning for at least 18 reactors in the range 200–220 MWe before increasing the unit size to 500 MWe.

In developing countries, weakness of industry and scarcity of capital may be more important factors than unit energy cost. Hence such countries may prefer to build smaller reactors if by so doing their own industries can supply the major components and the requirements for working capital are reduced. If a country needs energy to develop its economy, availability of nuclear power could be the dominant factor. As stated by H. Bhabha at the first Geneva Conference on the Peaceful Uses of Atomic Energy in 1955, "There is no power as expensive as no power".

Large nuclear power plants require hundreds of kilometers of transmission lines to form an electrical grid system. Because of the large size of reactor units, planning for expansion of the grid on a ten-year time scale can easily result in costly errors in matching load demand, transmission capacity and generating capacity. Small mass-produced reactors with a one-year delivery time could meet changing growth patterns in a more flexible manner. If the rate of growth unexpectedly declined, factories could reduce production; if the demand increased, factories could operate 24 hours a day or could expand their production facilities.

Decentralizing nuclear power would also spread the risk. It is now well known that people will accept frequent small disasters more readily than rare catastrophes. For example, according to the International Air Transport Association in Geneva, 22 fatal crashes of commercial aircraft in 1980 caused 812 deaths. There has been no public outcry against commercial flying and no crisis of confidence. Contrast that situation with the Three Mile Island accident, which caused no deaths but gave rise to extreme reactions from journalists and the public. Although we may have to endure the legacy of Three Mile Island for many years, a decentralized system of small reactors which effectively eliminates the possibility of a single big accident should have a significant advantage in terms of licensing and insurance, and in gaining public acceptance.

When nuclear-generated electricity is used for space heating less than one-third of the original heat energy is actually delivered to the consumer. If the consumer needs heat only it should be three times more efficient to obtain the heat directly from the source and not indirectly in the form of electricity. However, it is not that simple; if consumers are widely dispersed the factor-of-3 advantage in fuel efficiency may not be enough to balance the cost of providing insulated pipelines to carry steam or hot water from a central heating plant. Whether electrical heat or small-reactor heat is the best choice for a particular application will depend on many factors besides the fuel cost.

If both heat and electricity are required for local consumption, the most efficient arrangement is the production of electricity by a back-pressure turbine operating at a selected outlet temperature such that the outlet steam from the turbine can be used for industrial processes or space heating. To achieve peak efficiency from cogeneration of heat and electricity, all the capital equipment should be fully utilized. In practice of course this never happens.

In cold countries like Canada the annual consumption of low-temperature energy for space heating is at least double the annual consumption of electrical energy. Furthermore, the seasonal variation in outside air temperature results in a low annual load factor for space heating, generally less than 30%. Therefore if there were a massive conversion to all-electric space heating the winter heating load would dominate electrical demand, causing a severe and costly reduction in the overall electrical load factor. Small heating factors with a low capital cost per thermal kilowatt could alleviate this situation as indicated in Table 1. In practice we would probably use hybrid combinations of nuclear heat and oil heat to improve the load factors of the nuclear units and thereby to reduce the total cost of energy.

TABLE 1 The capital contribution to unit heating cost (in mills per kilowatt-hour).

Annual load factor (%)	Small heating reactor ($500 kW_{th}^{-1})[a]	Large nuclear plant ($2000 kW_e^{-1})[a]
30	28	114
40	21	86
50	17	68
60	14	57

[a] The assumed capital costs include all distribution costs. The annual charge on capital is assumed to be 15%.

In the phrase "Small is Beautiful" Schumacher (1973) expressed a popular and widespread reaction to the doctrine that bigger is better. However, he emphasized the duality of the human requirement when it comes to the question of size: "There is no single answer. For his different purposes man needs many different structures, both small ones and large ones, some exclusive and some comprehensive. Yet people find it most difficult to keep two seemingly opposite necessities of truth in their minds at the same time. They always tend to clamour for a final solution . . . " We are reminded that the most vocal critics of nuclear power are demanding that we follow the "soft" path exclusively and abandon the "hard" path. They refuse to accept Schumacher's duality.

3 ATOMS FOR WAR AND PEACE

The world reactor list to the end of 1980 shows that in recent years only India and the Soviet Union have built commercial reactors in unit sizes less than 250 MWe. India is committed to six pressurized heavy-water reactors by 1984 and is planning ten more of the same standard 220-MWe design before increasing the unit size to 500 MWe (Nuclear News, 1981).

In the Soviet Union, four small reactors are supplying energy to the remote mining town of Bilibino (Baturov et al., 1977). Each of the four units can provide 12 MWe and 25 MW of heat. It has also been reported that a 5-MW heating reactor is operating in the city of Dimitrovgrad (Nucleonics Week, 1980). During the past 15 years the Soviet Union has tested four different types of small reactors for remote regions, but beyond the four-unit plant at Bilibino future Soviet plans are unknown.

The most successful application of small reactors has been in military submarines, starting with the *Nautilus* in 1955. There are now 297 nuclear submarines in the navies of four nations: the United States, 115; the Soviet Union, 162; the United Kingdom, 15; France, 5 (International Institute for Strategic Studies, 1980). For comparison there are 245 commercial power reactors (> 30 MWe) in operation throughout the world, with an average capacity of 600 MWe. The safety and reliability of submarine reactors over a time span of 25 years proves beyond doubt that the technical infrastructure for building and operating large decentralized nuclear energy systems is highly developed in the United States and the Soviet Union. Whether or not such systems would be commercially viable is uncertain.

Reactor manufacturers in the UK and France have designed commercial land-based versions of the submarine reactor but have not yet made a sale. In the deep-sea environment the small reactor has a clear advantage over its air-breathing competitors but as soon as it surfaces it becomes just another energy option.

Small reactors are being used in surface warships to gain both tactical and strategic advantage, but attempts to introduce nuclear power to merchant shipping have not been successful. Despite outstanding technical performances by the American vessel *Savannah* and the German vessel *Otto Hahn,* both of these merchant ships were taken out of service because of high operating costs and a world surplus of shipping capacity. The Soviet Union is operating nuclear-powered icebreakers but it is difficult to judge the true economic advantages. Canada has been weighing the costs and benefits of a nuclear icebreaker for the past 5 years but has finally decided to build a conventional diesel-powered vessel.

Between 1954 and 1976 the United States built and operated seven small land-based reactors and one barge-mounted reactor for military purposes (Bennett and Mann, 1978). Most of these were dual-purpose plants producing both heat and electricity for army bases; one of them also produced fresh water by desalination. They ranged in size from 300 kWe to 10 MWe, and all except the largest used highly enriched uranium fuel. In 1967 when five reactors were in operation, with a total installed capacity of 95 MWth and equipped for 18 MWe of electrical power production, the total field crew was 173 men. They were supported by 57 men at central command, and in addition there were 50 students in training plus 30 instructors, for an overall total of 310 men. The program was terminated in 1976 because the high cost of maintaining and operating small reactors in remote locations could not be justified in terms of military objectives. In other words, these small reactors could not compete with diesel plants.

One of the most significant small-reactor demonstrations took place in Sweden from 1963 to 1973. The 70-MWth Agesta plant supplied heat and electricity to a Stockholm suburb of 40,000 people for 10 years until high operating costs forced it to close down (Hannerz and Larsson, 1977). The decision to close down was made just before the 1973 oil crisis, and there was no possibility of reversing the decision because the entire heavy-water inventory had been sold to Canada. As the surrounding community had become accustomed to smokeless nuclear heat there were many protests from the public when Agesta was shut down.

To summarize, the safety and reliability of small reactors have been well-proven in a developing country, in remote locations, in a densely populated city, and in submarines, icebreakers, and merchant ships. However, small reactors have not come into widespread commercial use because of the availability of cheap oil.

4 NEW DIRECTIONS

During the past 5 years several new small-reactor designs have emerged. In the larger unit sizes up to 300 MWe, reactor manufacturers in France, the UK, the Federal Republic of Germany, and the United States are offering pressure-vessel designs based on existing reactors used for ship propulsion. The French and British designs are adaptations of nuclear submarine power plants; the FRG and US entries lean heavily on experience with the nuclear merchant ship *Otto Hahn.* Although these designs have been actively marketed, commercial negotiations with several countries have not yet resulted in any firm orders.

India remains the only nation committed to a significant number of power reactors in the range 200–500 MWe. The 20-year plan is to have four units of 200 MWe, 14 units of 220 MW, and 12 units of 500 MWe, for a total installed capacity of 9900 MWe. India is demonstrating that the pressurized heavy-water reactor, based on Canadian technology, is a viable route to energy self-sufficiency for a developing country. India is also convinced that the 200-MWe unit size is a good choice for the initial phase of their nuclear program.

In the category of low-temperature reactors for heat only, four new designs have appeared during the past 5 years. They are the 200-MWth SECURE reactor (Nilsson and Hannus, 1977), the 100-MWth THERMOS reactor (Dupuy et al., 1977), the 500-MWth DHAPP reactor (Skvortsov et al., 1977), and the 2-MWth SLOWPOKE reactor (Hilborn and Glen, 1981).

In all four cases there has been great emphasis on safety features which would permit them to be located in populated areas. Not surprisingly, the four concepts incorporate many of the same design features. The reactors are all designed for low-temperature low-pressure operation, which reduces both the probability and potential consequences of a pressure-component failure. The reactor containment structures are earthquake resistant and protected against terrorists. In all cases the reactor cores are submerged in underground water-filled pools and in three of the designs the heat exchangers are also submerged. Two of the designs rely on convective circulation instead of mechanical pumps. Three of the reactors do not require control rods in the core region and two of them do not require safety rods.

An overview of the safety features of the four designs reveals two common characteristics which stand out above all others: (1) a simplicity of concept which should be highly credible to nontechnical policy makers and the public and (2) a high degree of inherent protection against mechanical and operator errors. To further illustrate these characteristics the Canadian SLOWPOKE concept will be described more fully.

5 SIMPLE IS SAFE

Instead of attempting to scale down a large power reactor, our approach is to uprate a 20-kWth research reactor called SLOWPOKE. Developed at the Chalk River Nuclear Laboratories in 1970 (Kay et al., 1973), it is a low-temperature pool-type reactor used for teaching and research at four Canadian universities and for industrial applications at two other Canadian laboratories.

Because of its inherent safety the SLOWPOKE reactor has no mechanical or chemical automatic shutdown devices and is licensed to operate without an operator in the reactor room. The reactor power is regulated by a single motor-driven control rod responding to a signal from a self-powered neutron detector. There are no safety rods, ionization chambers, start-up instruments, or coolant pumps. The reactor is turned on with a switch and reaches full power within a few minutes.

If a fault in the regulating system causes full withdrawal of the control rod, the reactor power is safely limited by the negative temperature coefficient of the core. From 1971 to the present this kind of fault has occurred once in 25 reactor-years of SLOWPOKE operation.

The proposed 2-MWth heating reactor shown in Figure 2 is intended for large buildings and institutions in urban areas and for remote northern communities. It is similar in concept to the SLOWPOKE research reactor and is being designed to have similar safety characteristics, based on limited reactivity additions and a large negative void coefficient. Full-time reactor operators would not be required but alarms for fire, intrusion, and radiation would be continuously monitored at a location away from the reactor site. In an urban area, a number of reactors would be monitored at a single control center.

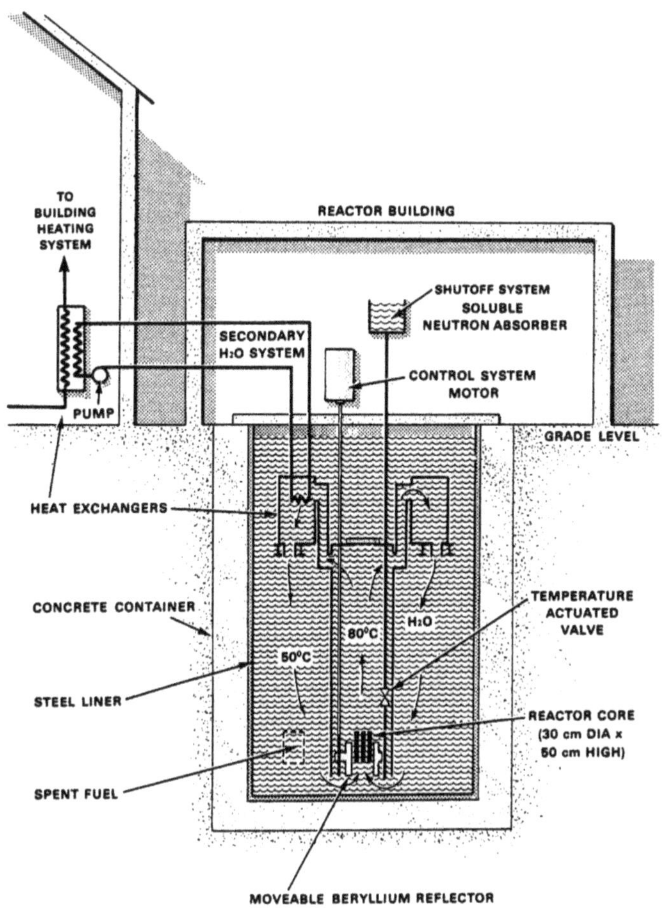

FIGURE 2 A 2-MW SLOWPOKE heating reactor.

The reactor is doubly contained in a water-filled pool and a concrete vault. The core consists of 200 uranium oxide fuel elements of the type used in CANDU power reactors, but using 5% enriched uranium instead of natural uranium. Each fuel element is 0.5 m long and contains 0.5 kg of uranium.

Reactor power is controlled by a motor-driven beryllium annulus surrounding the core and responding to a signal from a temperature sensor. The coolant temperature is normally maintained at 80° C. Core cooling is by natural convection and the pool water is separated from the hot water delivered to the consumer by heat exchangers. For an extended shutdown soluble poison can be added to the pool.

Our present proposal for a 2-MWth heating reactor is a starting point for an assessment of the SLOWPOKE concept over the range 2–20 MWth. If our analysis confirms technical feasibility and indicates economic potential we plan to build a prototype unit at Chalk River in 1984–1985.

6 THE BOTTOM LINE

A preliminary estimate of the present cost of energy from the 2-MWth SLOWPOKE reactor using 5% enriched uranium oxide indicates that nuclear heat is approximately competitive with electric heat and oil heat for an annual load factor of 50% (Hilborn and Glen, 1981).

To estimate small-reactor energy costs for the long term it is assumed that the fuel elements would contain plutonium instead of uranium. Other assumptions are as follows.

(1) Capital cost of small reactors, \$500 $kWth^{-1}$.
(2) Capital cost of oil furnaces, \$50 $kWth^{-1}$ (combustion efficiency, 75%).
(3) Operating and maintenance costs, 0.4 cents $kWth\,h^{-1}$ for both nuclear and oil heating.
(4) World price of heavy oil at the point of use, 26¢ l^{-1}.
(5) Cost of plutonium in fabricated fuel elements, \$200 g^{-1}.
(6) In-reactor plutonium inventory, 2 kg MW^{-1}.
(7) In-reactor residence time of fuel, 1 full-power year.
(8) Out-of-reactor residence time of fuel, 1 calendar year.
(9) Fuel-consumption, 1.6 g Pu MWd^{-1}.
(10) Annual charge on capital and inventory, 15%.
(11) Annual load factors, 30% and 60%.

The estimated costs of nuclear and oil heating are compared in Table 2. The bottom line of Table 2 shows that small plutonium-fueled reactors would begin to be competitive if the price of oil doubled. The most uncertain factor is the total cost of plutonium fuel, which depends on the production cost in fission reactors, fusion reactors, or spallation accelerators, plus the cost of chemical processing and fuel fabrication. The total fuel cost in dollars per gram of plutonium in fabricated fuel was arbitrarily taken to be four times the 1980 US price of highly enriched uranium in the form of uranium hexafluoride.

A debatable question is how inflation should be accounted for over the 20-year lifetime of capital equipment. If the "constant dollar" approach is taken, the annual

TABLE 2 Nuclear and oil heating costs (in mills per thermal kilowatt-hour).

Item	30% load factor		60% load factor	
	Nuclear	Oil	Nuclear	Oil
Capital	28	3	14	1
Operating	4	4	4	4
Fuel inventory	30	0	18	0
Fuel consumption	13	30	13	30
Total	75	37	49	35

charge on capital is then derived from the difference between the prevailing borrowing rate and the inflation rate. Typically a 4% difference yields an annual charge of 7.4% for 20 years, which is one-half of the annual charge rate assumed in Table 2. As a basis for an investment decision, private industry might assume 100% depreciation over 5 years while a nonprofit government utility would probably adopt a much longer period, corresponding more closely to the expected lifetime of the equipment.

7 THE SECOND NUCLEAR ERA?

In a recent article Weinberg (1981) observed that the "first nuclear era" was already 30–50% complete and would run its course when easily mined uranium was exhausted. Because present-day power reactors convert less than 1% of mined uranium to heat energy, world resources of uranium will support a limited number of these light-water and heavy-water reactors. Weinberg suggested 1000 units of 1000 MWe for 50 years, assuming a world uranium resource base of 10 million tons.

The second nuclear era, if there is going to be one, requires an abundance of neutrons to provide an abundance of fissile material. However, the second nuclear era may never begin if the public cannot accept the possibility, however remote, of a catastrophic reactor accident.

The underlying difficulty is that the public understands consequences but not probabilities. The possibility of 3000 deaths and $14 billion in property damage as documented by Rasmussen (1975) is specific and frightening, but an accident probability of 10^{-6} per reactor year is an abstract concept beyond the understanding of the average person. When a serious accident actually happens it is no consolation to the victims that the event was statistically expected or unexpected.

8 JUST A FRIENDLY LITTLE REACTOR

Both the real and perceived consequences of the worst accident to a small reactor should be several orders of magnitude less severe than those postulated for a large power reactor. At the same time, as the number of small reactors increases the probability of an accident increases, and we can imagine a situation where the overall risk to the public would be approximately the same for a small number of large reactors as for a large number of small reactors. However, a public which overestimates consequences and ignores probabilities may tend to be less apprehensive of small reactors than large reactors.

Eventually the public may accept accidents to small reactors to the same extent that they accept fires, explosions, and airplane crashes, as long as the consequences are not obviously worse. There is no need to invoke theoretical probability arguments. Reactor designers, however, will continue to assess the overall nuclear risks in terms of the probabilities and consequences of accidents, according to national regulatory standards.

In Canada we have already had some public reaction to our small-reactor development program even though we are still at the laboratory stage and have not committed a prototype for construction. Popular articles in newspapers and magazines have been factual and for the most part favorable. People were surprised to discover that research reactors have been operating in seven Canadian cities for many years. Journalists quickly seized on the concept of an unattended minireactor in the basement which would replace the conventional oil furnace. Preliminary discussions have been held with national and provincial regulatory agencies concerning the siting of small reactors in cities but it is too early to draw any useful conclusions.

One of the most encouraging responses was from a mass-circulation "back-to-nature" magazine that is well known for its antinuclear bias. While generally negative in outlook, the article presented antinuclear and pronuclear arguments in a reasonably balanced manner. The title of the article was completely satisfactory to both sides: "Just a Friendly Little Reactor".

9 THE PATH AHEAD

For the near term the Soviet Union has demonstrated dual-purpose reactors (12 MWe + 25 MWth) in a remote community and India has demonstrated 200-MWe units for electricity production. Neither of these reactors requires a reactor pressure vessel and the Indian reactor does not require enriched uranium. Dual-purpose pressurized-water reactors have been proposed by France, the FRG, the UK, and the United States. Low-temperature reactors for heat only have been proposed by France, Sweden and Finland, the Soviet Union, and Canada. Unit sizes range up to 300 MWe for the dual-purpose reactors and 500 MWth for the heat-only designs. None of the proposed reactors has yet been built, and it could take 10–15 years before one or more of these new designs is well proven enough for commercial applications.

For the long term we envisage regional systems of small reactors supported by a large nuclear fuel center encompassing facilities for fissile-fuel production, chemical reprocessing, remote fabrication, waste management, and nuclear safeguards. Looking 25–50 years ahead, will the public accept the siting of small reactors in cities and will they regard a nuclear fuel center as just another industrial complex? Perhaps a politician's reply is the most appropriate: "Man will occasionally stumble over the truth, but most of the time he will pick himself up and continue on" (Winston Churchill).

REFERENCES

Baturov, B.B., Melentjev, L.A., Bulkin, Yu.M., Galaktionov, I.V., Svereva, G.A., Tokarev, Yu.I., Chernyaev, V.A., and Smirnova, E.S. (1977). The use of nuclear reactors for central heat supply and district heating. In the Proceedings of an International Conference on Nuclear Power and its Fuel Cycle, Salzburg, May 1977. International Atomic Energy Agency, Vienna.

Bennett, L.G.I. and Mann, R.F. (1978). Analysis of the operating experience of small, military nuclear reactors. Final Report Task 25B02A. Defence Research Establishment, Ottawa, Ontario.

Dupuy, G., Labrousse, M., Lerouge, B., and Schwartz, J.P. (1977). THERMOS reactors. In the Report of the Topical Meeting on Low-Temperature Nuclear Heat, August 1977. Technical Research Centre of Finland, Otaniemi, Finland.

Hannerz, K. and Larsson, Y. (1977). Reactor waste heat utilization and district heating reactors. In the Proceedings of an International Conference on Nuclear Power and its Fuel Cycle, Salzburg, May 1977. International Atomic Energy Agency, Vienna.

Harms, A.A. and Häfele, W. (1981). Nuclear synergism: an emerging framework for energy systems. American Scientist, 69:3.

Harms, A.A., Krenciglowa, E.M., Sassin, W.W., and Hilborn, J.W. (1981). Accelerator-supported regionally distributed nuclear heat systems with nuclear waste abatement capacity. Annals of Nuclear Energy, 8:9.

Hilborn, J.W. and Glen, J.S. (1981). Small reactors for low temperature heating. In the Proceedings of the Canadian Nuclear Society, 2:87.

International Institute for Strategic Studies (1980). The Military Balance 1980–81. Bartholomew Press, Dorking, UK.

Jassby, D.K. (1981). The fusion-supported decentralized nuclear energy system. Journal of Fusion Energy, 1:1.

Kay, R.E., Stevens-Guille, P.D., Hilborn, J.W., and Jervis, R.E. (1973). SLOWPOKE: a new low-cost laboratory reactor. International Journal of Applied Radiation and Isotopes, 24:509.

Mehta, S.K., Kakodkar, A., Balakrishnan, M.R., Ray, R.N., Murthy, L.G.K., Chamany, B.F., and Kati, S.L. (1977). Innovations in PHWR design, integration of nuclear power stations into power systems and the role of small size nuclear power plants in a developing country. In the Proceedings of an International Conference on Nuclear Power and its Fuel Cycle, Salzburg, May 1977. International Atomic Energy Agency, Vienna.

Nilsson, L. and Hannus, M. (1977). SECURE nuclear district heating plant. In the Report of the Topical Meeting on Low-Temperature Nuclear Heat, August 1977. Technical Research Centre of Finland, Otaniemi, Finland.

Nuclear News (1981). World list of nuclear power plants. Nuclear News, American Nuclear Society, 24:2.

Nucleonics Week (1980). The Soviets' first mini-nuclear plant. Nucleonics Week, 21:1.

Rasmussen, N.C. (1975). Reactor safety study: an assessment of accident risks in U.S. commercial nuclear power plants. WASH-1400, NUREG-75/014. US Nuclear Regulatory Commission, Washington, D.C.

Schumacher, E.F. (1973). Small is Beautiful. Harper and Row, New York, pp. 61–62.

Skvortsov, S.A., Sokolov, I.N., Krauze, L.V., Nikiporetz, Yu.G., and Philimonov, Yu.V. (1977). The low-temperature water–water reactor for the district heating atomic power plant. Report of the Topical Meeting on Low-Temperature Nuclear Heat, August 1977. Technical Research Centre of Finland, Otaniemi, Finland.

Weinberg, A.M. (1981). The future of nuclear energy. Physics Today, 34:3.

ON THE ROLE AND TECHNOLOGICAL READINESS OF FAST BREEDERS AND FUSION–FISSION HYBRIDS IN THE WORLD NUCLEAR FUTURE

S.I. Abdel-Khalik and G.L. Kulcinski
Nuclear Engineering Department, University of Wisconsin, Madison, Wisconsin 53706 (USA)

G. Kessler
Nuclear Research Center, D-7500 Karlsruhe (FRG)

ABSTRACT

This study focuses on the future role of hybrids and fast breeders and the possible time windows for their introduction. The technological readiness of both reactors and associated fuel-cycle facilities during these time windows is examined. Four possible options allowing world nuclear demand to be met within the known uranium resource base are identified: (1) breeders alone rapidly introduced in the year 2000; (2) hybrids alone rapidly introduced in 2000; (3) hybrids and breeders simultaneously introduced in 2000 following traditional market-penetration rates; and (4) hybrids and breeders rapidly introduced simultaneously in 2020. It is shown that breeders can be commercially available in 2000 but that the rapid market penetration required in option (1) would require exceptional efforts. Options (2) and (3) are ruled out since it is impossible for hybrids to be commercially available by 2000. An ambitious fusion development plan would allow hybrids to be commercially available by 2020. Option (4) represents a fall-back position in case the introduction of breeders is delayed beyond 2000 or the rapid market penetration rate required in option (1) cannot be achieved.

1 INTRODUCTION

The idea of utilizing nonfission neutron sources to produce fissile material by transmutation of the fertile isotopes of thorium and uranium dates back to the early 1950s (Lewis, 1952; Imhoff et al., 1953, 1954; Lawson, 1955). It was recognized at that early date that the natural fissile content of uranium ore resources is insufficient to support a large expanding economy of thermal fission reactors. Fusion–fission hybrids and accelerator breeders were offered as potential competitors to fission fast breeders. However, interest in these concepts waned as rich deposits of uranium were discovered in the 1950s, providing a large inexpensive source of U-235. The interest was rekindled in the late 1960s as a result of the significant increase in demand for U-235 by the commercial Light-Water Reactors (LWRs) then operating and planned (Jung, 1969; Lidsky, 1969). Further interest in these options was generated by the recent slowdown in the development of fast breeders in several key countries.

This study focuses on fusion–fission hybrids as sources of fissile material and vehicles for the early introduction of fusion. The idea is to surround the fusion reaction region with a blanket of fertile material so that the fusion neutrons will convert the fertile isotopes Th-232 or U-238 to U-233 or Pu-239, respectively. For a D–T fusion reactor the blanket should also allow for tritium breeding to make the reactor self-sufficient.

The fissile material produced in the hybrid can be burned in fission reactors or it can be partially burned in situ to release considerably more energy than is generated by fusion alone. Thus hybrids can nicely couple the "fast neutron-rich but energy-poor" D–T fusion process with the "energy-rich but neutron-poor" fission process. Neutron multiplication in the hybrid blanket through (n, 2n), (n, 3n), and (n, fission) reactions makes it possible for the total number of breeding captures per D–T fusion neutron to be considerably larger than unity. This means that, even if hybrids are made to breed their own tritium fuel, large quantities of fissile materials can still be produced per unit of fusion energy (Maniscalco, 1976).

The most attractive feature of the hybrid concept is that it may allow fusion to make an early and significant contribution to world energy needs (Bethe, 1978; Teller, 1979). The fission energy produced in the supported fission reactors and in the hybrid blanket itself makes it possible to relax the fusion gain requirements in the hybrid. Reduced gain and plasma-confinement parameters in magnetic fusion devices and low driver efficiency or target gain in inertial-confinement fusion may be tolerable. Hence it is reasonable to expect that hybrids could be deployed much earlier than pure fusion devices and may possibly open the way for their introduction.

Fusion–fission hybrids have potentially much higher fuel-production rates per unit thermal power than fast breeders (Fortescue, 1977). Neutronic analyses and conceptual reactor studies have shown that a U–Pu hybrid can produce enough plutonium to fuel as many as six LWRs of equivalent thermal power on a steady-state basis (Maniscalco et al., 1978). The support ratio is even higher for Th–U hybrids because U-233 is a more efficient fuel for thermal fission reactors and because in the hybrid blanket Th-232 has a much lower fast-fission cross section than U-238. This is significant not only from an economic viewpoint but also from the standpoint of concern about hybrid ownership and possible proliferation.

A study was recently conducted to determine the impact of fusion–fission hybrids on world uranium demand and to identify the preferred hybrid concept from a resource standpoint (Abdel-Khalik et al., 1981). The following conclusions were reached.

(1) From a resource standpoint, hybrids which breed their own tritium fuel and have a low blanket energy multiplication are preferable.
(2) Hybrids have the potential to lower the cumulative uranium demand to values well below the established resource base. However, the time window for hybrid introduction is quite near and narrow (2000–2020).
(3) If breeders or plutonium high converters are not used, hybrids must be introduced early (2000) and must penetrate the market rapidly if the projected nuclear component of the energy demand is to be met within the known resource base. If

delayed till 2020, the demand can be met only if hybrids are coupled to the breeders and if both reactor types are allowed to penetrate the market rapidly.

(4) Traditional market-penetration constraints are too restrictive so that hybrids will not "do the job" unless they are simultaneously introduced with the breeders in the year 2000.

(5) The use of hybrids results in a significant reduction in the maximum annual uranium demand. The demand disappears entirely 35–40 years after hybrid introduction. This means that uranium accessibility for large consumers with few resources of their own will not be the main problem; adequacy of the resource base will remain the primary issue.

This study focuses on the role of hybrids and fast breeders in the world nuclear future and the time window(s) for their introduction. The "technological readiness" of both reactors and associated fuel-cycle facilities during these time windows will be examined. The aim is to determine whether the introduction dates and market-penetration rates, identified by Abdel-Khalik et al. (1981) as necessary to meet the projected demand within the known resource base, can be achieved.

We begin by summarizing the methodology of, and results obtained by, Abdel-Khalik et al. (1981) with regard to how the cumulative uranium consumption of the world is affected by the introduction dates and market-penetration rates of both hybrids and breeders. Secondly, current and planned facilities for the development of fast breeders are examined to assess the potential for the timely and rapid market penetration of breeders during the time window identified by Abdel-Khalik et al. (1981). Finally, the state of the art of fusion physics and technology is examined to determine the earliest possible introduction date for commercial hybrids.

2 TIME WINDOWS FOR HYBRID AND BREEDER INTRODUCTION

The time period between the present and the year 2075 is examined. It is assumed that pure fusion reactors will not contribute significantly to world energy needs till the end of that period. The low demand projection of world nuclear capacity of the International Nuclear Fuel Cycle Evaluation (INFCE, 1979) extrapolated to the year 2075 has been used (Figure 1). This forecast is the most recent available and is based on estimates made by individual countries of their projected energy needs. The use of such a low forecast will yield optimistically low estimates of the cumulative uranium consumption.

Figure 2 is a schematic diagram of the different scenarios examined by Abdel-Khalik et al. (1981). I and II are reference cases where LWRs are coupled to either plutonium high converters (HC) or fast breeders (FBR). Scenarios III and IV are for Th–U hybrids (HYB) coupled to either light-water or heavy-water denatured reactors (LWRD, HWRD), while scenarios V and VI are for U–Pu hybrids coupled either to high converters or fast breeders along with LWRs. For the Th–U systems the LWRs built before hybrid introduction are assumed to operate in a once-through mode (OT) and may be converted to light-water denatured reactors when U-233 bred by the hybrids becomes available. In all the hybrid scenarios the tritium required to start the hybrid is produced in power-generating reactors of the Savannah River type (SR). The hybrids are assumed to breed their own

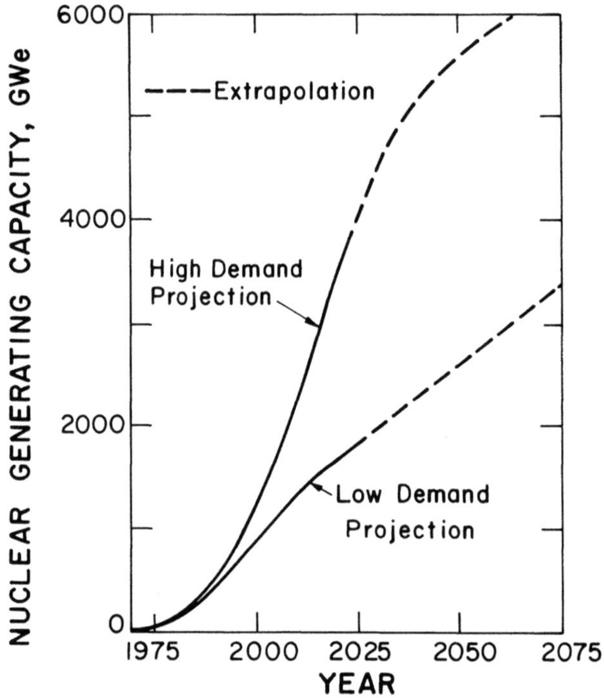

FIGURE 1 Projected world nuclear capacity (excluding areas with centrally planned economies) (INFCE, 1979); the low demand projection was used in this study.

tritium fuel and to have a low blanket energy multiplication (suppressed-fission type); these have been shown to be preferable from a uranium-resource standpoint. Design parameters for the different reactors shown in Figure 2 are given by Abdel-Khalik et al. (1981).

The optimum time-dependent shares of the different reactor types in all the scenarios have been determined using the strategy-optimization code SOP-KA (Klumpp, 1979). The code was designed to minimize the cumulative uranium consumption over the period of interest. The analyses have been performed for different introduction dates and market-penetration constraints. Two market-penetration constraints have been examined (Figure 3). Constraint A allows full market penetration within ten years of the introduction date. This is clearly optimistic and is used to determine a lower bound on the uranium requirements. Constraint B was obtained using a logistic substitution model based on historical market-penetration data for different energy sources.

The annual and cumulative uranium requirements for the scenarios examined with different introduction dates and market-penetration constraints have been obtained. These are used to identify the time window(s) for hybrid and breeder introduction that are necessary to meet the demand within the resource base.

Figure 4 shows the cumulative uranium consumption as a function of time for the different scenarios examined. Here the hybrids, breeders, high converters, and tritium

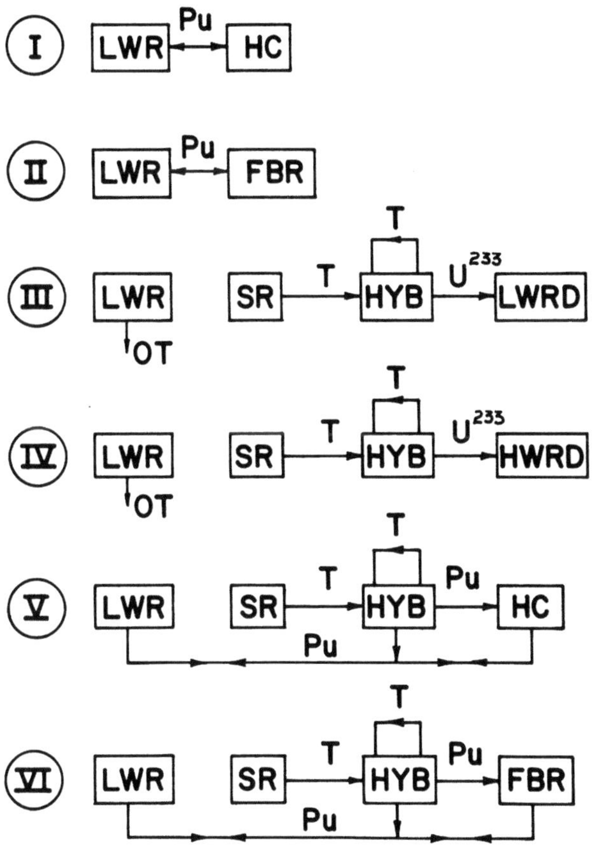

FIGURE 2 A schematic diagram of the different scenarios examined.

producers are assumed to enter the market in the year 2000. Market-penetration constraint A has been used so that these results represent the lowest possible uranium-demand values for the different scenarios. The hatched band in Figure 4 represents the sum of the reasonably assured (RAR) and estimated additional (EAR) uranium resources with recovery costs up to $130 per kilogram of uranium. The band width is ± 20% of EAR (OECD–IAEA, 1978). It is clear from the results shown in Figure 4 that hybrids have the potential to lower the cumulative uranium demand to values well below the resource base. These results also show that the breeder alone (scenario II) can "do the job" if it is introduced early (year 2000) and is allowed to penetrate the market rapidly.

Results similar to those in Figure 4 for an introduction date of 2020 with the rapid market-penetration constraint A are shown in Figure 5. These results indicate that delaying hybrid introduction till 2020 would make it impossible to meet the nuclear component of the demand within the uranium resource base. For comparison, the results for scenario II with breeder introduction in the year 2000 are superposed on Figure 5.

The results presented in Figures 4 and 5 represent lower bounds on the cumulative uranium consumption for the different scenarios because of the relatively unrestricted

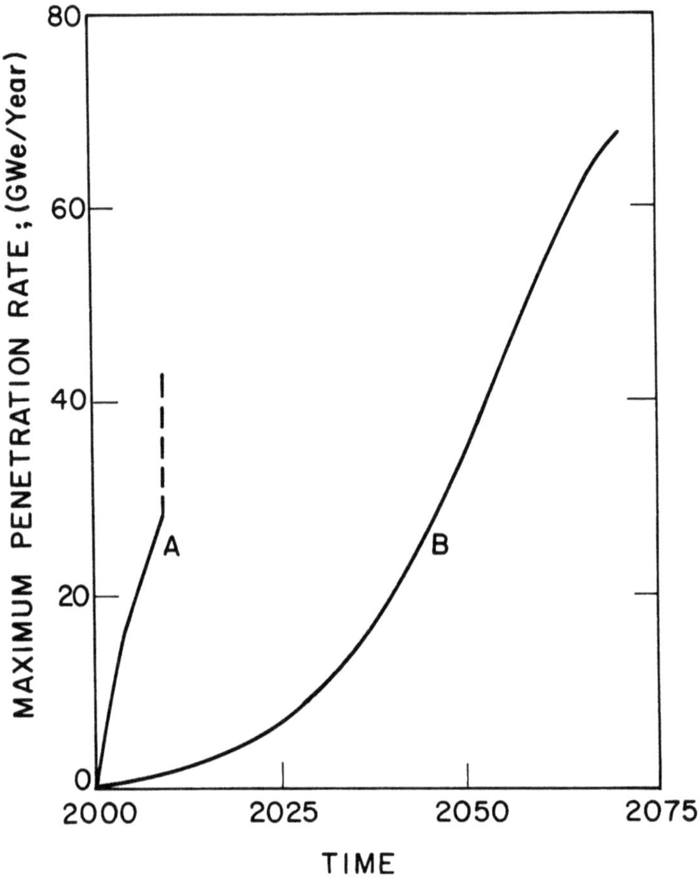

FIGURE 3 Market-penetration constraints for an introduction date of 2000.

market penetration assumed (constraint A). The computations have been repeated for traditional market-penetration constraints (constraint B). The results are shown in Figures 6 and 7 for introduction dates of 2000 and 2020, respectively.

Figure 6 shows that if hybrids and advanced reactors are introduced in the year 2000 and are allowed to penetrate the market under historical constraints (constraint B) the demand can be met within the resource base only if hybrids are coupled to plutonium high converters or fast breeders (scenarios V and VI). If the introduction date is delayed till 2020 (Figure 7) the cumulative uranium demand for all hybrid scenarios will be considerably larger than the resource base. The results shown in Figures 6 and 7 show that traditional market-penetration constraints will be too restrictive if breeders alone are introduced in the year 2000 or if hybrids and breeders are simultaneously introduced in 2020.

The results presented in Figures 4–7 show that the time window for breeder and hybrid introduction is relatively near and narrow (2000–2020). There are four possible

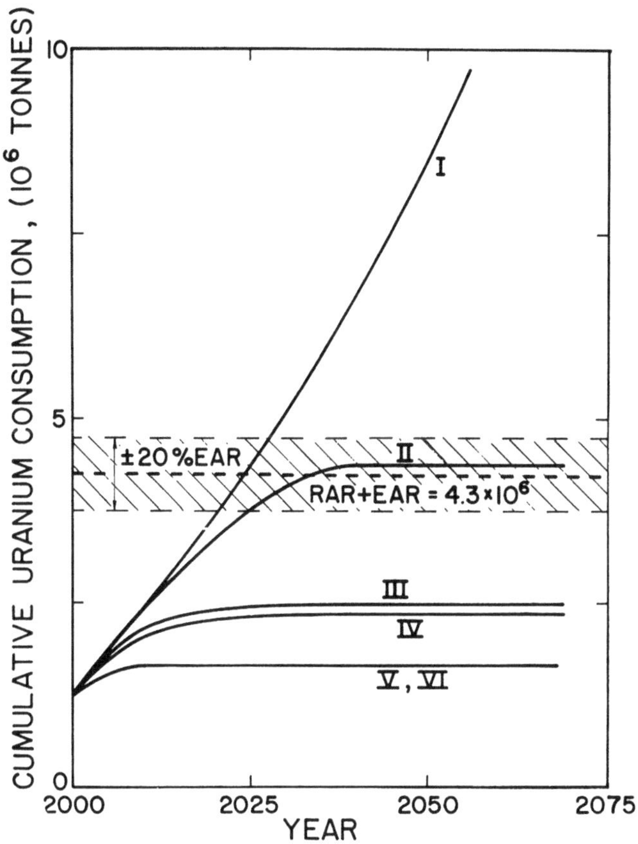

FIGURE 4 The variation of cumulative uranium consumption with time for the different scenarios shown in Figure 2 (these represent the lowest possible values since constraint A is used with an introduction date of 2000).

scenarios under which the world demand can be met within the resource base, as shown in Table 1: (1) breeders only are rapidly introduced in the year 2000 (scenario II with constraint A); (2) hybrids are rapidly introduced in the year 2000 (scenarios III–VI with constraint A); (3) hybrids and breeders (or high converters) are simultaneously introduced in the year 2000 (scenario V or VI with constraint B); (4) hybrids and breeders (or high converters) are rapidly introduced simultaneously in the year 2020. In the remainder of this paper we assess the technological readiness of hybrids and breeders to determine which, if any, of these four possibilities can be realistically achieved.

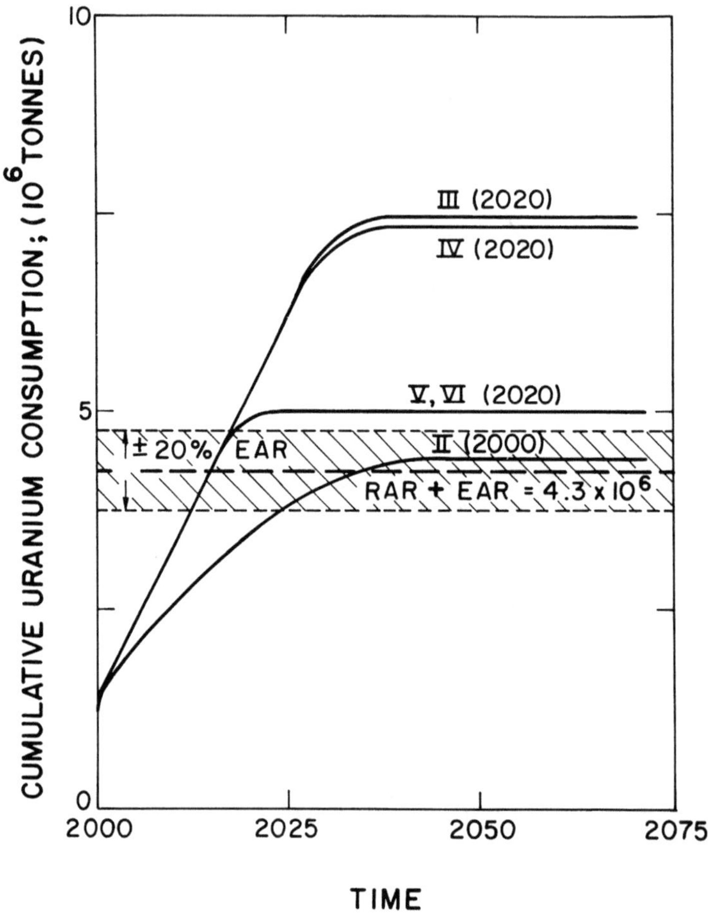

FIGURE 5 A comparison between breeder scenario II and hybrid scenarios III–VI for the market-penetration constraint A with introduction dates of 2000 and 2020, respectively.

3 THE TECHNOLOGICAL READINESS OF FAST BREEDERS

The purpose of this section is to determine whether the first option identified in Table 1 as necessary for the world demand to be met within the resource base can be realistically achieved. Specifically, there are two questions to be addressed: (1) Can the liquid-metal-cooled fast breeder reactor (LMFBR) be commercially available by the year 2000? (2) Can the breeder be rapidly introduced into the market (constraint A)?

Current plans for the development of fast breeders and associated fuel-cycle facilities are shown in Figure 8. Solid lines indicate facilities in operation or under construction while broken lines indicate plants under consideration. At present, LMFBR development is well beyond the stage of theoretical physics and engineering testing.

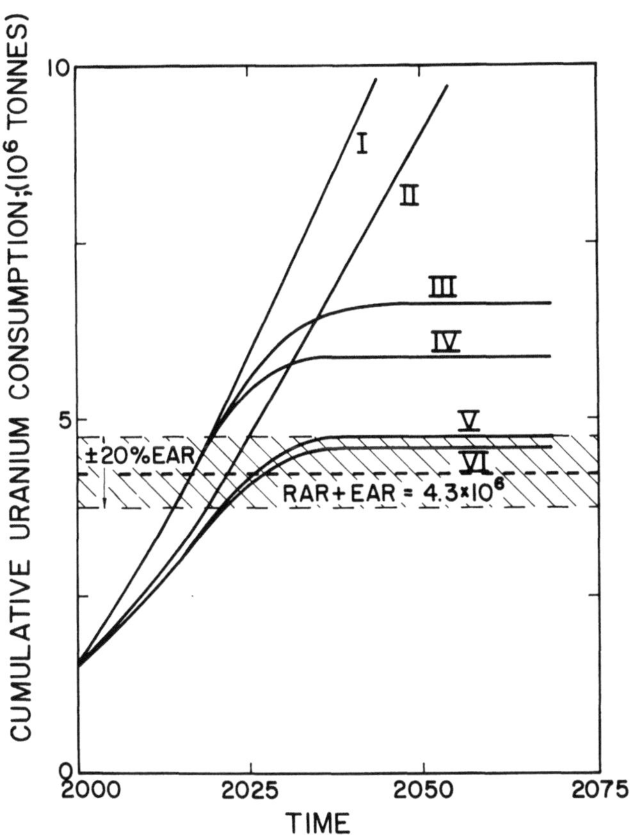

FIGURE 6 Cumulative uranium consumption for the different scenarios shown in Figure 2 for the traditional market-penetration constraint B with an introduction date of 2000.

Intermediate-size (200–300 MWe) LMFBRs are in operation while the first commercial-size plants are either beginning operation or under construction (Figure 8). Three intermediate-size prototype reactors have been operating for several years, namely the BN-350 in the Soviet Union started in 1973, Phenix in France (1973), and the PFR in Scotland (1975). Three other intermediate-size prototype reactors are planned, namely the SNR-300 in the Federal Republic of Germany (FRG) which will start around 1986 and the CRBR in the United States and MONJU in Japan which are scheduled to start somewhat later. As for commercial-size LMFBRs, the BN-600 at Sverdlovsk in the Soviet Union has been in operation since 1980 while the 1200-MWe Super-Phenix is now under construction in France and is scheduled to begin operation in 1983. Construction of two similar reactors, Super-Phenix 2 and 3, is scheduled to begin about one year after Super-Phenix begins operation. Construction of two 800-MWe plants (BN-800) is planned in the Soviet Union. In addition, design studies for large commercial reactors are continuing in the FRG (SNR-1200), the United Kingdom (CDFR-1300), and the Soviet Union (BN-1600).

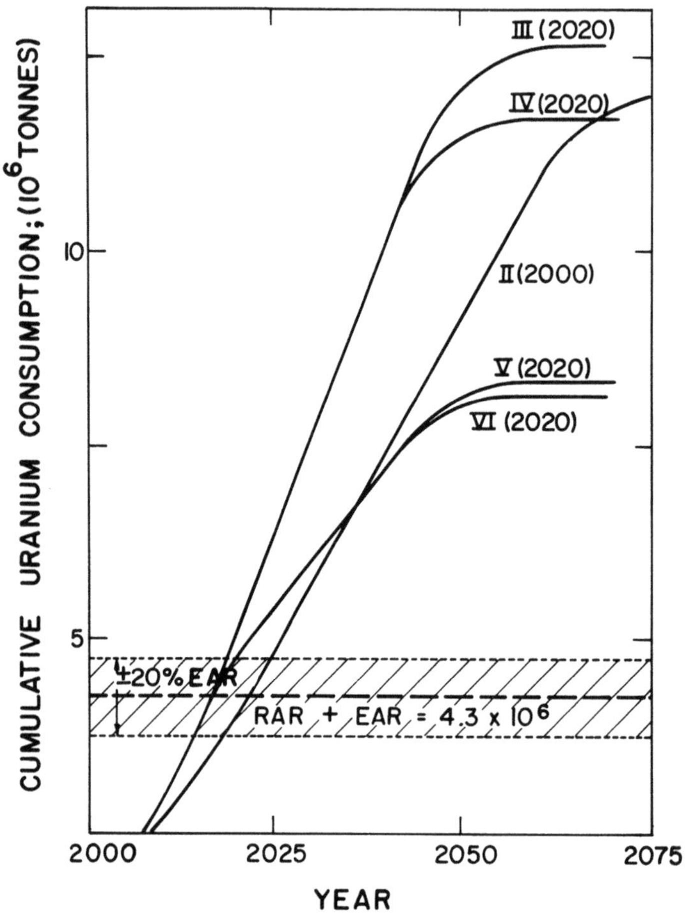

FIGURE 7 A comparison between breeder scenario II and hybrid scenarios III–VI for the market-penetration constraint B with introduction dates of 2000 and 2020, respectively.

Fast breeders require a closed fuel cycle where the reactor itself represents only one-third of the total system. Hence in order to assess the technological readiness of fast breeders it is necessary to examine the status of reprocessing and refabrication plants. The largest facility in operation at present is an 8 t yr^{-1} pilot plant at Dounreay in the UK which has been used to reprocess the fuel for the PFR. A 5 t yr^{-1} reprocessing plant, named TOR, at Marcoule, France, has been used to reprocess most of the fuel for Phenix. Small experimental facilities are currently operated in the United States, the FRG, and Japan. Medium-sized plants with capacities of 30–60 t yr^{-1} and 50–100 t yr^{-1} are currently under discussion for the CDFR and Super-Phenix 1, 2, and 3, respectively. The CDFR plant will be built after the decision to build Super-Phenix 2 and 3 is made. Commercial-size fast-breeder reprocessing plants will have capacities in the range of 200–300 t yr^{-1} and will each be able to support 10–15 GWe of LMFBR plants.

TABLE 1 Possible scenarios for meeting world nuclear demand within the uranium resource base[a].

Scenario	Constraint	Introduction date	Result
FBR (II)	A	2000	Yes
	A	2020	No
	B	2000	No
	B	2020	No
HYB (III–VI)	A	2000	Yes
	A	2020	Possible (only with FBRs, HCs)
	B	2000	Possible (only with FBRs, HCs)
	B	2020	No

[a] FBR, fast breeder; HYB, hybrid; HC, high converter.

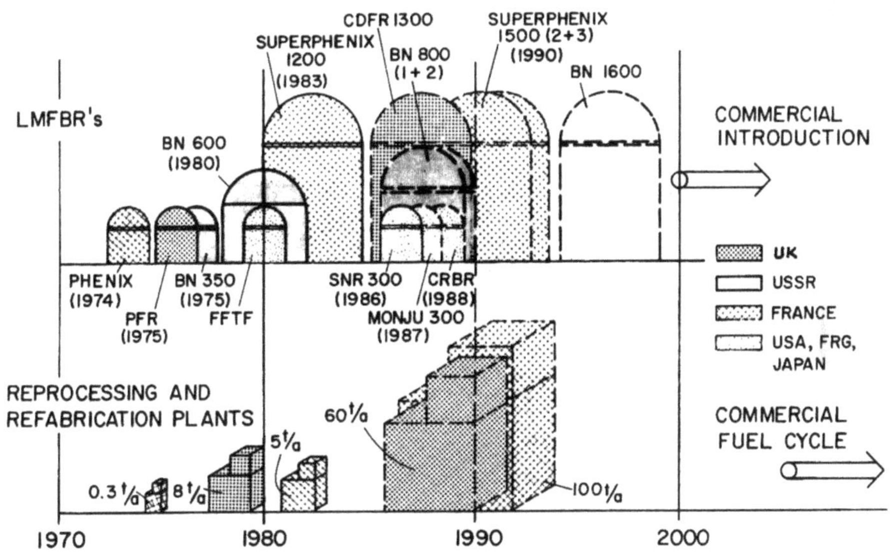

FIGURE 8 The current status and planned facilities for LMFBRs and associated fuel-cycle plants.

On the basis of the foregoing discussion, we conclude that LMFBRs can be commercially available by the year 2000. The complete fuel cycle would operate commercially around 2005 or somewhat later since each reprocessing plant supports at least 10 GWe of LMFBR plants. The ability of fast breeders to penetrate the market rapidly is difficult to ascertain. However, it is clear that exceptional efforts would be needed to approach the fast penetration scenario required (constraint A). Justification of such exceptional efforts would naturally depend on the urgency of the decision to introduce the breeder commercially, which in turn depends on other factors including the nuclear capacity requirement, the availability of natural uranium, the availability of capital, and the availability of industrial capacity for breeder construction.

4 THE TECHNOLOGICAL READINESS OF FUSION–FISSION HYBRIDS

The purpose of this section is to determine whether or not it is possible for hybrids to be commercially introduced within the time window (2000–2020) identified earlier as necessary for the world demand to be met within the known resource base. It should be emphasized at this point that, unlike fast breeders, fusion technology is in its infancy with most of the effort being currently devoted to physics experiments. Hence estimates of commercial introduction dates and market-penetration potential are far from certain. Here we examine the "best-case" scenario which would lead to the earliest possible commercial introduction.

Examination of the current status of fusion physics and technology reveals that there are only three fusion devices which might realistically achieve the necessary performance and be commercially available at such an early date: namely tokamaks, tandem-mirror devices, and light-ion beam-driven inertial-confinement fusion systems. The current status and projected progress milestones in fusion physics are shown in Figure 9 where the product $n\tau_E$ of the plasma ion density and the confinement time is plotted together with the average ion temperature T_i.

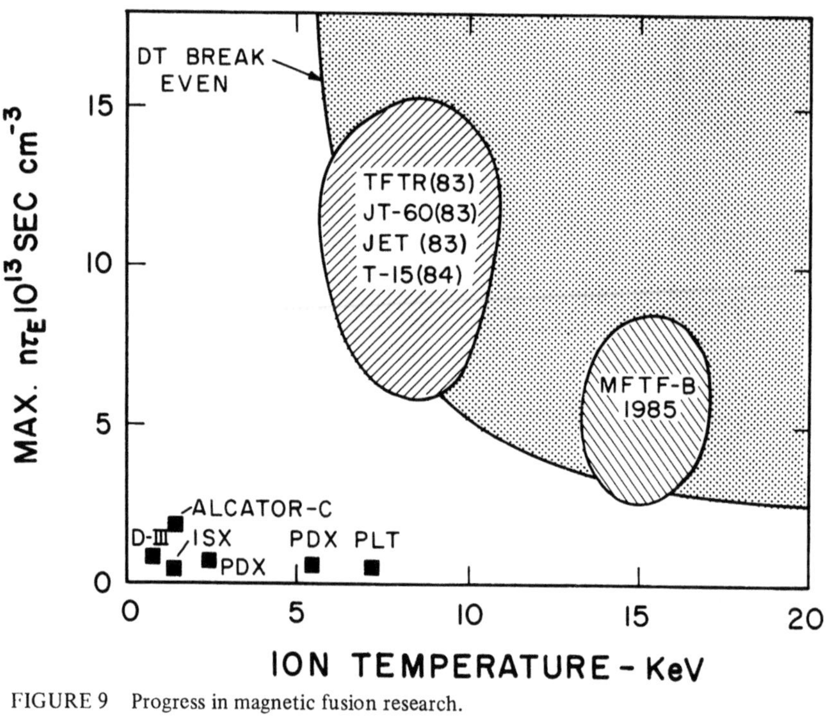

FIGURE 9 Progress in magnetic fusion research.

Temperatures sufficient for ignition have already been reached in the PDX and PLT while the $n\tau_E$ product has reached 2×10^{13} s cm^{-3} in Alcator-C. The next generation of tokamaks (JET, T-15, TFTR, and JT-60) will approach or exceed the energy break-even point within the period 1983–1985. Both the TFTR and JET will produce D–T neutrons, albeit on a short (≈ 1 s) time scale. Similarly the MFTF-B tandem-mirror device should produce plasma conditions in its central cell region which exceed break-even conditions in the 1985–1986 period. Therefore it appears that by the mid-1980s at least one magnetic-confinement device will reach energy break-even and produce D–T neutrons.

Progress in inertial confinement is somewhat more difficult to predict; the best measure of success is the gain on target defined as the ratio between the thermonuclear energy produced and the driver energy on target. Figure 10 shows the relation between the gain and driver energy on target (Bodner, 1980) based on a conservative set of assumptions for the coupling efficiency, the symmetry of implosion, the degree of preheat, etc. Superposed on Figure 10 are the present US plans for neodymium glass lasers (Shiva and Nova), CO_2 lasers (Antares), and light-ion beam systems (PBFA). Without any firm experimental data on hand, one may speculate that inertial-confinement fusion drivers that are large enough to reach energy break-even will be available in the mid-1980s.

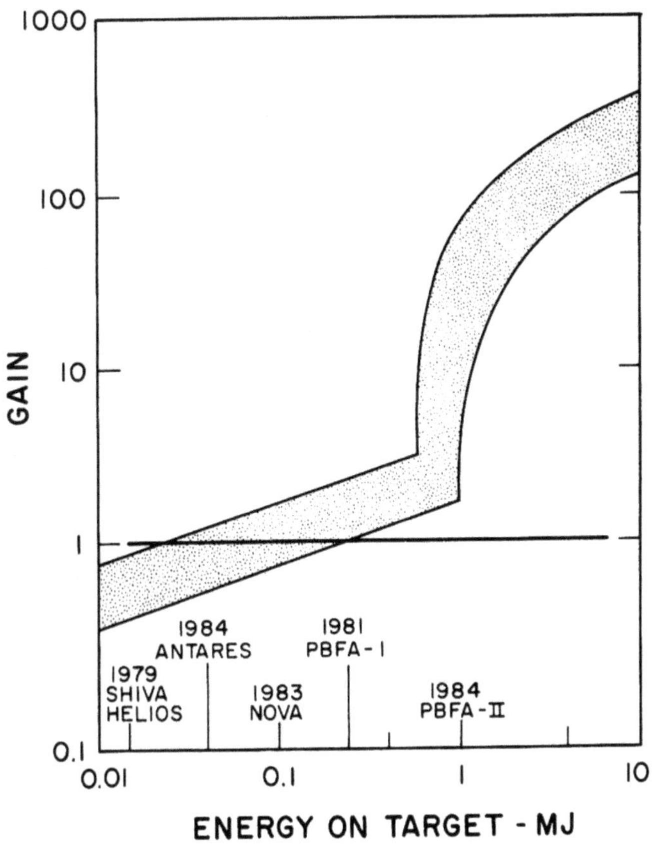

FIGURE 10 Energy–gain predictions for inertial confinement.

If one accepts the premise that fusion will reach the break-even point by the mid-1980s, the question remains: What needs to be done between energy break-even and commercial introduction of hybrids? First we examine the transition from physics to engineering test facilities. Table 2 outlines the possibilities for significant (> 10 kg yr^{-1}) fissile-fuel production in engineering test facilities.

TABLE 2 Possibilities for significant (> 10 kg yr^{-1}) fissile-fuel production in engineering test facilities.

Device	Earliest date for significant fissile-fuel discharge
Tokamak	
INTOR (IAEA)	1995
FED (United States)	Not possible
Tandem-mirror reactor	
TASKA (FRG–University of Wisconsin)	1995
Light-ion beam	Not planned yet

The most ambitious program for a tokamak facility is the INTOR project currently being coordinated by the International Atomic Energy Agency (IAEA, 1980). Examination of its potential operating schedule reveals that if it were to contain fertile material then the earliest date that fissile fuel could be produced at a rate of 10 kg yr^{-1} would be 1995. A lower-power lower-duty-cycle device, FED (ORNL, 1981), is currently being studied in the United States but it is not capable of producing fissile fuel in the desired quantities.

On the tandem-mirror reactor side, a device named TASKA (Badger et al., 1982) is being jointly designed in collaboration between laboratories in the FRG and the University of Wisconsin; this would produce 10 kg of fissile fuel per year by 1995 (at the present time there are no plans to include fertile material in the blanket).

There are presently no plans for inertial-confinement fusion devices which could produce fissile fuel on such a large scale but it is possible that one based on PBFA-II results could be in operation by the late 1990s.

The foregoing discussion indicates that if break-even is reached by the mid-1980s it will be possible to have an engineering test facility producing about 10 kg of fissile fuel per year in the mid-1990s. The next question is then: Can one proceed directly from an engineering test facility like INTOR or TASKA to a commercial hybrid reactor producing tons of fissile material per year? Most likely the answer to this question will be negative because of a multitude of technological problems which make it necessary to design, build, and operate a demonstration prototype hybrid reactor before commercialization. Figures 11 and 12 outline the chronology for commercial hybrid introduction without and with a demonstration reactor. It should be emphasized that these schedules are quite ambitious and represent the earliest possible introduction dates.

Figures 11 and 12 indicate that it is *impossible* for hybrids to be commercially available by the year 2000. If devices of the INTOR/TASKA level are built by 1990 and

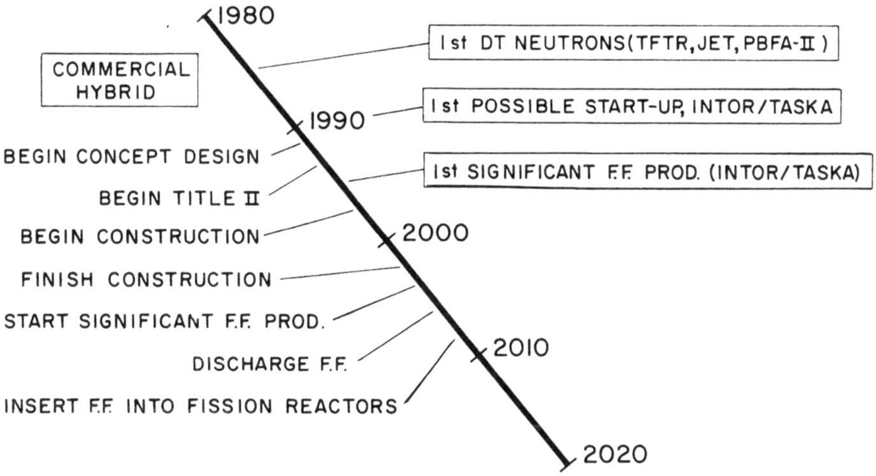

FIGURE 11 The shortest possible route to a commercial D–T hybrid. (FF, fissile fuel.)

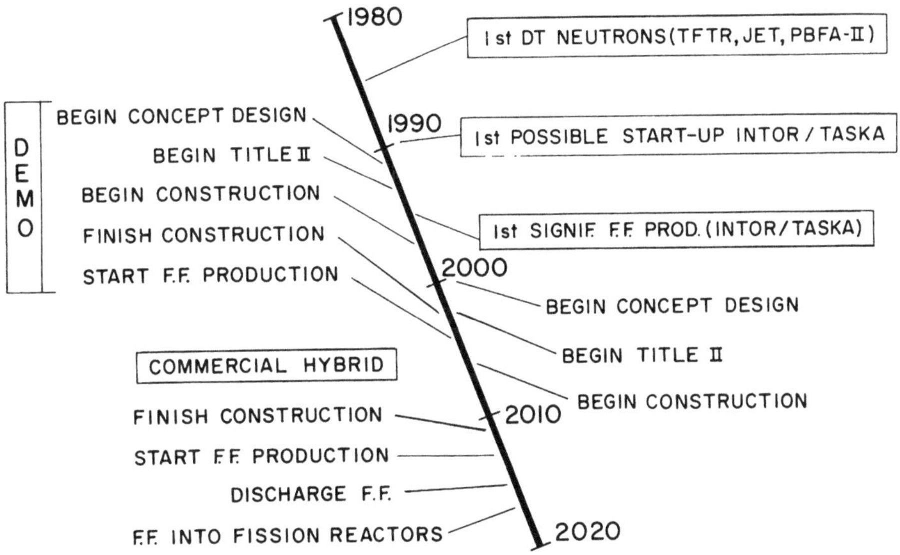

FIGURE 12 A more likely route to a commercial D–T hybrid. (FF, fissile fuel.)

if no demonstration prototype reactor is required, fissile fuel from the first commercial hybrid would be available around 2010. However, if a demonstration reactor is required, the commercial hybrid introduction date shifts to around 2020.

Considering the level of progress in physics and technology which has to be achieved to make the foregoing ambitious schedules feasible, it seems unreasonable to expect hybrids to be commercially available before 2020.

5 DISCUSSION AND CONCLUSIONS

We have focused on the role of hybrids and fast breeders in the world nuclear future and the time windows for their introduction. The technological readiness of both reactors and associated fuel-cycle facilities during these time windows has been examined. Four possible options for the world nuclear demand to be met within the uranium resource base have been identified (Table 1): (1) breeders alone are rapidly introduced in the year 2000 (constraint A, Figure 4); (2) hybrids alone are rapidly introduced in the year 2000 (constraint A, Figure 4); (3) hybrids and breeders (or high converters) are simultaneously introduced in 2000 following traditional market-penetration constraints (constraint B, Figure 6); (4) hybrids and breeders (or high converters) are rapidly introduced simultaneously in 2020 (constraint A, Figure 5).

It has been shown that fast breeders can be commercially available by the year 2000. However, the rapid market penetration required in option (1) would require exceptional efforts. Options (2) and (3) have been ruled out since it is impossible for hybrids to be commercially available by the year 2000. It has been shown that, following an ambitious fusion-development plan, hybrids can be commercially available by 2020; however, the rapid market penetration required in option (4) would again require exceptional efforts. This option would then represent a "fall-back" position in case breeder introduction is delayed beyond 2000 or if the rapid market penetration required in option (1) cannot be achieved. The issue then is whether this "insurance policy" is worth the development costs for the hybrid option. Such a question is beyond the scope of this paper.

REFERENCES

Abdel-Khalik, S.I., Jansen, P., Kessler, G., and Klumpp, P. (1981). Impact of fusion–fission hybrids on world nuclear future. Atomkernenergie Kerntechnik, 38: 1–11.

Badger, B., et al. (1982). TASKA: A Tandem Mirror Fusion Engineering Test Facility. Nuclear Engineering Department Report UWFDM 500. University of Wisconsin, Madison, Wisconsin.

Bethe, H.A. (1978). The fusion hybrid. Nuclear News, 21:41–44.

Bodner, S.E. (1980). Critical elements of high gain laser fusion. NRL-Memorandum Report 4453. Naval Research Laboratory, Washington, D.C.

Fortescue, P. (1977). Comparative breeding characteristics of fusion and fast reactors. Science, 196:1326–1329.

IAEA (1980). INTOR – International Tokamak Reactor – Zero Phase. International Atomic Energy Agency, Vienna.

Imhoff, D.H., et al. (1953). Proposal for a driven thermonuclear reactor. USAEC Report LWS-24920. California Research and Development Corp.

Imhoff, D.H., et al. (1954). A driven thermonuclear power breeder. Report CR-6. California Research and Development Corp.

INFCE (1979). Working Group V on Fast Breeders, Final Report. International Atomic Energy Agency, Vienna.

Jung, H. (1969). Two suggestions regarding controlled fusion: approximation equations for fusion–fission reactors and fusion-reaction-tube reactors. Nuclear Fusion, 9:169.

Klumpp, P. (1979). Die Bedeutung von Leichtwasserreaktor-Hochkonvertern für den Natururabedarf der Bundesrepublik Deutschland-Analyse möglicher Strategien des Übergangs von Leichtwasserreaktoren zu Schnellen Brütern. Dissertation, University of Karlsruhe, Karlsruhe.

Lawson, J.D. (1955). A survey of some suggested methods of realizing fusion reactors. AERE-GP/M, No. 185. Atomic Energy Research Establishment. Harwell, UK.
Lewis, W.B. (1952). The significance of the yield of neutrons from heavy elements excited to high energies. DR-24 Chalk River Nuclear Laboratories, Chalk River, Ontario, Canada. Reprinted as AECL-968 (1960), Atomic Energy of Canada Limited, Chalk River, Ontario.
Lidsky, L.M. (1969). Fusion–fission symbiosis: general considerations and a specific example. In Proceedings of the Nuclear Fusion Reactors Conference, Culham Laboratory. British Nuclear Energy Society, Culham, UK.
Maniscalco, J.A. (1976). Fusion–fission hybrid concepts for laser-induced fusion. Nuclear Technology, 28:98.
Maniscalco, J.A., Hansen, L.F., and Allen, W.O. (1978). Scoping studies of ^{233}U breeding fusion–fission hybrid. In Proceedings of the ANS Topical Meeting on the Technology of Controlled Nuclear Fusion, 3rd, Santa Fe, New Mexico. CONF-780508. American Nuclear Society, LaGrange Park, Vol. 1, pp. 159–169.
OECD–IAEA (1978). World uranium potential – an international evaluation. OECD Report. Organization for Economic Cooperation and Development, Paris.
ORNL (1981). Initial trade and design studies for the fusion engineering device. ORNL/TM-7777. Oak Ridge National Laboratory, Oak Ridge, Tennessee.
Teller, E. (1979). Fission or fusion? American Nuclear Society, Transactions, 33:214.

Section Three

Fusion and Dense Plasma Neutron Sources

THE TOKAMAK AS A CANDIDATE FOR A D–T FUSION REACTOR

G.H. Wolf
Institut für Plasmaphysik der Kernforschungsanlage Jülich GmbH, Association EURATOM–KFA, D-5170 Jülich, Postfach 1913 (FRG)

ABSTRACT

Based on current conceptions, the most important features of the tokamak fusion reactor are presented. The characteristic values which would have to be attained for successful operation of the reactor are then compared with the present status of tokamak physics. The comparison forms the basis for an assessment of those areas of plasma physics and reactor technology where critical problems remain to be solved.

1 INTRODUCTION

The source of usable fusion power closest to realization is the D–T reaction yielding 17.6 MeV of energy, of which 80% is carried by a neutron. Including secondary (breeding) reactions in the blanket, about 20 MeV per fusion event may be gained. Thus the fusion flame of a 3000-MWth unit emits about 10^{21} neutrons (14.1 MeV) per second. A blanket system surrounding the burning chamber is required to absorb these neutrons; in this blanket the power of the neutrons is transformed into heat which can be used for electricity generation or – when the necessary high-temperature technology has been mastered – for power-consuming chemical reactions. Eventually most of the neutrons are required to breed tritium from lithium with or perhaps without the use of additional neutron multiplication. Alternatively the neutrons may also be applied to breed fissile materials (e.g. from U-238) using surplus neutron reactions for the tritium production.

The favored mode of reactor operation, which it is believed will become possible with toroidal magnetic confinement, is thermonuclear burning, the heat losses from the plasma being replenished by the α-particle energy (3.5 MeV) from the fusion processes themselves. Thus it is essential that these α particles remain magnetically confined until they have transmitted their energy to the fusion plasma; thereafter they should be pumped away. This mode of operation is preferable to other schemes in which the heat losses from the plasma are mainly replaced by heating from external sources throughout the operating period. Particularly in view of their low efficiency, such external heating devices would place an additional burden on the global power balance and would endanger the chances of obtaining a useful net energy gain. Furthermore, if we view the fusion reactor as a neutron source, this mode of "thermonuclear burning" gives it a specific advantage over any of the neutron sources which require high-power beams in order to maintain neutron-producing reactions in a target.

This paper is intended to give a short description of the characteristic features of a tokamak reactor and – from the present status of tokamak technology – to outline some of the problems which remain to be solved.

2 PRESENT CONCEPTIONS OF THE REACTOR (SEE TABLES 1 AND 2)

(1) A conventional tokamak reactor works in a pulsed mode. During the quasi-stationary "burning" phase no input of external power into the plasma is required in order to maintain the characteristic plasma parameters (temperature $T = 10$–20 keV, density $n = (1$–$3) \times 10^{14}$ cm^{-3}). Heating the plasma towards "ignition" requires a net heating power of the order of 100 MW (or more) to be deposited in the plasma from external sources; the duration of this heating phase is several seconds.

TABLE 1 Some characteristics of a tokamak reactor.

Parameter	Experimental status	Goal	Unit
$n\tau_E$ (ignition)	10^{12}–3×10^{13}	$(2$–$4) \times 10^{14}$	s cm^{-3}
Pulse duration	0.1 $(10)^a$	100 +	s
Heating power density	0.3 $(0.05)^a$ $(1)^b$	0.4–1	W cm^{-3}
Plasma temperature T	$7(0.5$–$1)^a$	10–25	keV
Particle density n	$10^{14\,d}$ $(6 \times 10^{14})^b$	10^{14}	cm^{-3}
Magnetic field B	3.5 $(9)^b$	5–6	T
Relative plasma pressure β	0.01–0.03^e	0.06	–
Fusion power density p_f	Negligible^f	2–5	W cm^{-3}
Average surface heat load	10	40–100	W cm^{-2}
Average heat load per unit volume inside blanket	–	4–10	W cm^{-3}
Plasma current I_p	1^g	10	MA
Minor plasma radius a	0.5^g	2.5	m
Major radius R	1.5^g	10	m

[a] For ohmic heating only.
[b] For specific high-field tokamaks with ohmic heating.
[c] Superconducting.
[d] There are indications of an $n \propto B/R$ relationship.
[e] The required improvement is expected from an elongated cross section.
[f] Expected to become significant with TFTR and JET. TFTR is the Toroidal Fusion Test Reactor (Princeton, New Jersey) and JET is the Joint European Torus (Culham, UK).
[g] For JET $I_p = 3$–5, $R = 3$, and $a = 1.4$.

TABLE 2 A fusion reactor: the limitations on various parameters.

Ignition
Sufficient magnetic thermal insulation: $n\tau_E \propto n^2 a^2$?
or $n \propto B/R$?

Impurities (radiation, pressure)
Appropriate heating methods (efficiency, core)

Power density: $p_f \propto \beta^2 B^4 \langle \sigma v \rangle / T^2$
Magnetic field strength (superconduction, forces)
β = plasma pressure/magnetic pressure
Neutron wall load, plasma–wall interaction

Pulse duration
Transformer flux swing (volt-seconds)
Accumulation of impurities
Sufficient exhaust of helium (ashes)
Control of burning temperature

Coolant temperature
Tritium permeation into coolant
Degradation of material properties
Corrosion by lithium, coolant, and reaction partners

Reactor lifetime (replacement of components)
Erosion and corrosion of
  Elements of the first wall (plasma, neutrons)
  Blanket component
  Neutron multipliers
Burnup of breeding material
Unit size?

(2) The duration of the burning phase (essentially the pulse duration) is finally limited by the transformer flux swing driving the induced loop current I_p typical of tokamaks. Within this limitation the pulse duration could extend to several thousand seconds, provided that no critical accumulation of wall-released impurities or of ashes (helium) occurs since this would lead to much shorter burning times. Without an effective exhaust for helium the pulse duration is limited to 20–100 s; this is probably too short for a positive energy yield and would probably require a number of thermal cycles that is larger than can be tolerated.

(3) There are schemes which promise a transition to a genuine stationary mode of reactor operation. However, they require either the application of power-consuming auxiliary devices (radio-frequency transmitters, injectors) or plasma confinement by means of other magnetic configurations (e.g. those in stellarators) which do not need an induced loop current and which lead to a loss of toroidal symmetry.

(4) The fusion power density p_f averaged over the whole volume of the burning chamber is of the order of watts per cubic centimeter (i.e. MW m^{-3}), say typically 2–5 W cm^{-3}. This is derived from an average magnetic field strength B of 5–6 T, since $p_f \propto \beta^2 B^4$ where β is the ratio between plasma pressure (including some helium pressure) and magnetic pressure. One-fifth of p_f is the α-particle heating power p_α transmitted to the plasma, while four-fifths of p_f goes via the neutrons into the blanket. The maximum

value of p_f is about equally determined (i) by the maximum plasma pressure which can be tolerated by the magnetic field and (ii) by the amount of fast neutrons which penetrate the first wall system and which – integrated over the years – are expected to lead to significant material degradation effects (radiation damage). Material degradation by neutrons, however, is a problem common to all sources of fast neutrons where walls are needed to separate the loci of neutron production from those of neutron absorption; in fact the problem may become worse for more compact neutron sources as compared to the large-volume fusion plasma (see also point (7) in this section).

(5) The heat load per unit volume from the neutrons absorbed in the blanket is typically twice the value of p_f. The area power density related with the penetration of these neutrons through the first wall is typically 200–500 W cm^{-2} (2–5 MW m^{-2}); however, only a small fraction of it is absorbed there. By contrast, the average thermal load on the surfaces of the first wall caused by plasma–wall interaction processes is around 40–100 W cm^{-2} (0.4–1 MW m^{-2}); furthermore, it should be noted that a significant fraction of this surface heat load tends to be focused by magnetic field lines on small parts of the total surface which consequently may become exposed to local values that are much higher than that expressed by the average.

(6) The minimum dimensions of the burning chamber and the minimum total fusion power p_f result from the minimum value of $n\tau_E = 2 \times 10^{14}$ s cm^{-3} required for thermonuclear "ignition". $n\tau_E$ is a scale value describing the capability of the magnetic confinement system to provide sufficient thermal insulation. The minimum dimensions estimated from present scaling experience are as follows: minor plasma radius $a = 1.5$–2 m and total fusion power $p_f = 500$–1000 MWth. This may be reduced by a factor of 3–10 if the demand for ignition does not need to be maintained (e.g. for a fusion–fission hybrid system). For commercial reactors larger unit sizes are usually considered: typically major radius $R = 10$ m, minor plasma radius $a = 2.5$–3 m, and $p_f = 3000$ MWth (see Figure 1).

(7) The burning chamber is almost completely enveloped by blanket components for tritium breeding in order to obtain a sufficiently high breeding ratio (i.e. > 1). Depending on the neutron fluence through the layers of the first wall which separate the fusion plasma from the blanket volume and on the actual materials used, these elements of the first wall may need to be replaced several times during the total lifetime of the reactor.

(8) Obvious features of a fusion reactor are its large electrical components for producing the magnetic fields, inducing the plasma current, and controlling the magnetic confinement; thus fusion is based on large-scale electronuclear technology. The magnet system for both the toroidal magnetic field and probably also the poloidal magnetic field has to consist of superconductors; the maximum magnetic field strength at the toroidal field coils is in excess of 10 T. The stored magnetic energy is of the order of 100 gigajoules.

3 THE PRESENT STATUS OF TOKAMAK PHYSICS

3.1 With Ohmic Heating Only

(1) In tokamaks where the plasma is heated solely by the induced tokamak current I_p (i.e. without additional heating) and with values of I_p of typically 0.5 MA, the range

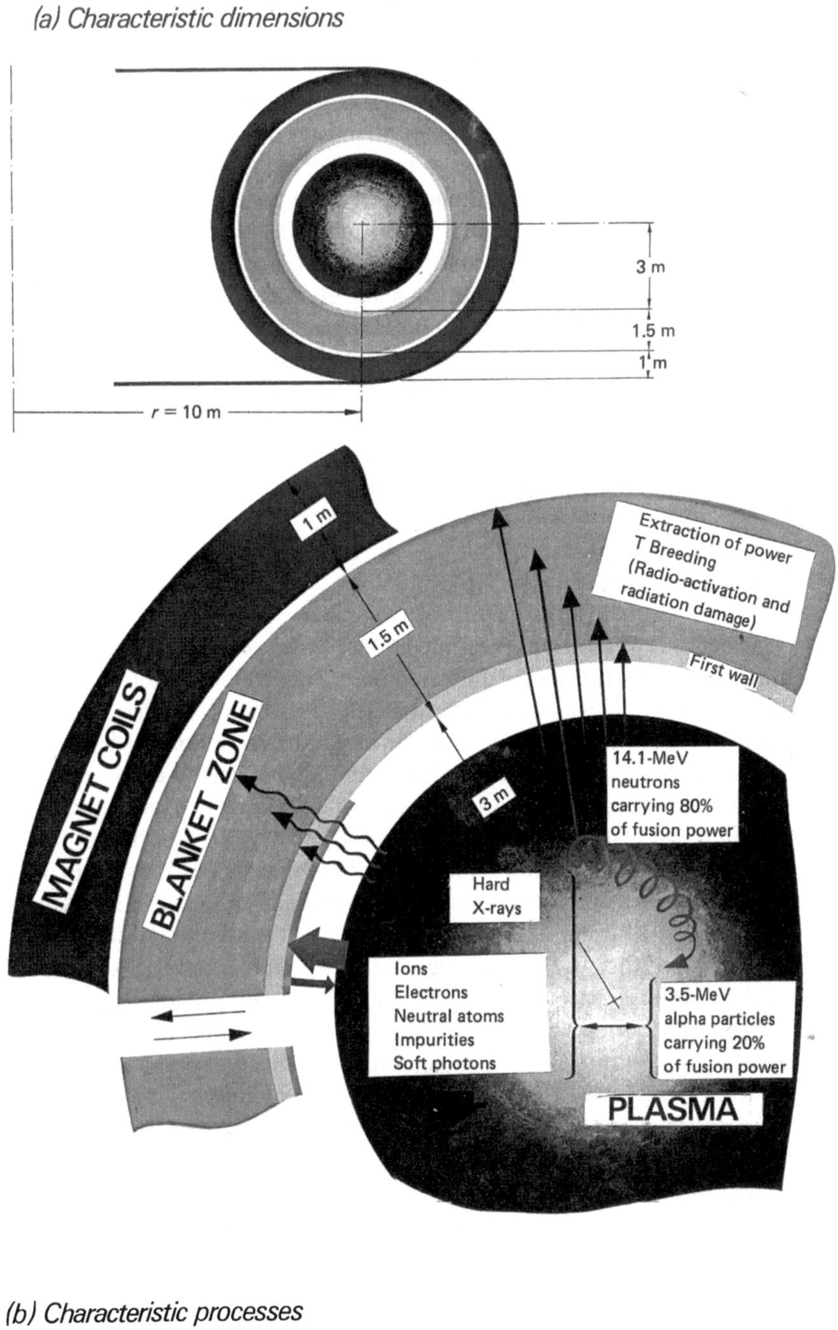

FIGURE 1 Cross-sectional diagrams of a toroidal D–T fusion reactor: (a) characteristic dimensions, (b) characteristic processes in each zone. The plasma–wall interaction has four main components: (1) heat transfer (20% of fusion power), (2) release of impurities from the wall, (3) erosion of the wall structure, (4) fuel cycle and exhaust (fuel, ashes, and impurities).

of magnetic field strength has been extended up to 9 T. The pulse duration has been extended beyond 10 s (ASDEX*). With typical plasma temperatures $T_{e,i}$ of 1 keV the value of $n\tau_E$ is around 10^{13} s cm^{-3}; for high-field devices it has even exceeded 4×10^{13} s cm^{-3} (FT). Using large magnetic fields an average plasma particle density of 6×10^{14} cm^{-3} has been achieved. Z_{eff}, a scale figure for plasma purity (ideally $Z_{eff} = 1$), could be reduced to between 1 and 1.5. Disruptive instabilities, which are not yet completely understood, are still a major concern although there is a growing belief that it may become possible to control or even suppress them (ASDEX).

(2)There is some dispute about whether there exists a general empirical scaling law which correlates $n\tau_E$ with the relevant plasma parameters and dimensions. A relationship like $n\tau_E \propto n^2 a^2$ (the ALCATOR scaling) has appeared to be a useful approximation over a wide range. However, the detailed transport processes which might lead to such a relationship are not understood. The ion heat conduction apparently scales neoclassically and the electrical resistivity (with respect to the tokamak current I_p) classically. The maximum tolerable plasma densities vary in proportion to B/R (the Murakami limit) which means that for reactor dimensions the resulting density values are too low.

3.2 With Additional Heating

(1) The methods of additional heating that have been successfully applied so far are (i) adiabatic compression, (ii) Neutral Injection (NI), and (iii) Ion-Cyclotron Resonance Heating (ICRH); some understanding of the detailed heating processes has been gained. The applied power levels have extended into the megawatt range; for smaller power levels electron-cyclotron resonance heating and "lower-hybrid" heating have also been used. For the regime of low-to-medium plasma densities, the achieved increase in plasma temperature ("heating rate") has been typically around 1 keV MW^{-1} (with NI and ICRH) and the maximum ion temperature T_i that has been obtained is 7 keV (NI in PLT). Differences in the heating rates observed in different tokamaks and variations in reproducibility are not yet sufficiently understood. Until now the duration of the heating pulses has been limited to about 0.2 s, mainly due to technical constraints (e.g. insufficient cooling) on the heating devices. The presence of impurities observed during the heating pulse, however, leads to some concern that when prolonged heating pulses are applied the buildup of impurities may become a major problem; this is caused partly by an increased load on specific wall elements (limiters), partly by injected impurities, and partly by incomplete confinement of the high-energy orbits and by plasma-antenna contact (see also point (5) of Section 4.1).

(2) The power densities (W cm^{-3}) attained in the plasma by means of additional heating have already reached the values which are needed in a reactor. This statement does not hold for the resulting power flux densities (W cm^{-2}) onto the first wall since the volume/surface ratio is still too small. The maximum plasma density n which has been achieved is larger than that obtained without additional heating. There are indica-

* Acronyms have the following meaning: ASDEX, the Axial-Symmetrisches Divertor Experiment (Garching, FRG); ALCATOR, the high-field tokamak at MIT (Cambridge, Massachusetts); FT, the Frascati Tokamak (Frascati, Italy); PLT, the Princeton Large Torus (Princeton, New Jersey); ISX-B, the Impurity Study Experiment B (Oak Ridge, Tennessee).

tions, however, that n is still dependent on B/R, although with a somewhat larger coefficient which improves the density extrapolation for larger devices.

(3) The relative plasma pressure β is important for the thermonuclear power density and thus also for ignition (e.g. in the Joint European Torus (JET)) and for the required heating power. For plasmas with a circular minor cross section, values of β were observed (e.g. those for ISX-B) which apparently have reached – perhaps even exceeded – the theoretically expected upper limit. However, both quantities (the theoretical upper limit and the experimental observations) are still the subject of dispute and concern, including also the predicted benefit of elongated (i.e. D-shaped) minor plasma cross sections.

(4) Until recently the magnetic fields for plasma confinement were generated mainly with "conventional" i.e. (non-superconducting) coils, though partly using cryotechniques (e.g. the 9-T field in ALCATOR). The magnets consisted of both band coils and Bitter coils (disc coils).

4 ASSESSMENT

Plasma physics and technology are closely interlinked. While fully appreciating this fact, I shall, however, attempt to subdivide this assessment into these two categories. Given the structure of this paper, some repetition of arguments or aspects is unavoidable.

An essential feature of tokamaks is that they are far ahead of the other systems in approaching those plasma parameters which are required for ignition, i.e. $Q = \infty$. Q is the ratio of the fusion power generated in the plasma to the power input from external sources into the plasma (at any time during the quasistationary phase of the pulse). (There is another definition of Q (see for instance Raeder et al., 1981, p. 270) which describes the total energy balance by integrating all power-consuming or power-producing terms over one pulse period. Obviously, using this definition, $Q = \infty$ can never be achieved.)

4.1 Plasma Physics

(1) Due to our insufficient knowledge of the relevant transport processes and scaling laws, the minimum dimensions (in particular the minor plasma radius a and the magnetic field B) which are required for ignition cannot be exactly predicted. Important information on these questions is expected from TFTR, JET, and JT-60*; the first two of these – when operating with tritium – should in addition demonstrate α-particle confinement and the generation of significant fusion power, and it is expected that the $Q = 1$ limit will be exceeded. We may be reasonably optimistic with respect to our chances of confining an ignited plasma in burning chambers of the large sizes (and magnetic field strengths) that are required for power reactors. The values of $n\tau_E$ encountered in large size power reactors may even yield so much extra heating power, over and above the balance between α-heating and heat conduction losses, that a significant fraction of the α power may be (or may *need* to be) emitted via impurity radiation even out of the core.

---

* JT-60, the JEARI Tokamak 60 (Japan).

(2) Since ignition has not been achieved yet there is uncertainty about the effects of the burning process on transport mechanisms, stability (including internal disruption phenomena), and temperature control. Various schemes have been proposed to solve the last task. However, final answers to these questions will not become available until relevant experiments have been done. The results of JET may be of relevance here, depending on how closely thermonuclear ignition can be approached. It is unlikely, however, that JET will give information on the question of how to control the burning temperature.

(3) According to present opinion, no disruptions of the plasma current (major disruptive instabilities) can be tolerated in a power reactor. It is unclear whether this condition also holds for a DEMO. For INTOR a limited number of such disruptions – determined, for example, by fatigue – may be tolerated. There is encouraging experimental evidence from the divertor experiment ASDEX and from the stellarator W VII that control of disruptions may in principle become possible. Whether this will also be the case for "conventional" tokamaks is as yet unclear; generally it is hoped that improved control of the plasma–wall interaction processes will also eliminate the disruptive instability.

(4) The possibility of achieving pulse durations of at least some hundreds of seconds is still uncertain. An elementary precondition for this is sufficient outward transport of the α particles (ashes) from the plasma core, through the action of transport mechanisms which may be influenced by α-particle heating. By contrast the extraction of these α particles from the plasma boundary layer into an efficient exhaust is a more technological than theoretical problem which is connected with the question of whether a magnetic divertor is required and, if so, which type of divertor. This question is itself closely interlinked with the problem area of plasma–wall interaction to be discussed under point (5). Present research programs are also examining how to prolong the induced tokamak current by external methods, with the final goal being d.c. operation.

(5) The concentration of helium (ashes) and impurities in the plasma core – originating from plasma–wall interaction – must remain below small critical values in order (i) not to endanger ignition or to quench burning prematurely and (ii) to limit the erosion and corrosion of the first wall as the source of these impurities, for example, by sputtering, microarcing, or evaporation. It still needs to be demonstrated that the respective criteria can be fulfilled and what will be the appropriate methods to achieve this goal. This means that the underlying processes of power transfer and ash exhaust are not yet sufficiently controlled or even understood. The experimental progress achieved so far needs to be extended towards plasmas that are subject to additional heating, taking into account, however, that the wall of a reactor will be exposed to flux densities (of both particles and energy) which will exceed those obtained in present experiments and that this will occur during as-yet unachieved pulse durations and via transport mechanisms which are codetermined by α-particle heating. Theoretical concepts have been developed for the desired properties of the plasma boundary (e.g. radiation cooling, convection cooling, and shielding) which suggest possible solutions. More experimental results are needed in order to make a choice between the various possibilities of plasma exhaust and boundary diversion (e.g. between mechanical and magnetic divertors).

(6) There is much discussion about the question of whether the achievable values of β will be sufficient for economic power production. In view of the maximum tolerable

neutron fluence through the first wall which also limits the power density, the β problem, although very important for reactor economy, should not be overrated at the present state of the art. Improvements in magnet technology and the acceptance of larger reactor unit sizes might still result in wall-loading being the limiting factor for the power density.

(7) There is still some uncertainty with respect to the required and achievable plasma density n (influencing both $n\tau_E$ and p_f). If the dependence of n on B/R suggested from present experiments cannot be overcome, we are faced with a serious difficulty, leading to a limit on the major radius R, additional demands for higher magnetic fields B (see also point (1) of Section 4.2), and operation at higher plasma temperatures T on the $\langle\sigma v\rangle/T^2$ curve. However, predictions from transport models give some cause for optimism.

(8) Developing the quite formidable apparatus that will be required for heating the plasma towards ignition poses considerable technological problems such as efficiencies and pulse duration and reliability. The adaptation of beam energy (efficiency!) or antenna size to the demands of heating the core of a reactor plasma are also part of the problem. One major limitation on performance stems from the need to cool those parts which are exposed to high thermal loads. There is a tendency to favor RF heating for future applications, one of the reasons being the expected efficiency of the process. This should not, however, cause proven methods and their potential for improvement (e.g. sources of negative ions) to be neglected. Refueling appears to become difficult only if magnetic divertors combined with heavily pumped divertor chambers need to be used; otherwise a moderate gas feed combined with strong "recycling" may suffice for a satisfactory particle balance.

4.2 Technology

Here only those questions will be mentioned which are specific for a tokamak system with respect to its potential as a reactor; thus, more general questions of fusion-reactor technology are excluded. The problems of heating, refueling and plasma–wall interaction have already been discussed under points (6)–(8) of Section 4.1 since they represent typical points of overlap between physics and technology.

(1) The general opinion is that the magnetic field of an electricity-producing D–T reactor can be economically sustained only by using superconducting magnets. This may also be true for a breeding D–D hybrid reactor, though not necessarily for a breeding D–T hybrid reactor. In view of the β limitation it is still unclear whether NbTi conductors will yield a toroidal field compatible with satisfactory thermonuclear power densities. It is assumed that further development of superconducting-magnet technology will lead to increased magnetic fields, that the side effects from unavoidable electromagnetic transients can be reduced to tolerable levels, and that solutions will be found for supporting the large tilting forces on the coils. By contrast, the geometrical arrangement of the poloidal magnet system poses great difficulties, particularly for those coils which one would like to place inside the toroidal magnet. The questions of access, assembly, and repair (reliability) are delicate key problems which have not yet been solved. It would be desirable to find a modular design at least for the inner parts of the poloidal field system (segments); however, no solution has yet been found.

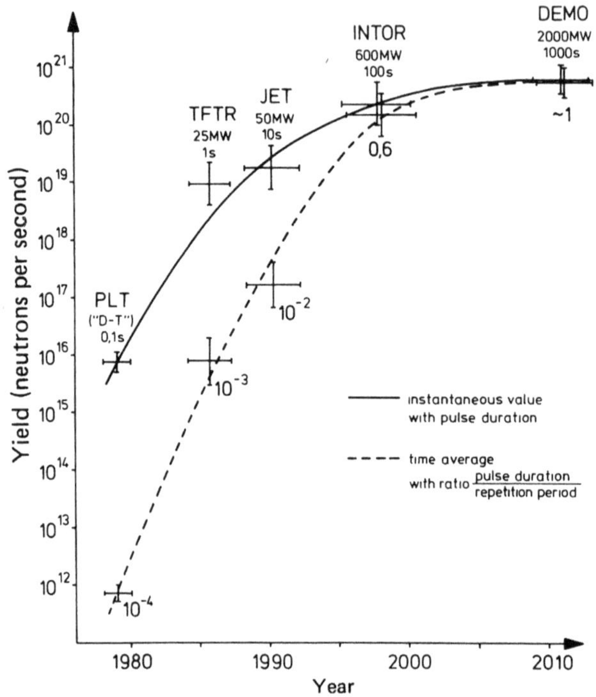

FIGURE 2 Expected neutron yield of future tokamaks (3000 MW $\simeq 10^{21}$ neutrons per second). (The values for PLT were obtained with deuterium only, but are scaled towards D–T reactions.)

(2) When exploiting the high neutron yield of future tokamaks (see Figure 2) for breeding purposes, the blanket system must cover the burning chamber almost completely, especially if fertile materials of a low breeding ratio are used. For tokamaks this leads to additional difficulties with respect to assembly and access which result from the toroidal geometry, particularly in the so-called inner areas (i.e. towards the center line of the machine). Another serious blanket problem typical of tokamaks is caused by the transients of the loads inherent in pulsed operation which lead to additional (e.g. fatigue) difficulties. However, not much effort has yet been invested in finding solutions for these problems, given that the power densities in the blanket are relatively low. It is still much too early to conclude that the foregoing problems will render the construction and operation of a tokamak reactor impossible.

(3) The satisfactory containment of tritium – for both normal operation and fault conditions – has not yet been demonstrated. This problem is fairly independent of the specific reactor concept and is mainly concerned with the transport of tritium through heat exchangers, etc. Some complications may arise, however, from the geometrical difficulties as discussed in point (2) of this section.

(4) The most desirable mode of operation would be to achieve a steady state. Significant efforts should be made to include in the research program schemes which promise eventual transition to this mode of operation.

5 REMARKS

For the first time in the history of technical science it is foreseeable that planned efforts oriented towards the development of a specific machine will need to be extended over several generations (see also Figure 3). In such a situation progress can only be defined by the achievement of intermediate steps.

Present experimental results are encouraging enough to warrant the statement that machines of the tokamak type (and probably also those of the stellarator type) have the potential to reach and to maintain thermonuclear burning, or at least $Q > 1$, within reasonable machine parameters; for all other fusion design concepts such a statement would be premature.

This first step on the long development path is commonly called the *demonstration of physical feasibility*. Anticipating the success of this step, preliminary efforts have now been initiated to attack the second step of the development which is the *demonstration of technical feasibility*. Recognition of the serious difficulties which become apparent in approaching this second goal, however, should not be taken as an excuse to concentrate current efforts on other concepts which, on the one hand, appear to be technically simpler but, on the other hand, inherently contain a much larger fraction of speculative extrapolation with respect to their ability to fulfill the physical requirements.

An assessment of the tokamak system with regard to the third and last step – the *demonstration of economic usefulness* – is particularly difficult since it requires knowledge of the general economic situation as it may develop for several decades in the future. Excluding here the more general, but certainly very important question of whether there will be a need for fusion and whether fusion will be able to fulfill this need, we have to compare the tokamak system with its possible alternatives. The tokamak appears to have no significant economic disadvantages.

The most apparent distinguishing feature of the tokamak (or its "cousin", the stellarator) is its potential for confining an ignited plasma whereas most of the alternative concepts may in principle only operate as power amplifiers. By contrast, the essential disadvantages which can be observed are caused by the toroidal magnetic field configuration which leads to the well-known difficulties of assembly, maintenance, and repair, particularly with respect to the remote handling procedures required. Experiments which will go into operation during this decade are expected to demonstrate whether this toroidal geometry is in fact necessary to confine thermonuclear plasmas.

Looking back on more than 20 years of research in magnetic fusion (see also Figure 3), the following conclusions stand out.

(a) Enormous progress has been achieved in the difficult art of mastering plasma confinement. Although our theoretical understanding of the significant physical mechanisms has also improved, several important phenomena are still unexplained. We have been able to shed some light on the road ahead of us by specifying more closely our conceptions about future reactors and our estimates of the demands we may wish to place upon them. The achievements made so far have resulted not from the introduction of new configurations and principles but rather from thorough, hard, and responsible work together with global cooperation: this approach has yielded success in a series of small but important steps.

(b) Magnetic confinement requires two rather complex systems: (i) the outside

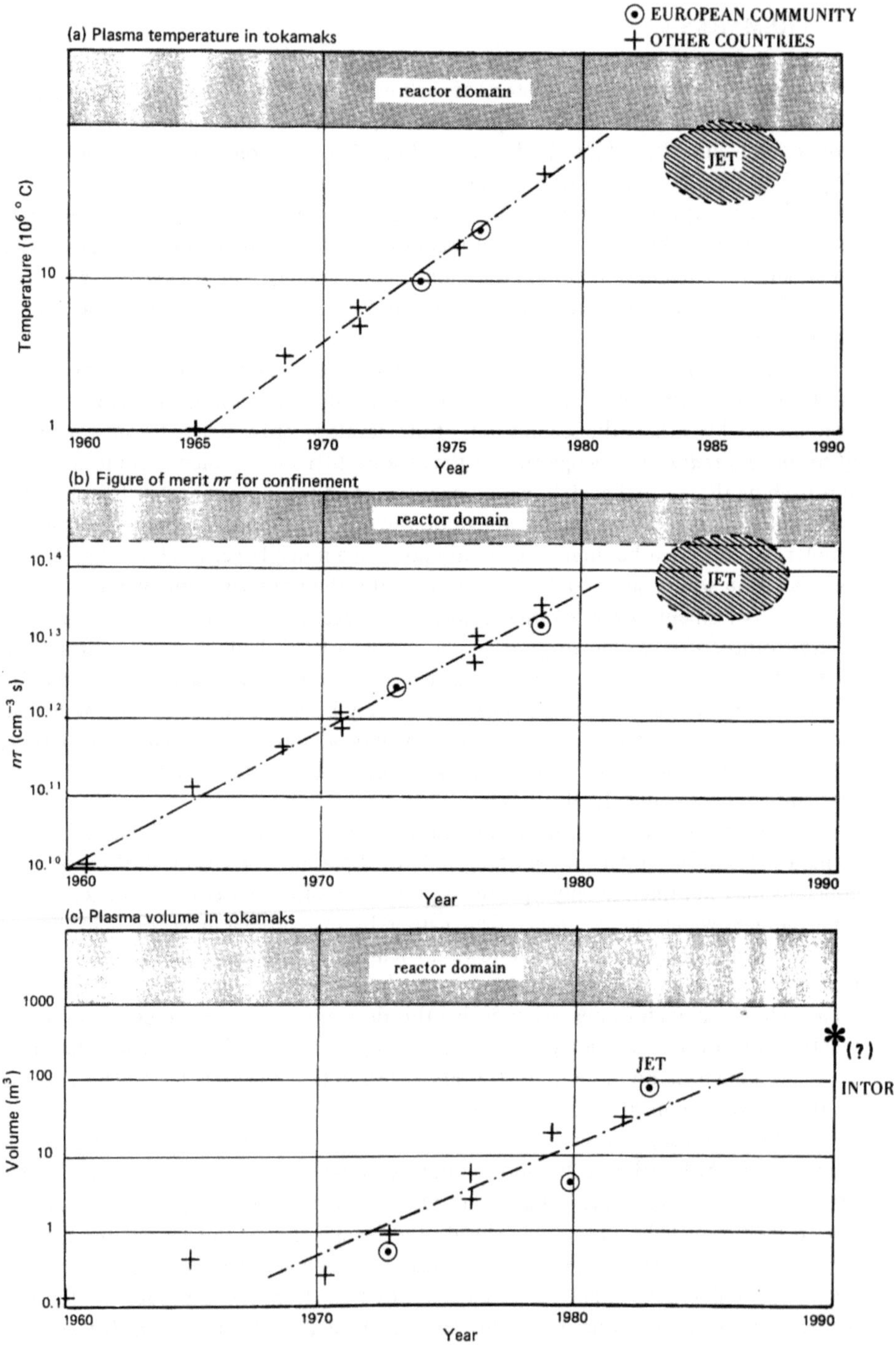

FIGURE 3 Progress in various significant parameters: (a) plasma temperature in tokamaks, (b) figure of merit $n\tau$ for confinement, (c) plasma volume in tokamaks. (By courtesy of the European Community Fusion Programme.)

arrangement of magnetic coils, transformer, energizing and feedback circuits, and heating equipment; (ii) plasma currents and plasma flows interacting with each other and with the outside circuits. There is a mutual complementarity with respect to the degree of complexity of these two systems; i.e. we have either a very complex and intrinsically coupled (stability!) plasma configuration enclosed by a relatively simple magnet arrangement or – at the other extreme – a highly controlled plasma configuration enclosed by a relatively complicated magnet arrangement. Though we do not know the optimum yet, it looks as if working tokamaks are not so far off.

ACKNOWLEDGMENTS

The author gratefully acknowledges helpful discussions with colleagues from Garching and Jülich, in particular J. Eidens, G. Grieger, E. Hintz, M. Kaufmann, M. Keilhacker, and J. Raeder.

BIBLIOGRAPHY

Because of the nature of this survey no references to specific original publications can be made within the restricted space available. Instead, a selection of books and proceedings is given where practically all the relevant facts on the state of the art may be found.

Argonne National Laboratory (1980). STARFIRE, a commercial fusion power plant study. ANL/FPP-80-1. Argonne National Laboratory, Argonne, Illinois.
Bulletin of the American Physical Society 25, 8 (October 1980).
Grieger, G. (Editor) (1980). Contributions to the IAEA Technical Committee Meeting on Large Tokamak Experiments, 4th, Tokyo, April. IPP, Garching, FRG.
Häfele, W., Holdren, J.P., Kessler, G., and Kulcinski, G.L. (1977). Fusion and fast breeder reactors. RR-77-8. International Institute for Applied Systems Analysis, Laxenburg, Austria.
IAEA (1979). Plasma Physics and Controlled Nuclear Fusion Research 1978, Vols. I–III. International Atomic Energy Agency, Vienna.
IAEA (1981). Plasma Physics and Controlled Nuclear Fusion Research 1980, Vols. I–III. International Atomic Energy Agency, Vienna.
IAEA (1981). INTOR International Tokamak Reactor, Conclusions of the IAEA Meeting, 6th, Vienna, April. EUR FU BRU/XII 2/81/EDV 24. International Atomic Energy Agency, Vienna.
JET (1979). Joint European Torus, Annual Report. EUR 6822 EN, Commission of the European Community, Brussels.
Köppendörfer, W. (1981). The ZEPHYR experiment. In the Proceedings of the Symposium on Fusion Technology, Oxford, September, 1980. Pergamon Press, Oxford, Vol. I, p. 77.
Plasma Surface Interactions in Controlled Fusion Devices, North-Holland, Amsterdam, 1980.
Raeder, J., Bünder, R., Dänner, W., Klingelhöfer, R., Lengyel, L., Leuterer, F., and Söll, M. (1981). Kontrollierte Kernfusion. B.G. Teubner, Stuttgart.

THE FUSION BREEDER: ITS POTENTIAL ROLE AND PROSPECTS*

J.D. Lee
Lawrence Livermore National Laboratory, Livermore, California (USA)

ABSTRACT

The fusion breeder concept utilizes 14-MeV neutrons from DT fusion reactions to produce more neutrons than are required to breed the tritium needed to sustain the fusion process; this excess is used to produce fissile fuel for subsequent use in fission reactors. The historical development of and motivation for the fusion breeder are described and a number of conclusions drawn about the possible future role of such a device. First, it could significantly influence the development of both fission and fusion plants by allowing fission to meet demand with a variety of reactor types and by acting as an early commercial fusion application. Second, if serious development work starts now, the device could be ready relatively quickly and in time to avert some of the energy/environment problems foreseen for the 21st century. Third, the concept seems economically favorable, would set a ceiling to U_3O_8 prices, and could show a very high benefit/cost ratio. Finally, the device would give power utilities the option of just buying fissile fuel as they do at present from enrichment plants.

1 INTRODUCTION

The fusion breeder is a concept that utilizes 14-MeV neutrons from $D + T \rightarrow n(14.1\ MeV) + \alpha(3.5\ MeV)$ fusion reactions to produce more neutrons than are needed to breed the tritium (T) required to sustain the fusion process. This excess capacity is used to produce fissile material (Pu-239 or U-233) for subsequent use in fission reactors. We concentrate on a class of blankets we call fission suppressed. The blanket is the region surrounding the fusion plasma in which fusion neutrons interact to produce fuel and heat.

The fission-suppressed blanket uses non-fission reactions (mainly (n, 2n) or (n, n′t)) to generate excess neutrons for the net production of fuel. This is in contrast to the fast fission class of blankets which use (n, fiss) reactions to generate excess neutrons. Fusion reactors with fast fission blankets are commonly known as fusion–fission hybrids because they combine fusion and fission in the same device.

* Work performed under the auspices of the US Department of Energy by the Lawrence Livermore National Laboratory under Contract W-7405-Eng-48.

2 MOTIVATION

The motivation for developing the fusion breeder is basically the same as for the fission breeder – to tap the enormous energy potential of the world's abundant U-238 and Th-232 resources. While the basic motive is the same there are important differences that make the fusion breeder concept most intriguing. For example:

1. The fusion breeder requires no fissile inventory and the doubling time for the fusile inventory (T) is measured in weeks, not years.
2. The fusion beeder can produce many times (up to ~ 30) more net fissile product per unit of power than a fast breeder reactor.
3. Power densities are 10 to 100 times less than those in the fission breeder.
4. The fusion breeder adds flexibility to the long-term fission energy option. All sorts of fission reactors can be considered, not just fast breeders, by virtue of an inexhaustible, prolific source of fissile material from fusion breeders.
5. The fuel-production and energy-production roles are performed in separate devices.

From fusion's viewpoint a motivation for the introduction of the fusion breeder is the potential for an earlier, competitive application of fusion that in turn would advance fusion technology towards its ultimate goal of being a stand-alone energy source. Conditions that must exist for fusion breeder deployment are as follows:

1. The fusion program must succeed and produce an engineering demonstration of fusion.
2. Fission, with reprocessing, must be accepted and have assumed its logical place in the energy mix.
3. Fusion-breeder-specific technology (blankets and fuel cycles) must be developed and demonstrated.

The fusion breeder could quickly become an important element in the energy system. This is possible because one fusion breeder could support the makeup fuel needs of many fission reactors, 15 or more times its own nuclear power. In addition to quick introduction another advantage of high support ratios is that system economics are fairly insensitive to performance and cost uncertainties concerning the fusion breeder.

The lower power density inherent in the fission-suppressed fusion breeder should simplify design and licensing compared to systems with high fission-power densities.

The timely development of the fusion breeder could be considered an insurance policy – insurance to protect against a uranium resource shortfall and uranium cartels, environmental limitations on coal, economic and other limitations on alternatives such as conservation, solar, or fusion. Preliminary economic analysis suggests the fusion breeder could be competitive with mined uranium when U_3O_8 reaches a price of about 100 \$ lb^{-1}, about twice the cost uranium was selling for a few years ago. The mere existence of a demonstrated fusion breeder technology should put a cap on uranium prices.

It is conceivable that the fusion breeder could be beneficial to fission breeder deployment by providing some or all of the initial fission inventories for the fission breeders and thus overcome their doubling-time constraints. But at the same time the fission breeder will then have to be competitive with non-breeders from both the economic and public-acceptance points of view for there to be an incentive for deployment. Thus the fusion breeder adds a new dimension to fission and fusion by providing the flexibility of making non-net breeding fission reactors renewable (long-term) energy options, and by allowing fission breeders to be deployed regardless of their doubling time. This in turn provides an application for fusion that is economically viable, earlier in its development when energy from fusion alone may be two to three times more expensive.

The US national magnetic fusion program is aiming for an engineering demonstration of fusion by the late 1990s. A successful demonstration of fusion coupled with development of fusion-breeder-specific technology would then allow a fusion breeder demonstration in the 2000–2010 period. This is fortuitous because this is about the period when it is predicted that a uranium shortfall may curtail commitment to further conventional nuclear construction.

We believe the fusion breeder could play an important role in our energy future and its economic benefits should far outweigh its overall additional development cost relative to the main line magnetic fusion development program.

3 HISTORY

The idea of marrying fusion and fission was first considered at the Lawrence Livermore Laboratory (Imhoff et al., 1954) in the early 1950s at the beginning of the fusion program. But the idea was not pursued because large uranium deposits were discovered and fusion was found to be far more difficult to achieve than at first perceived. In the early 1960s Weale performed integral experiments with a natural uranium pile and a DT neutron source and measured nuclear reaction rates of 14-MeV neutrons in uranium (Weale et al., 1961). In the late 1960s Lidsky at MIT did some studies on fusion–fission and presented the concept of fusion–fission "symbioses" (Lidsky, 1969).

At about the same time (1972) Lee at Livermore started looking at fusion neutron-induced fission of U-238 as a way to improve the power balance of low-Q fusion systems*. This work culminated with Livermore collaborating with General Atomic in the conceptual design of a "standard mirror" hybrid reactor (Bender, 1978). This reactor used a minimum-B mirror fusion driver and a fast fission U-238 blanket. The fusion driver operated at 400 MW fusion power, driving the blanket at 3400 MW. Net output from this reactor was 600 MWe and 2000 kg Pu-239 yr^{-1} (at a capacity factor of 0.74).

At this point in Livermore's fusion–fission program, emphasis switched from standard-mirror to higher performance tandem-mirror fusion drivers (Fowler and Logan, 1977) and we also began investigating the fission-suppressed class of fissile-breeding blankets (Lee, 1979a; Lee and Moir, 1981). In 1979 Livermore, Bechtel, General Atomic, and General Electric worked together on a conceptual design study of tandem-mirror hybrids. This was the first year of a proposed multiyear study. The first year of this study had two

---

* Q is the ratio of the fusion power generated to the power input into the plasma from external sources.

principal purposes. One was to examine the applicability of a low technology ($Q \sim 2$) tandem mirror as the driver for fusion–fission; the other was to develop and compare conceptual designs of U-233-producing blankets. The result was two conceptual designs, one based on fast fission of thorium, the other on the fission-suppression concept (Moir et al., 1980). Conclusions reached were that the tandem mirror should make a good driver and that both blanket types resulted in similarly attractive economics but had grossly different operating characteristics and feasibility issues. The fission-suppression blanket case produced three times more fissile fuel (U-233) per unit of nuclear power and its peak fission-power density and fission product after heat was ~100 times less for a given fusion neutron current. The high specific fissile production, low fission power density, and low fission product inventory made the fission-suppression blanket very attractive.

While the potential performance of fission-suppression blankets is attractive their feasibility is less certain. The specific blanket concept examined in the 1979 study used beryllium for neutron multiplication, and a molten salt ($LiF–BeF_2–ThF_4$) as a mobile fuel and heat transfer fluid from which the U-233 (as well as T) could be removed economically at low concentration. The feasibility issues of concern are how to protect the Be from the molten salt (MS), how to deal with Be swelling, and what structural material(s) to use to contain the MS. The structural material must also stand up to the fusion neutron environment and not overly reduce breeding. In this design graphite was used to contain sintered Be powder and TZM (a molybdenum alloy) was chosen for the structure. Graphite is known to be compatible with MS but its radiation resistance is of concern. TZM is known to be compatible with MS and is expected to be reasonably resistant to radiation damage, but there are feasibility problems with TZM fabrication.

Another concern is molten salt processing technology: it is thought to be feasible but a large-scale development effort will be required for commercialization. Tandem-mirror drivers and fission-suppressed blankets were also found to compare favorably to other drivers and blanket types in a 1980 EPRI-sponsored Westinghouse-led feasibility assessment of fusion–fission (Chapin, 1980).

The attractive performance of the 1979 fission-suppressed case, tempered somewhat by its feasibility and other concerns, set the stage for our 1981 work. The principle objective of this work is to develop fission-suppressed blanket design concepts that perform about as well as the 1979 case but that do not share its feasibility and development problems. A secondary objective is to look at higher performance tandem mirrors as drivers, since with the development of a new tandem concept (thermal barriers) higher performance could be achieved without pushing technology to extremes.

The history described above is intended to summarize how the Livermore fusion breeder concept has evolved. For a more general review of fusion–fission, the publications of Lee (1978, 1980) and of Maniscalco et al. (1981) are recommended.

4 FUSION–FISSION SYSTEMS

A fusion–fission system consists of fusion breeders providing fissile inventory and/or makeup fuel to fission reactors. The fusion breeder has three major elements:

1. The fusion driver (neutron source).
2. The blanket (for breeding).
3. The fuel cycle (reprocessing and refabrication).

The world-wide fusion research effort is pursuing both magnetic and inertial confinement approaches to commercialize fusion. At present the tokamak and the tandem mirror, both magnetic approaches, are considered the top contenders. All are potential candidates for drivers for fusion–fission. The latest in a series of tokamak commercial reactor design studies is STARFIRE (Baker et al., 1980). The 3500-MW fusion STARFIRE design (shown in Figure 1) is representative of how a tokamak fusion breeder might look. The tandem mirror has an open-ended axial geometry as opposed to tokamak's closed toroid. The conceptual 3000-MW fusion tandem-mirror driver in Figure 2 shows the tandem's major features (Carlson, 1980).

Blankets for fusion breeders (or fusion–fission hybrids) can have nuclear characteristics that range from fast fission of U-238 for neutron multiplication to fission-suppressed cases that use Be or Pb (n, 2n) or ^{7}Li(n, n$'$T) reactions for the generation of excess neutrons. The breeding ratio and energy release from 14-MeV neutrons with various multipliers is given in Table 1. These are idealized cases in that no blanket engineering requirements (such as structure or coolant, etc.) are included. U-238 is the most prolific multiplier per 14-MeV neutron while Be gives the highest net fissile production per unit power. Performance estimates for three blanket types, where conceptual engineering design requirements are included, are given in Table 2. As an example, the mechanical layout of a conceptual central cell module for a tandem mirror with a beryllium/molten salt (Be/MS) blanket is shown in Figure 3.

TABLE 1 Ideal breeding ratio and energy release with various neutron multipliers.

Breeding			Fissile atoms per MeV[c]
Multiplier[a]	Ratio[b]	Energy (MeV)[b]	
U-238	4.2	190	0.017
Th-232	2.4	50	0.028
Be	2.7	22	0.077
Li	1.8	16	0.050
Pb	1.7	19	0.037

[a] Plus fertile materials.
[b] Atoms or energy per 14-MeV DT fusion neutron.
[c] (Breeding ratio — 1)/energy.

At this point in the evolution of the fusion breeder concept we believe the fission-suppressed class of blankets has the most potential. This belief is based on the fact that the fission-suppressed blanket has the lowest fission power density and after-heat, and the lowest fission product inventory which should simplify design and licensing. It also gives the highest support ratio which should facilitate deployment. Utilities could buy fissile fuel from fusion breeders as they do today from enrichment plants.

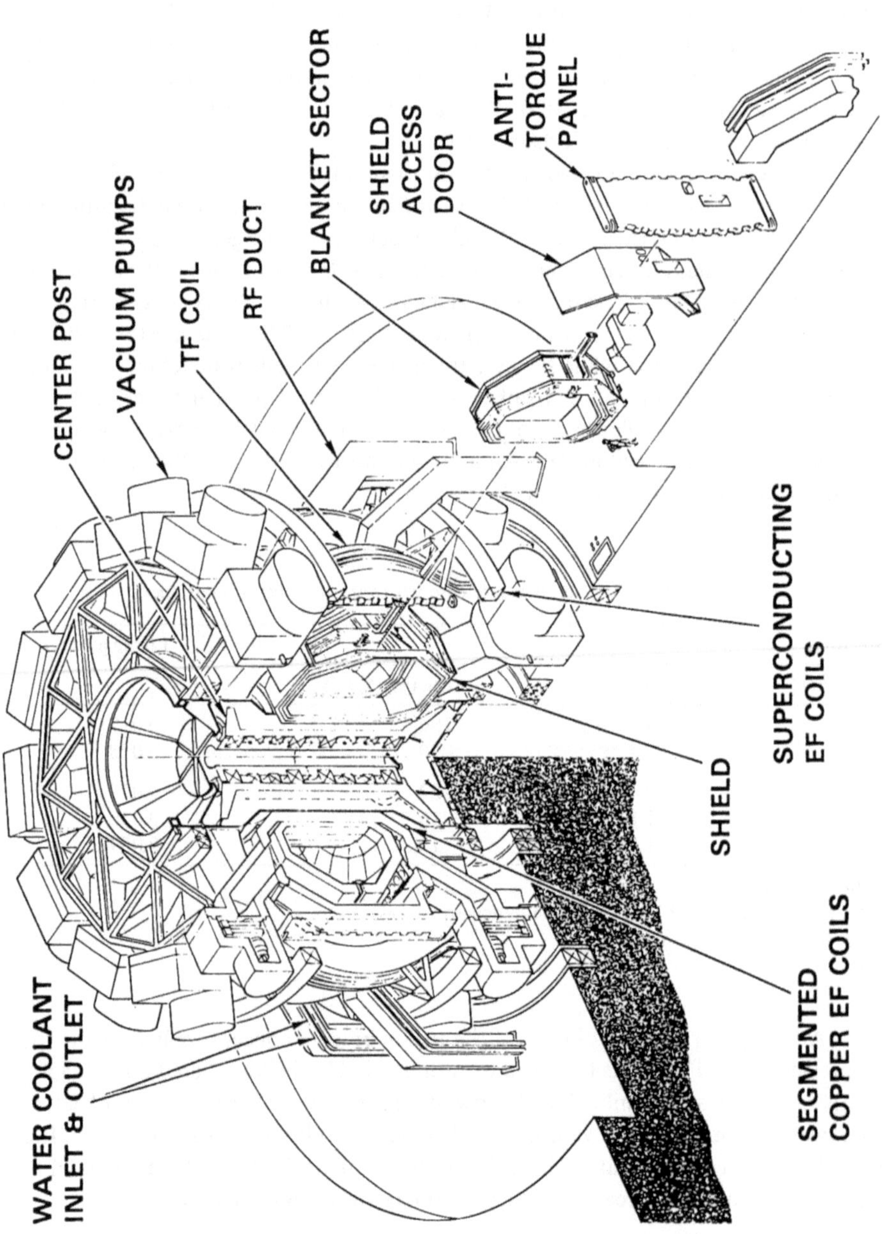

FIGURE 1 Reference design for the STARFIRE tokamak reactor.

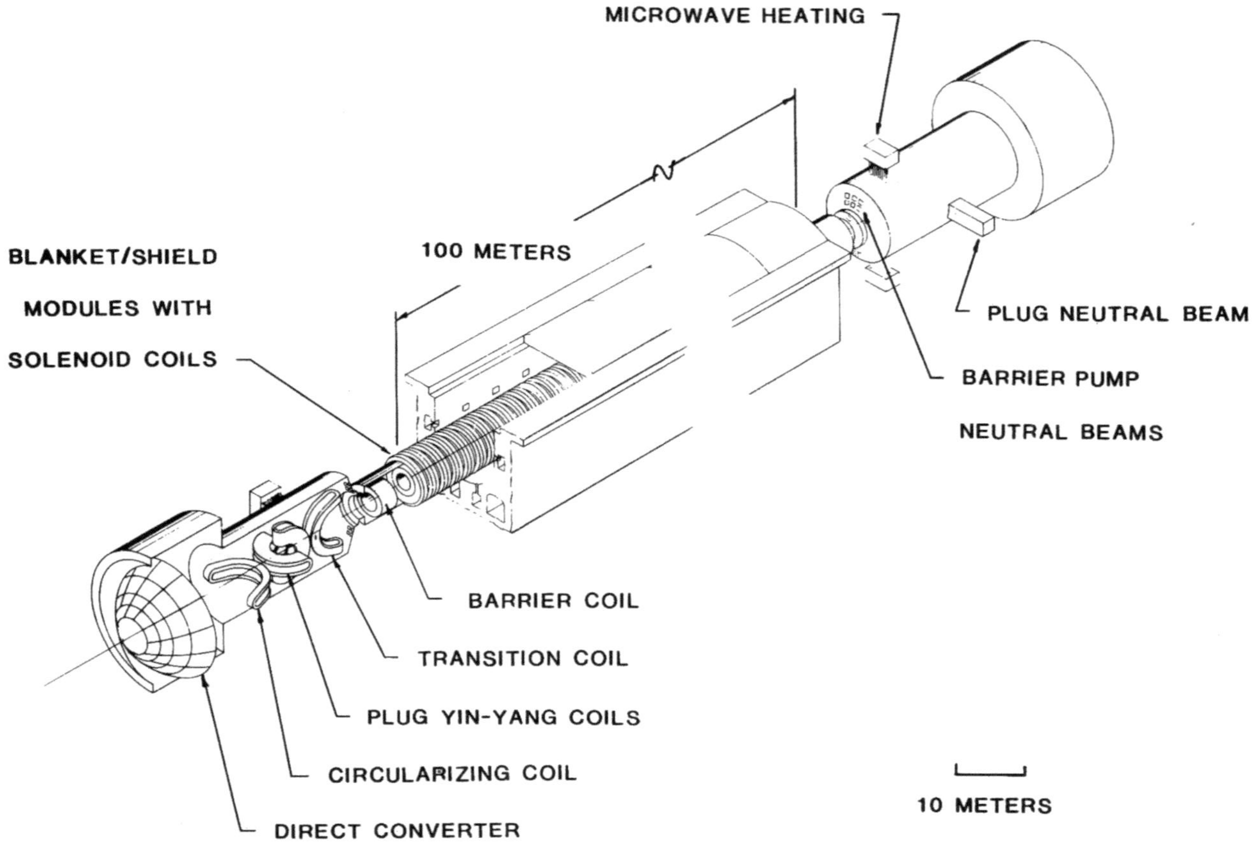

FIGURE 2 Major features of a 3000-MW tandem-mirror driver.

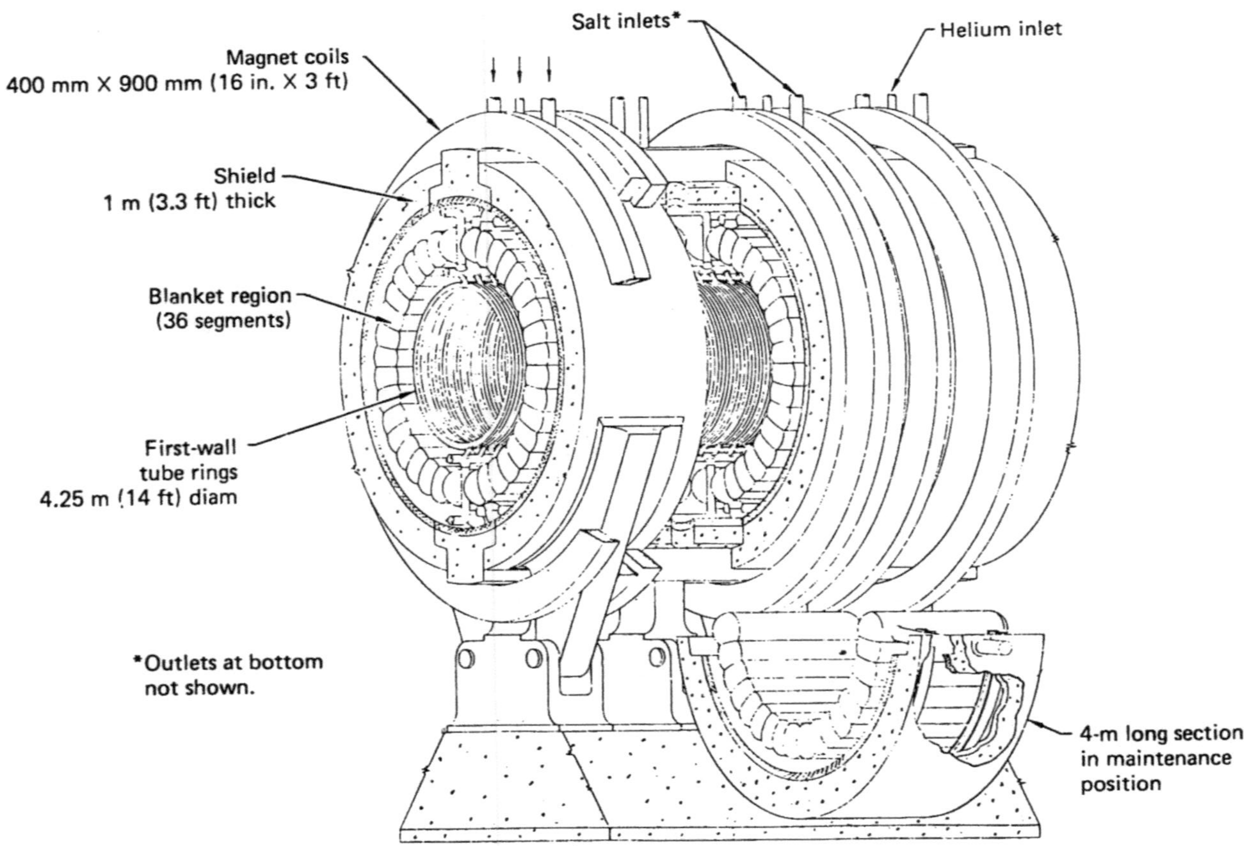

FIGURE 3 Mechanical layout of a central cell module for a tandem-mirror device with a beryllium/molten salt blanket.

TABLE 2 Performance estimates for engineered blankets.

	Fast fission	Fast fission	Fission suppressed
Blanket type	U-238	Th	Th
Tritium breeding	1.0	1.0	1.0
Fissile breeding ratio[a]	1.5	0.8	0.8
Energy multiplication (M)[b]	11	5	1.6
Fission power density (peak, W cm^{-3})	350	100	10
Fissile production rate (kg yr^{-1})[c]	2700	2900	9600
Fission reactor power supported (MW)[d]	19,000	30,000	99,000
Nuclear support ratio[e]	5	9	25

[a] Net atoms per DT fusion.
[b] M = energy/14 MeV.
[c] Breeder power = 4000 MW nuclear.
[d] LWR makeup = 142 kg Pu or 97 kg U-233 $GWyr^{-1}$.
[e] Ratio of fission reactor power to fusion breeder power.

A successful fission-suppressed blanket requires that:

1. Fast fission of fertile material must be suppressed by limiting its exposure to fast neutrons.
2. An effective non-fission neutron multiplier must be used to generate significant numbers of neutrons in excess of those needed to breed tritium to fuel the DT fusion reaction.
3. Breeding must not be compromised by parasitic neutron capture.
4. Fissile material must be removed at low concentrations to inhibit in-situ fission reactions.
5. Fuel reprocessing and fabrication must be at low cost.
6. Tritium inventory must be kept low to limit its loss due to radioactive decay.
7. Wherever possible, conventional technology should be used to minimize the cost, risk, and time for the necessary development program.

The types of fission reactors that could be fueled (initial inventory and/or make-up) by fusion breeders span the entire spectrum from conventional thermal reactors (LWR, HWR, etc.) to advanced thermal reactors (HTGR, LWBR, etc.) and fast breeders. Conventional fuel cycles can be used as they are or modified to make better use of fissile materials. The fusion breeder could provide the U-233 needed for large-scale use of the Th/U fuel cycle in conventional or advanced reactors.

Now for a specific example of a fusion–fission system. The fusion breeder part of the system has a tandem mirror for the fusion driver, and a beryllium fission-suppressed blanket with a molten salt fuel cycle. This fusion breeder supplies makeup fuel to LWRs operating on Th/U-233 with thorex reprocessing and recycle. This fusion breeder has a fusion power of 2700 MW resulting in a total nuclear power of 4000 MW. Net electrical and fissile productions by this fusion breeder are 1000 MW and 6200 kg U-233 yr^{-1} at 70% capacity factor. The U-233 produced supports 24,000 MWe of LWR capacity. System power levels are listed in Table 3. Note that 96% of the system's net electric production is provided by the fission reactors.

TABLE 3 Examples of predicted fusion–fission system power levels.

Fusion breeder	
Fusion power	2700 MW
Nuclear power	4000 MW
Electric power, gross	2400 MW
Electric power, net	1000 MW
Fissile material (U-233) production, net[a]	6200 kg yr^{-1}
Fission reactor (LWR) power supported[b]	24,000 MWe

[a] At a 70% capacity factor.
[b] Annual U-233 makeup = 255 kg GWe^{-1}, at a 70% capacity factor.

Results of our economic evaluation of this example of a fusion–fission system are given in Table 4. While our ability to estimate fusion breeder economics is very limited, economics is still an important aspect of our studies. Note that in this case the fusion breeder cost is quite high (4.4 billion dollars) but when divided by the number of LWRs supported, the system capital cost is only 20% higher than those of the LWRs alone. The resulting system electricity cost is 4.3 cents $kWhr^{-1}$ which is only ~15% higher than if the LWRs were using U–Pu recycle and U_3O_8 at 90 \$ kg^{-1}. We estimate that this fusion breeder would be competitive with U_3O_8 at a price of about 180 \$ kg^{-1}.

TABLE 4 An example of fusion–fission system economics[a] (1980 US dollars).

Total capital cost of fusion breeder	4.4 billion \$
Total capital cost of LWRs	870 \$ kWe^{-1}
Capital cost ratio (LWRs + FB/LWRs)	1.20
System electricity cost	4.3 ¢ $kWhr^{-1}$
Fissile material (U-233) transfer cost	45 \$ g^{-1}
Equivalent U_3O_8 cost	180 \$ kg^{-1}

[a] Assuming capacity factors of 70% and capital charges of 15% yr^{-1}.

The uncertainty in our economic estimate is large but because the fusion breeder can support many fission reactors, the uncertainty in overall system electricity cost is much smaller. For example, a 100% increase in U-233 cost leads to only about a 20% increase in electricity cost, while a 50% decrease in U-233 cost results in a 10% decrease in electricity cost. This insensitivity to uncertainties in fusion breeder economics has been referred to as its "robustness" (Dreyfuss et al., 1978). For more details on this example of a fusion–fission system, see Lee et al. (1980).

What type of blanket is best? While we at Livermore believe in the fission-suppressed blanket, this is still really an unanswered question. One of the goals of EPRI's assessment

(Chapin, 1980) was to compare blanket types. System economics for a tandem-mirror fusion breeder with three blanket types are compared with net breeder power in Figure 4. The three blanket types are thorium metal–fast fission (Th–FF), uranium carbide–fast fission (UC–FF), and the Be fission-suppressed blanket designated here as Th–FS. The Th–FS blanket was found to have some economic advantage over the other two but this advantage is reduced as breeder net electric power increases. On the other hand when other characteristics are compared, at 800 MWe, for example (Table 5), we see striking differences in LWR power supported (21,000 down to 6000 MWe), fusion power (2400 down to 700 MW), and breeder capital cost (4.1 down to 2.7 billion dollars). The highest values are with the fission-suppressed blanket, the lowest with the uranium fast-fission blanket.

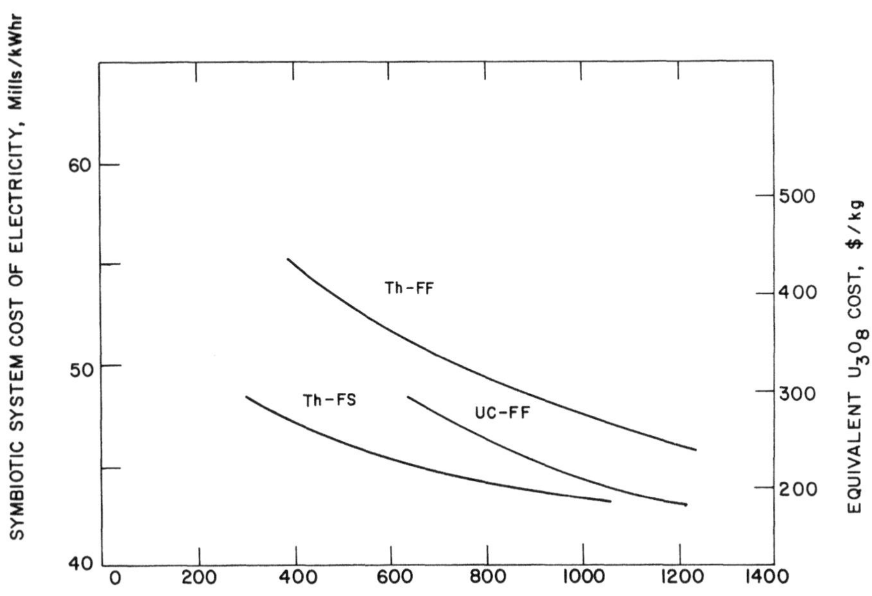

FIGURE 4 System power cost versus fusion breeder net power for three blanket designs.

What type of fusion driver is best? This question was also addressed in the EPRI study (Chapin, 1980). As a representative economic comparison, Figure 5 shows that all three major fusion approaches could make economically competitive drivers.

Major conclusions and recommendations of the 1980 EPRI study are listed below:

1. U_3O_8 resource limitations are only a question of time, and fusion–fission could eliminate the shortfall.
2. Fusion breeders could provide a ceiling on U_3O_8 real cost escalation in the 100 to 400 \$ kg^{-1} range.

TABLE 5 Parameters for an 800 MWe net breeder with three blanket types.

Parameter	Blanket type		
	UC–FF	Th–FF	Th–FS
Fusion power (MW)	655	980	2350
Central cell length (m)	22.5	34	81
Net fissile material production (kg yr^{-1})	2120 (Pu)	2600 (U)	5410 (U)
LWR supported (GWe)	6.0	10.3	21.1
Total capital cost of breeder (billion $)	2.69	2.76	4.1
Symbiotic system cost of electricity (¢ $kWhr^{-1}$)	4.65	4.90	4.34
Transfer cost of fissile material ($ g^{-1})	52.9	87.5	51.0
Equivalent cost of U_3O_8 ($ kg^{-1})	260	310	200

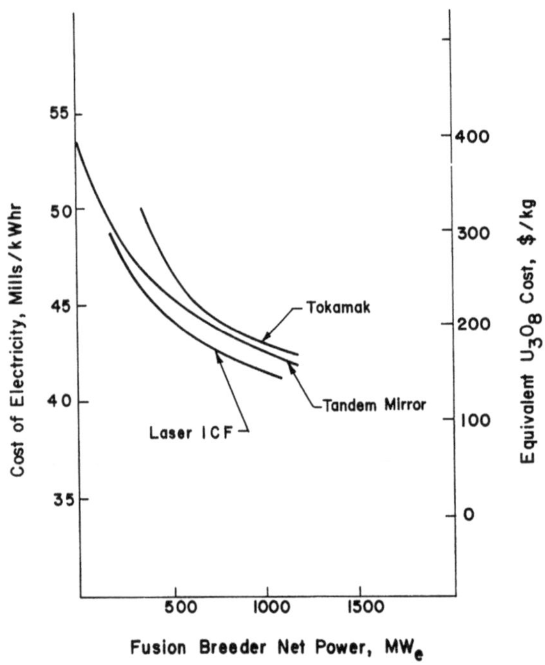

FIGURE 5 Estimated economics of various types of fusion driver (with thorium fission-suppressed blanket).

3. Commercial introduction will be needed in the period 2000–2020.
4. The fusion–fission concept has potential for near-term impact on utility decision making.
5. The three confinement approaches considered – tokamak, tandem mirror, and laser ICF (inertial confinement) – resulted in similar economics.
6. Cost and performance differences were due more to blanket/fuel-cycle choices.

7. A tokamak commercial demonstration model could be operational by 2000 with the tandem mirror and inertial-confinement demonstrations following after five and ten years, respectively.
8. Blanket development may be the pacing factor.
9. Fusion–fission should be made a viable energy option for the time when a U_3O_8 shortfall is anticipated.
10. Increasing conceptual design activity is needed.
11. Increased attention should be devoted to fission-suppressed blanket concepts.
12. In-depth development scenarios are needed.

The US funding for fusion–fission research and development in 1981 was approximately 1 million dollars, 0.7 million from the Department of Energy (DOE), and 0.3 million from the Electric Power Research Institute (EPRI). The DOE funding supports Livermore and four subcontractors (TRW, General Atomic Co., Westinghouse, and ORNL) working on the Tandem Mirror Hybrid (Breeder) project with major emphasis on development of fission-suppressed blanket concepts. The EPRI funding supports Combustion Engineering Co. on a project appraising past blanket designs.

Fusion breeder development is dependent upon the success and timetable of fusion development. Fusion development in the United States is scheduled to: demonstrate physics feasibility in the 1980s with TFTR (the Tokamak Fusion Test Reactor) and MFTF-B (the Mirror Fusion Test Facility-B); demonstrate engineering feasibility in the 1990s with the Fusion Engineering Device (FED); and have a demonstration electric power plant operational by 2000. The objectives and timetable for the US fusion program are specified in the Magnetic Fusion Energy Act of 1980. The FED could be a test bed for fusion breeder blankets as well as fusion driver technology. The demonstration device could be a fusion breeder, assuming that blanket and fuel cycles are sufficiently developed by then.

5 DEPLOYMENT

A major advantage we see for fusion–fission is its ability to support a much faster and more timely growth in nuclear capacity than either fission or fusion could alone. Conventional fission reactors (LWRs, etc.) will run out of fuel. The fission breeder (LMFBRs) will be limited first by the historical limitations on the introduction of a new technology and then by its doubling time when the Pu stockpile is exhausted. Fusion electric deployment will also be limited by introduction constraints.

Therefore, even with the successful commercial introduction of fission breeders and fusion electric devices, an energy shortfall is very likely as oil, gas, and uranium resources are exhausted. The fusion breeder, by virtue of its high support ratio, could provide the stepping stone needed to bridge the energy gap (shortfall) while the fission breeder and/or fusion electric reactors are being introduced.

For example, consider the following US scenario:

Nuclear generation goals
250 GWe from LWRs in 2000 (21% of demand);
1900 GWe from all nuclear sources in 2060 (50% of demand);
Requiring nuclear growth of $3.4\%\,yr^{-1}$.

LWR assumptions
U_3O_8 resource limit of 3 million tonnes;
Utilities must be convinced of availability of 30-year fuel supply to commit themselves to a new LWR;
Plutonium is recovered and stockpiled for use in LMFBRs.

LMFBR assumptions
First commercial plant in 2005, 8 GWe by 2020;
25% per annum growth after 2020 until a limit of 20 plants per annum is reached;
20 new plants per annum until growth = 10% yr^{-1}.
10% per annum growth until plutonium stockpile consumed, then limited to 18-year doubling time.

Fusion breeder assumptions
First commercial plant in 2015;
Initial introduction rate same as LMFBR;
30 plants on line by 2037 (each producing 4000 MW-nuclear and 6200 kg fissile fuel yr^{-1}).

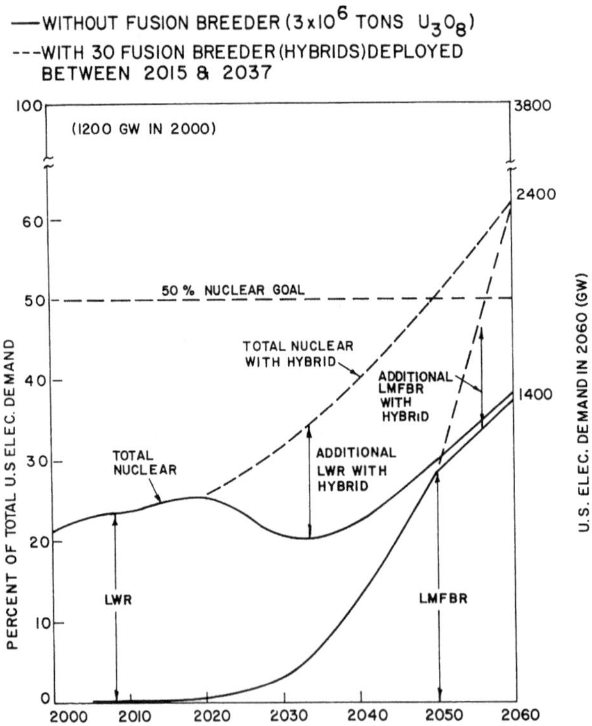

FIGURE 6 Projections of the amounts and proportions of US electricity demand that could be supplied using various strategies.

The projected effect of 30 fusion breeders on US nuclear capacity is substantial. As shown in Figure 6, without these fusion breeders, LWR expansion would stop in 2020 and LMFBR construction is not sufficient to keep nuclear's market share from dropping back to ~20% in ~2035. By 2060 LMFBR construction has increased nuclear's market share to ~37% but this is still ~500 GW short of the 50% nuclear goal.

With the fusion breeder, LWR expansion could continue until LMFBR expansion could catch up and continue to expand independent of its doubling-time constraints. For this example, 30 fusion breeders would allow nuclear to meet 60% of the demand in 2060, 1000 GW more than fission could alone. If fission breeders were not deployed, additional fusion breeders could be used to provide fuel for LWRs to meet the nuclear goal.

If fusion development is successful and deployment is similar to that of the LMFBR except ten years later, but if the fusion breeder option is not exercised, fusion reactors plus fission breeders could meet the long-term nuclear goal (after ~ 2055). But in the shorter term nuclear energy would miss its overall goal by a much wider margin. For this example, in 2040, there is ~ 700 GWe less nuclear power with fusion but without the fusion breeder.

While the deployment scenario just described is of course speculative, it is thought to be a good example of the fusion breeder's potential deployment advantages.

6 SUMMARY

To recapitulate, we feel that the fusion breeder has five major points in its favor. First, it could have a significant role in nuclear power development and deployment, for both fission and fusion plants, for it would allow fission to meet the demand with a variety of reactor types and is an earlier commercial mission for the fusion concept. Second, it could be ready in time to help avert serious energy or environmental problems in the 21st century if fusion's development timetable is met. Third, it appears to show favorable economics, with an increase in energy costs of $<$ 30%, based on preliminary studies, and is very robust; it would also put a cap on U_3O_8 prices. Fourth, it would give utilities the option of just buying fissile fuel, as they do now from enrichment plants. Fifth, it could have a very high benefit/cost ratio.

Serious development should start now.

ACKNOWLEDGMENT

My thanks are due to Berwald of TRW for helping with the deployment example.

REFERENCES

Baker, C.C. et al. (1980). STARFIRE, A Commercial Tokamak Fusion Power Plant Study. ANL/PPP-80-1; see also, An overview in inertial confinement fusion. In the Proceedings of the ANS Topical Meeting on the Technology of Controlled Nuclear Fusion, 4th, King of Prussia, Pennsylvania, October 1980, p. 1074, CONF801011.

Bender, D.J. (Editor) (1978). Reference Design for the Standard Mirror Hybrid Reactor. A joint report by Lawrence Livermore National Laboratory and General Atomic Company. Lawrence Livermore National Laboratory Report UCRL 52478/GA-A14796.

Carlson, G.A. (1980). The design of tandem mirror reactors with thermal barriers. In the Proceedings of the ANS Topical Meeting on the Technology of Controlled Nuclear Fusion, 4th, King of Prussia, Pennsyslvania, October 1980, p. 815.

Chapin, D.L. (Editor) (1980). Preliminary Feasibility Assessment of Fusion–Fission Hybrids. EPRI Project RP-1463, WFPS:TME-81-003.

Dreyfuss, D.J., Augenstein, B.W., Mooz, W.E., and Sher, P.A. (1978). An Examination of Alternative Nuclear Breeding Methods. RAND Report R-2267-DOE. The RAND Corporation, Santa Monica, California.

Fowler, T.K. and Logan, B.G. (1977). The tandem mirror reactor. Comments on Plasma Physics and Controlled Fusion Research, 2:167.

Imhoff, D.H. et al. (1954). A Driven Thermonuclear Power Breeder. CR-6. California Research and Development Corporation.

Lee, J.D. (1972). Neutronics of sub-critical fast fission blankets for D–T fusion reactors. In the Proceedings of the Intersociety Energy Conversion Engineering Conference, 7th, San Diego, California, 1972, p. 1294, American Chemical Society; see also, Lawrence Livermore National Laboratory Report UCRL-73952, 1972.

Lee, J.D. (1978). Mirror fusion–fission hybrids. Atomkernenergie/Kerntechnik (ATKE), 32.

Lee, J.D. (1979a). The Beryllium/Molten Salt Blanket. Lawrence Livermore National Laboratory Report UCRL-82663; also published in the Proceedings of the US/USSR Symposium on Fusion–Fission, 3rd, Princeton, New Jersey, 1979.

Lee, J.D. (1979b). Nuclear design of the LLL–GA U_3Si blanket. In the Proceeedings of the ANS Topical Meeting on Technology of Controlled Nuclear Fusion, 3rd, Santa Fe, New Mexico, May 1979, p. 219, CONF780508.

Lee, J.D. (1980). Tandem Mirror Fusion–Fission Hybrid Studies. Lawrence Livermore National Laboratory Report UCRL-84108; see also, Atomkernenergie/Kerntechnik (ATKE), 36.

Lee, J.D. and Moir, R.W. (1981). Fission-suppressed blankets for fissile fuel breeding fusion reactors. Journal of Fusion Energy, 1:3.

Lee, J.D. et al. (1980). Parametric systems analysis for tandem mirror hybrids. In the Proceedings of the ANS Topical Meeting on the Technology of Controlled Nuclear Fusion, 4th, King of Prussia, Pennsylvania, October 1980, p. 1455.

Lidsky, L.M. (1969). Fission–fusion symbiosis: general considerations and a specific example. In the Proceedings of the British Nuclear Energy Society Nuclear Fusion Reactor Conference, Culham, 1969, pp. 41–53. Culham Laboratory Report CLM-MFE.

Maniscalco, J.A. et al. (1981). Recent progress in fusion–fission hybrid reactor design studies. Nuclear Technology/Fusion, 1:4.

Moir, R.W. et al. (1980). Tandem Mirror Hybrid Reactor Design Study Annual Report. Lawrence Livermore National Laboratory Report UCRL-18808.

Weale, J.W., Goodfellow, H., McDaggart, M.H., and Mullender, M.L. (1961). Journal of Nuclear Energy A/B, 14:91.

PROSPECTS OF HYDROGEN PRODUCTION THROUGH FUSION

Edwin E. Kintner
Office of Energy Research, US Department of Energy, Washington, D.C. (USA)

ABSTRACT

With approximately two-thirds of world energy needs supplied by petroleum and natural gas, the expected future shortage of these resources will lead to a need for substitute fuels early in the next century. Synthetic fuels produced by conversion of the relatively abundant resources of coal can be expected to play a major role in filling this need. These resources can be substantially extended, and the environmental impacts of the conversion process considerably reduced, if external sources of hydrogen and process heat are provided. Nuclear fusion is a future inexhaustible energy source which can provide both hydrogen and high-temperature process heat. Moreover, certain unique features of fusion suggest the possibility of significant increases in overall process efficiencies through operation at very high temperatures. Studies are underway on the feasibility of coupling high-temperature electrolysis and thermochemical cycle processes with a fusion reactor and thus achieving splitting of water by fusion energy.

1 INTRODUCTION

Approximately two-thirds of the world's energy is at present supplied by petroleum and natural gas, and it is expected that this proportional usage will continue. As a result, sometime before the year 2030 – perhaps much sooner – natural petroleum supplies will not be able to meet the total demand and synthetic fuels, most probably made from coal, will be required to make up the deficit.

It is in the manufacture of these synthetic fuels that hydrogen and process heat from an "inexhaustible" source can significantly affect the world's energy economy after the next 30–50 years. The depletion of a dwindling and irreplaceable resource – coal – and the release of effluents to the environment from the production of synthetic fuels can be ameliorated by the integration of coal, process heat, and hydrogen into a symbiotic system. Practical fusion energy could be available within the next 30 years, and it could then provide an inexhaustible source of both hydrogen and heat. Therefore, in addition to developing fusion for the production of electricity, we in the United States are currently investigating alternative applications for fusion, especially the production of hydrogen and high-temperature process heat.

2 SYNTHETIC PRODUCTION OF GASEOUS AND LIQUID FUELS FROM COAL

The processes presently used for the conversion of coal to synthetic gaseous and liquid fuels are summarized in Figure 1. For Substitute Natural Gas (SNG), coal is first gasified with steam to produce CO and H_2 by the endothermic water-gas reaction, with the energy supplied by oxidation of additional coal. Some of the CO is then reacted with more steam (water-gas shift reaction) to form H_2 and CO_2. After removal of the acid gases (CO_2 and H_2S), the H_2/CO molar ratio is adjusted to 3/1 and the gases are catalytically combined to form CH_4 (SNG). More advanced methods catalytically gasify coal with steam to produce methane and CO_2 or directly hydrogenate coal to give methane (hydrane and flash hydropyrolysis). Further, coal can be directly catalytically hydrogenated to produce aromatic and aliphatic hydrocarbons (H coal). A third, indirect, route is to gasify coal and to combine the resultants CO and H_2 to give methanol, which is then catalytically dehydrated to form gasoline (the Mobil process).

3 NATURAL-RESOURCE AND ENVIRONMENTAL IMPLICATIONS OF COAL-CONVERSION PROCESSES

The wholesale use of coal as a major source of primary energy for meeting world demands for liquid and gaseous fuels over the next 50 years should be approached carefully since this would not only rapidly deplete coal resources but also create significant environmental and public-health effects by releasing large quantities of CO_2 into the earth's atmosphere. CO_2 is a major waste product of all the coal-conversion processes. Coal utilization efficiencies are only 30–45% in the production of synthetic gases and only 40–60% in the production of synthetic liquids. Thus 60–70% of the coal feed is used to produce steam and hydrogen and leaves the process as waste CO_2. Moreover, this is on top of the CO_2 released when the fuel is burned, thus tripling the "inevitable" CO_2 load on the environment.

Until now the vast quantities of CO_2 generated by combustion of hydrocarbons have been largely ignored since CO_2 is not known to have any adverse health effects at the concentrations at which it is presently found in the atmosphere. Recently, however, hypotheses have been proposed that CO_2 has produced a warming trend by causing a greenhouse effect in the earth's atmosphere. This is one environmental question which could pose a serious threat to life on earth. A recent IIASA report (Flohn, 1981) states that if fossil fuels supply all the world's energy needs then the average global temperature increase will be 2°C by the year 2030 and 4°C by the year 2050. To provide some perspective on how large the possible climatic effects of such a change would be a comparison with those changes that have occurred over historical and prehistoric periods is useful. An increase of 1°C would be equivalent to the increase in the earth's temperature over the last glacial period, some 5500–6500 years ago; a 2–5°C increase would be equivalent to the temperature increase during the last interglacial period, about 125,000 years ago. A warming of 4°C could lead to melting of much of the polar ice caps with a resulting increase in sea level which would flood most of the world's coastlines.

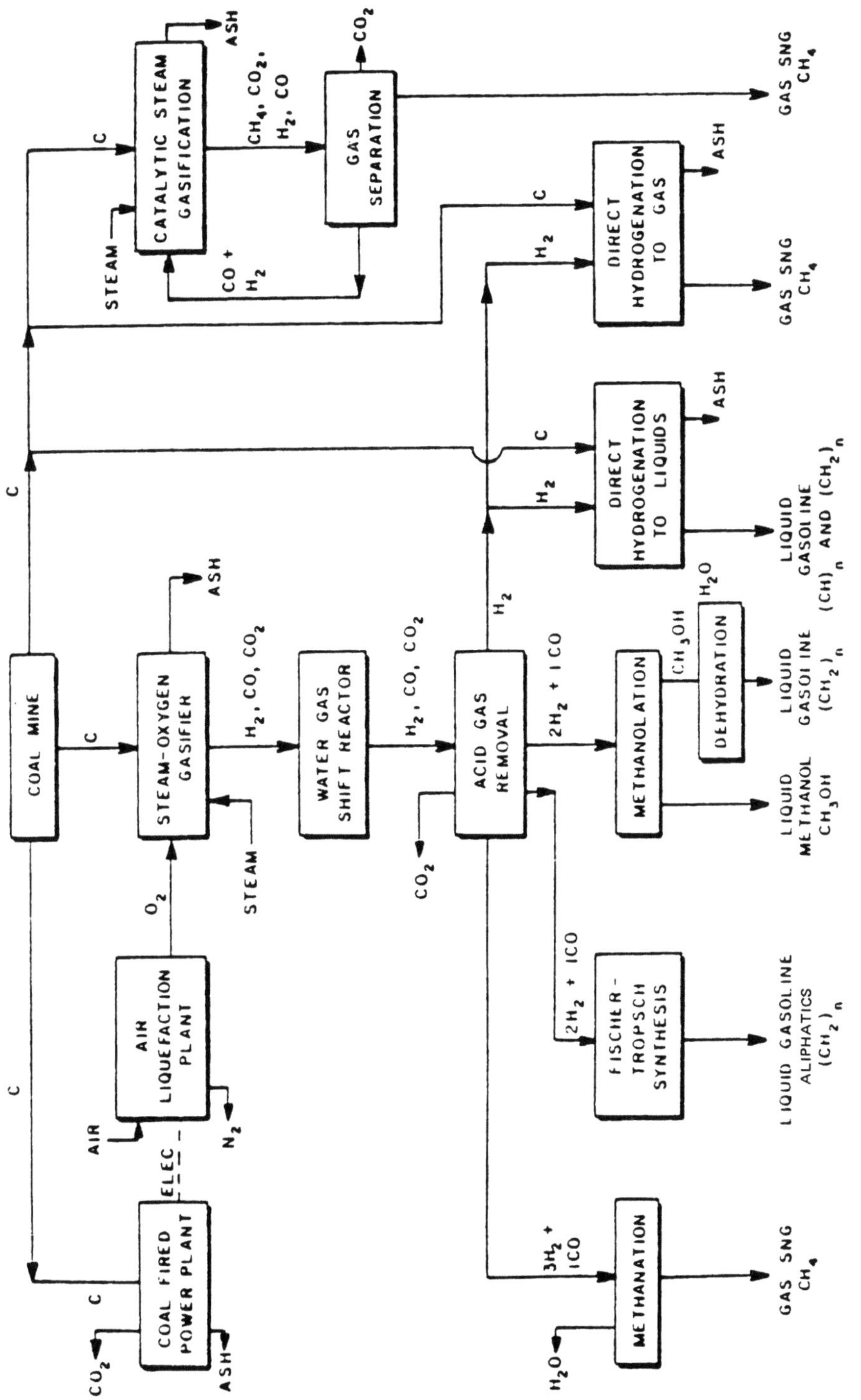

FIGURE 1 Present processes for conversion of coal to synthetic gaseous and liquid fuels.

4 HYDROGEN AND PROCESS-HEAT ENHANCEMENT OF COAL-CONVERSION PROCESSES

Thus, although it appears that coal-conversion processes may be badly needed to supply a large share of the world's demand for liquid and gaseous fuel in the next 30–50 years, serious problems of depletion of natural resources and release of CO_2 are inherent in these processes. It is for this reason that independent supplies of hydrogen and process heat could be most useful as additions in coal-conversion processes. An external source of hydrogen and process heat would eliminate (1) the need to burn a portion of the coal to produce heat to drive the process and (2) the need to react a portion of the coal with steam to produce hydrogen as an intermediate feed product.

Specifically, if an external source of hydrogen is used, the water-gas shift reaction ($CO + H_2O \rightarrow H_2 + CO_2$) would be eliminated, with a resultant SNG production-factor increase of 2.0 per unit input of coal. In methanol production followed by dehydration to gasoline the factor would be 2.4. The gasification process is driven toward completion at temperatures above 1000°C (i.e. coal and steam will react to form CO and H_2 with very small quantities of CO_2 resulting). This eliminates the need for an air-liquefaction plant and increases the capacity of the gasifier. Thus the maximum reduction in coal requirements results when external hydrogen and high-temperature process heat are utilized, and it amounts to a saving of approximately 70% in coal with almost no release of CO_2.

5 SOURCES OF HYDROGEN

Water-splitting processes are the most effective means for the production of hydrogen. From a resource perspective, hydrogen production by this method is, of course, unlimited since H_2O is the combustion product of H_2. The current technologies for the production of hydrogen are hydrocarbon reforming and coal gasification (see Table 1), which do not appear as attractive for the future as water-splitting processes. Specifically, the feedstock in hydrocarbon reforming is natural gas (methane), which is becoming prohibitively expensive to use as a feedstock for hydrogen production.

TABLE 1 Current hydrogen-production methods.

Hydrocarbon reforming	
(30 atm pressure, 1500°F temperature, nickel catalyst)	
Reform:	$CH_4 + H_2O \rightarrow CO + 3H_2$ (endothermic)
Shift:	$CO + H_2O \rightarrow CO_2 + H_2$ (exothermic)
Scrub:	$2NaOH + CO_2 \rightarrow Na_2CO_3 + H_2O$
Coal gasification	
Steam gasification:	$CH_{0.8} + H_2O \rightarrow CO + 1.4H_2$
Oxygen gasification:	$CH_{0.8} + 1.2O_2 \rightarrow CO_2 + 0.4H_2O$ + energy
Water-gas shift:	$CO + H_2O \rightarrow CO_2 + H_2$
The overall reaction for producing hydrogen from coal can be expressed as	
	$CH_{0.8} + 0.7O_2 + 0.6H_2O \rightarrow CO_2 + H_2$

In water-splitting processes the hydrogen–oxygen bond can be broken by the addition of energy in electrical and/or thermal form. When only electrical energy is provided the process is called electrolysis. The main impediment to large-scale commercialization of electrolysis is the availability and cost of input electric power. With conventional steam power cycles, electrical generation efficiencies are of the order of 40%. Since conventional electrolyzers now operate with an efficiency of 70%, although advanced electrolyzers may improve this to 95%, direct electrolysis is thus limited to overall efficiencies in the range 28–38%.

Fortunately, to raise this efficiency a portion of the prerequisite energy to break the bond can be provided as thermal energy through either electrochemical decomposition or High-Temperature Electrolysis (HTE).

Since the thermal component for electrochemical water decomposition is used directly at essentially 100% efficiency there is an advantage in making the ratio of the direct heat input to the electrical energy input as large as possible. At a temperature T the input thermal energy equals $T\Delta S$ where ΔS is the entropy change for the reaction. The electrical energy input equals the Gibbs free energy change ΔF and the sum of these energy changes equals the reaction enthalpy ΔH. The energy splits are shown as a function of temperature in Figure 2. As the temperature increases, the reaction enthalpy remains virtually constant. The Gibbs free energy for electrical input, however, decreases with increasing temperature and the thermal energy input $T\Delta S$ increases. The increasing fraction of thermal energy that is absorbed by the water molecule as the electrolysis temperature increases results in a higher process efficiency, so that more hydrogen can be produced for a given total energy input.

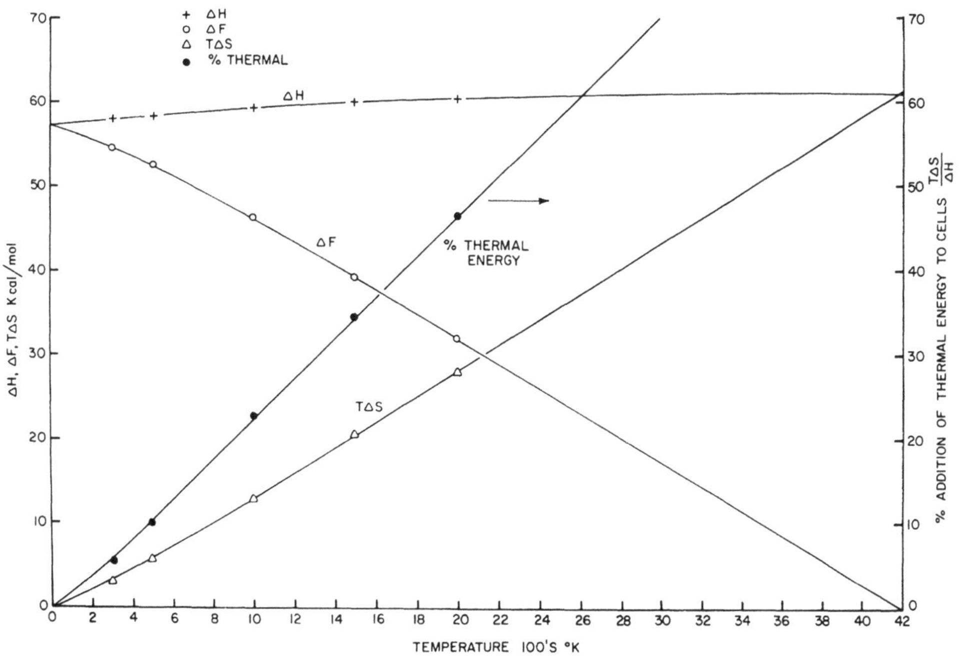

FIGURE 2 Fusion–HTE–synthetic-fuels thermodynamics of water decomposition (steam) $H_2O(g) = H_2(g) + \frac{1}{2}O_2(g)$.

Thus the splitting of a water molecule can be accomplished using only thermal energy, at a temperature exceeding 3000°C. Reasonable extrapolations of current materials technology place an upper limit on process temperatures far below this value. As it does not appear to be technically practical to carry out the process at such high temperatures, several laboratories have initiated research on thermochemical water-splitting processes which rely partly on input of heat and partly on the special properties of certain chemical catalysts that allow the breaking of the hydrogen–oxygen bond at temperatures between 600 and 1000°C. Thus by carefully designing the order in which heat is added and the critical chemical reactions take place it is possible to produce hydrogen with good efficiency at temperatures well below those needed for pure thermal decomposition. All the practical thermochemical cycles, as currently envisaged, require input temperatures of approximately 1000°C for the highest-temperature chemical step. Of the approximately 30 cycles under study worldwide, only three cycles have been developed to the stage where closed-loop tabletop models have been built and tested in the laboratory.

TABLE 2 Thermochemical cycles whose chemistry and closed-loop operation have been verified in the laboratory.

Sulfur–iodine cycle (General Atomic Company)

$$2H_2O + SO_2 + xI_2 \xrightarrow{\text{aqueous}} H_2SO_4 + 2HI_x$$

$$2HI_x \xrightarrow{\lesssim 573\,K} xI_2 + H_2$$

$$H_2SO_4 \xrightarrow{\lesssim 1144\,K} H_2O + SO_2 + \tfrac{1}{2}O_2$$

Sulfur cycle (part electrochemical) (Westinghouse Electric Corporation)

$$2H_2O + SO_2 \xrightarrow[\text{electrolysis}]{\text{aqueous}} H_2 + H_2SO_4$$

$$H_2SO_4 \xrightarrow[\text{temperature}]{\text{high}} H_2O + SO_2 + \tfrac{1}{2}O_2$$

Sulfur–bromine cycle (part electrochemical) (Ispra)

$$2H_2O + SO_2 + Br_2 \xrightarrow[320\text{–}370\,K]{\text{aqueous}} H_2SO_4 + 2HBr$$

$$2HBr \xrightarrow[\text{electrolysis}]{\text{aqueous}} Br_2 + H_2$$

$$H_2SO_4 \xrightarrow{1000\text{–}1100\,K} H_2O + SO_2 + \tfrac{1}{2}O_2$$

They are illustrated in terms of their principal chemical steps and reaction temperatures in Table 2. The main effort on the development of the sulfur–iodine cycle is at the General Atomic Company; for the sulfur cycle, at the Westinghouse Electric Corporation; and for the sulfur–bromine cycle, at the European Communities Joint Research Centre at Ispra.

6 ROLE OF FUSION ENERGY IN SYNTHETIC-FUELS PRODUCTION

It appears that fusion reactors can provide a better match than other inexhaustible energy sources with HTE and/or thermochemical cycles for the production of hydrogen and process heat at the requisite temperatures for synthetic-fuels production.

An inherent unique property of the deuterium–tritium (D–T) fusion reaction is that most of its energy is released in a very-high-energy neutron (i.e. 14.1 MeV). Such high-energy neutrons can penetrate deeply into blankets outside the critical components of the fusion reactor and can thus allow the generation of very high temperatures. Fission neutrons deposit most of their kinetic energy within the reactor core where considerations of fuel and cladding integrity limits the temperatures that are possible in practice to approximately 850°C (high-temperature gas-cooled reactor), but 80% of the energy of the fusion reaction is transported out of the plasma by high-energy neutrons and is deposited in the blanket where less critical engineering critieria make temperatures in the range 1000–1200°C appear possible.

Since hydrogen process efficiencies increase exponentially with temperature, the incrementally higher temperature which fusion reactors appear capable of producing can be used to increase significantly the overall efficiency of the HTE and thermochemical processes for hydrogen-production and coal-conversion processes for synthetic-fuels production. For example, in the SO_3-decomposition step of thermochemical cycles, no decomposition (without catalyst) of SO_3 is expected to occur in the 450–800°C temperature range. However, as can be seen from Figure 3, the conversion of SO_3 increases exponentially to 84% as the temperature approaches 1000°C.

7 US FUSION HYDROGEN PRODUCTION RESEARCH

In 1977 the US Fusion Energy Program funded several studies to identify and perform preliminary assessments of concepts that could be integrated with a fusion reactor to produce chemical fuels. By 1979, because of scientific and technological considerations with the other concepts, the HTE and thermochemical concepts for hydrogen production were selected for further evaluation and the development of scoping reactor designs. The HTE design is being performed at Brookhaven National Laboratory; the thermochemical reactor design is proceeding at Lawrence Livermore National Laboratory. Subsequent to the completion of these scoping designs, bench-scale feasibility experiments will be conducted for each concept and then, at the end of 1982, a reference concept will be selected for a detailed conceptual design study.

8 THE BROOKHAVEN NATIONAL LABORATORY FUSION REACTOR + HTE PLANT CONCEPT

8.1 Overview

Brookhaven National Laboratory has carried out a comprehensive scoping design study, called HYFIRE, of a large tokamak fusion reactor coupled to an HTE system to

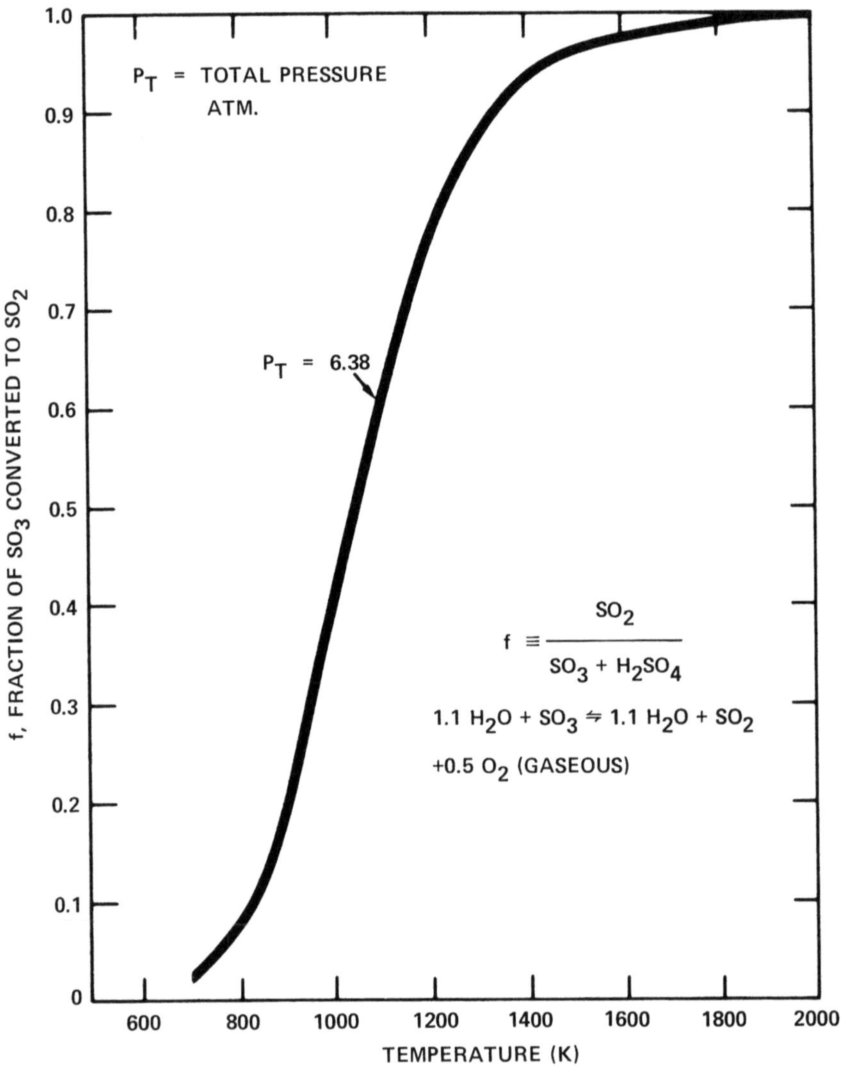

FIGURE 3 The SO_3–SO_2 equilibrium curve.

provide hydrogen and oxygen. The HYFIRE reactor design is based on the tokamak commercial electric power reactor, STARFIRE. The primary differences between the two reactors are in the blanket and the HTE process system. In addition to breeding tritium and supplying thermal energy for electric power, the HYFIRE blanket system must supply high-temperature process heat in the form of steam at temperatures in excess of 1000°C.

Figure 4 shows a simplified flow sheet of HYFIRE. It can be operated in either a hydrogen-production or an electrical-production mode.

Like STARFIRE, HYFIRE operates in a steady-state mode with the plasma current driven by radio-frequency power. The reactor has a major radius of 7 m and

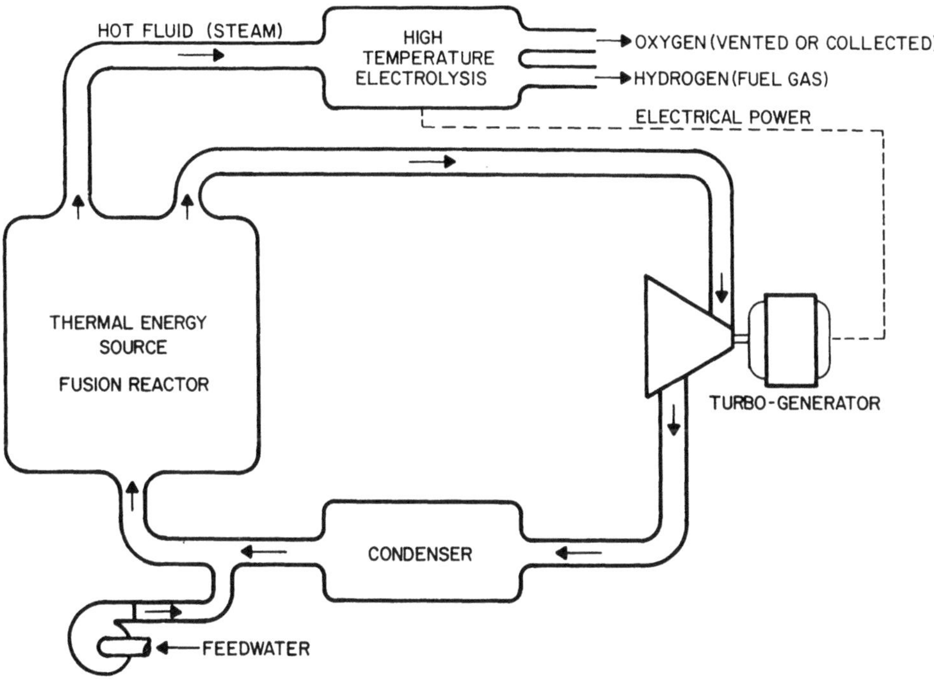

FIGURE 4 The general HTE process with a conventional power cycle.

produces 4000 MWth with an average neutron wall load of 3.6 MW m^{-2}. 1050 metric tons of hydrogen could be produced per day.

8.2 HYFIRE Blanket Design

The HYFIRE blanket must perform three functions: (a) provide high-temperature ($\geqslant$1000°C) process steam at moderate pressures (in the range 10–30 atm) to the HTE units; (b) provide high-temperature (700–800°C) heat to a thermal power cycle for generation of electricity for the HTE units; (c) breed enough tritium to sustain the D–T fuel cycle.

The deep penetration of the high-energy primary neutrons makes blankets with two temperature regions feasible. In this concept a relatively low-temperature (300°C) metallic structure is the vacuum–coolant pressure boundary, while the interior of the blanket, which is a simple packed bed of nonstructural material, operates at very high temperature (1400°C). Separate coolant circuits are required for the two temperature regions, as well as a thermal insulator to separate the regions. A representative HTE blanket module is shown in Figure 5.

For the HTE modules, refractory oxides (e.g. ZrO_2 or Al_2O_3) form the high-temperature region of the blanket. Such materials fill the interior of the blanket as solid rods or balls and in addition are used as a low-density solid block or fibrous thermal insulation between the high-temperature interior and the structural shell.

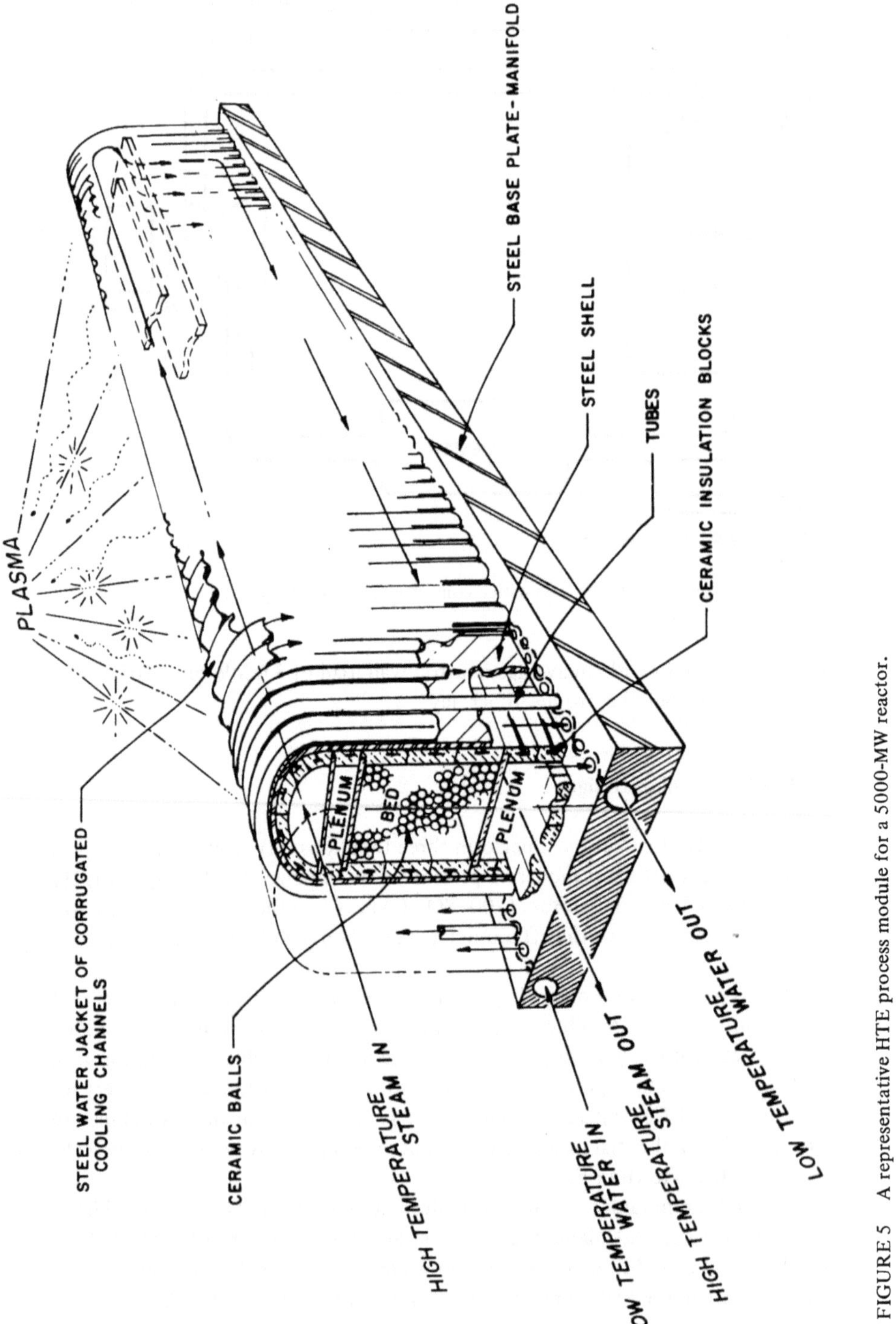

FIGURE 5 A representative HTE process module for a 5000-MW reactor.

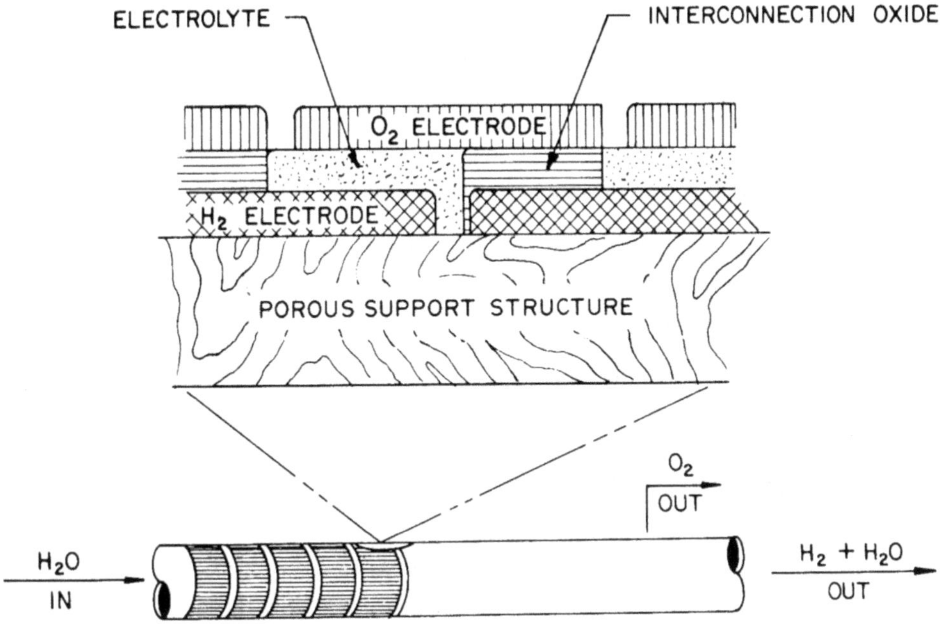

FIGURE 6 HTE cell design (the Westinghouse fuel cell).

The coolant for the hot interior can be a process fluid like steam or CO_2. This direct-heating feature eliminates the transfer of high-temperature heat across a metallic primary heat exchanger (which could severely limit the maximum temperature and choice of coolant) and is the mode of generating process heat for the HTE process.

8.3 HTE Design

A schematic diagram of a fuel cell is shown in Figure 6. This design serves as the basis for the high-temperature electrolyzer since an electrolyzer is a fuel cell in reverse. HTE uses arrays of tubes of relatively small diameter (1 cm), thick-walled porous ceramic (e.g. stabilized ZrO_2) on which a succession of thin electrode layers of suitably doped ceramics are deposited. The H_2 and O_2 ceramic electrodes are separated by a thin (a few thousandths of an inch) electrolyte layer of yttria-stabilized ZrO_2.

The major limitation with solid oxide electrochemical cells has been the difficulty with the connection between cells. However, recent developments have made available mixed oxides with all the required properties for the interconnectors.

8.4 The HYFIRE HTE Process

Two HTE process options have been identified during the scoping design study: (1) high-fractional conversion of steam to hydrogen (90% H_2 at the exit of the electrolyzer string); (2) low-fractional conversion of steam to hydrogen (10% H_2 at the exit of the first or second electrolyzer). Low-fractional conversion has a number of attractive features. Most important is the fact that a higher percentage of energy is supplied as thermal energy to decompose water than at the higher conversion.

8.5 The HYFIRE Plant-Process and Power-Conversion Design

A detailed process flow sheet for the conversion of fusion energy (thermal) to gaseous hydrogen via water electrolysis at high temperature is shown in Figure 7.

With regard to the steam circuit, a steam generator not unlike that of a pressurized-water fission reactor is used. The existing 1000-psia saturated steam is superheated to about 760°C by means of the helium. Steam then enters the high-pressure steam turbine and the high-temperature zone of the blanket. It leaves the blanket at 1400°C and goes directly to the high-temperature electrolyzers where approximately 10% is converted to hydrogen and oxygen in a single pass. The hydrogen together with the unreacted steam is passed on to the hydrogen-recovery portion of the flow sheet.

Preliminary studies and calculations indicate that a gross power-cycle efficiency in the 40–45% range appears to be achievable in HYFIRE using STARFIRE power recirculating parameters and power requirements. The corresponding H_2-production efficiency (the ratio of total fusion energy to the chemical energy of the hydrogen produced) is in the 50–55% range.

8.6 Requirements for Future Development of HTE

The two key technical areas requiring further development are the high-temperature blanket and the electrolyzer. With regard to the materials, the key blanket issues are (a) the integrity of the oxide in a steam atmosphere with radiation exposure and thermal cycling, (b) the long-term stability of the thermal insulator separating the structural shell and the high-temperature interior, and (c) the integrity of stainless steel or some other metallic structure, an issue common to all fusion blankets. An experiment on a high-temperature two-zone blanket must be designed and constructed to qualify such a blanket for testing in future fusion reactors.

Areas requiring further study in HTE are the manufacture of HTE cells, impurities and cation migration, and electrochemistry at elevated temperatures. Large-scale HTE technology could be developed by the time that the first commercial fusion reactors are operating.

9 THE LAWRENCE LIVERMORE NATIONAL LABORATORY FUSION REACTOR + THERMOCHEMICAL PLANT CONCEPT

9.1 Overview

Lawrence Livermore National Laboratory is currently performing a scoping design study of a commercial tandem-mirror reactor coupled to a thermochemical cycle to

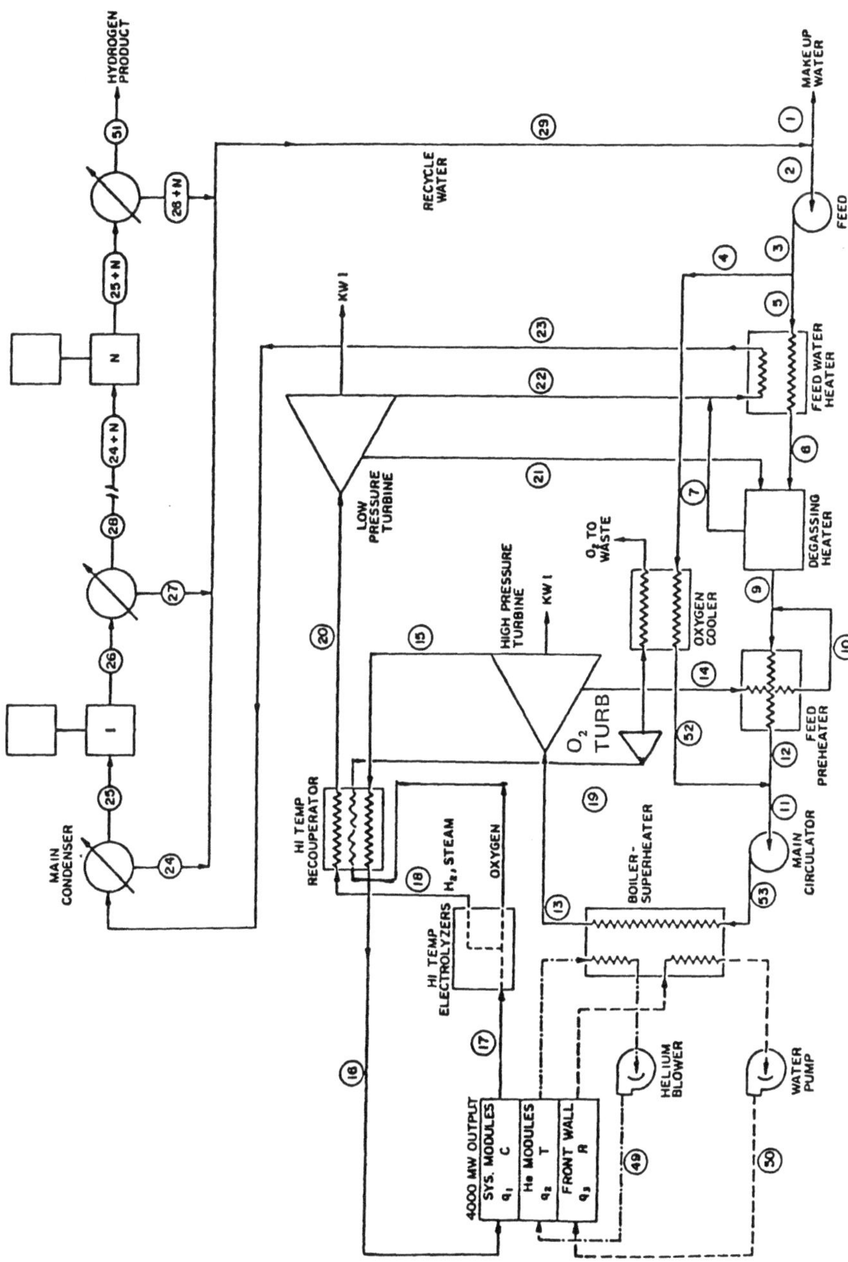

FIGURE 7 A flow sheet of the HYFIRE process.

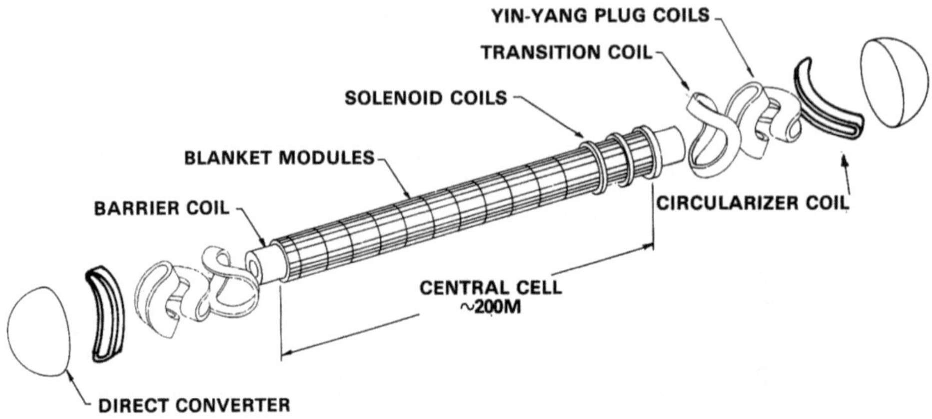

FIGURE 8 The tandem-mirror reactor.

produce hydrogen. The purpose of the study is to provide a basis for a technical feasibility assessment of such a system.

A primary reason for selecting a tandem mirror is its simpler reactor configuration. As shown in Figure 8, the reactor consists of a long central cell in which the power-producing D–T plasma is confined by straight magnetic field lines produced by simple circular Nb–Ti superconducting coil modules. These coils are about 4 m apart and the central-cell length is 200 m.

9.2 Description of the Reference Thermochemical Cycle

At the beginning of this study, three candidate thermochemical cycles were considered: the sulfur–iodine cycle (General Atomic), the sulfur cycle (Westinghouse), and the sulfur–bromine cycle (Ispra). The sulfur–iodine cycle was selected as the reference for the scoping design (see Table 2).

This cycle is a pure thermochemical cycle operating in an all liquid–gaseous environment. None of the reactions goes to completion; therefore it is necessary to separate reaction products from the reactants. Major parts of the process are associated with the separation and purification of the reaction products. A key to the successful operation of the process is the liquid–liquid phase separation of the lower-density phase containing H_2SO_4 and H_2O and a higher-density phase containing HI, I_2, and H_2O.

From an engineering perspective, the SO_3-decomposition step is a difficult problem because it requires that highly corrosive reactants be heated to 930°C. Moreover, it is common to all the most likely thermochemical cycles.

9.3 The "Joule-Boosted" Decomposer

The "Joule-Boosted" decomposer uses electrical instead of thermal energy to heat the reactants of the SO_3-decomposition step. The technical implications are the

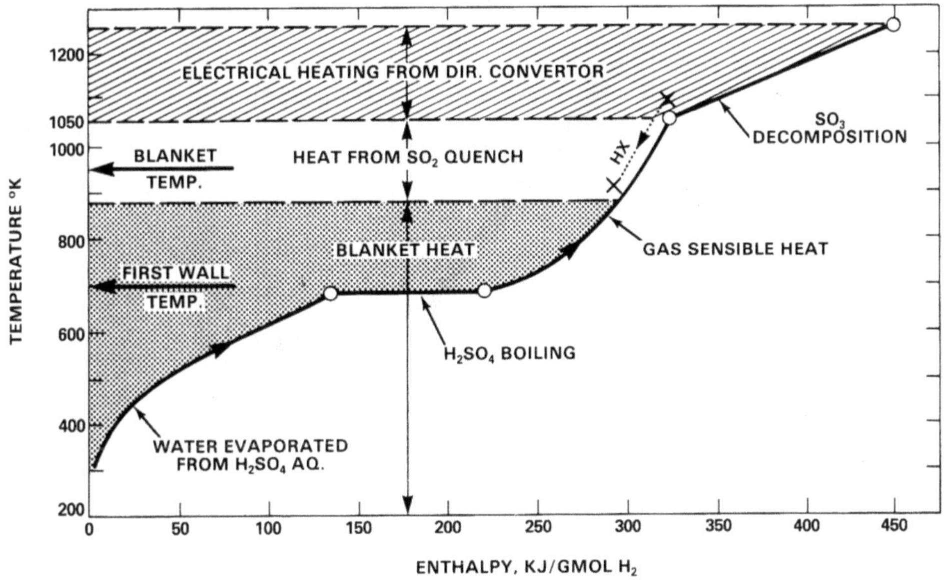

FIGURE 9 The heating curve for the H_2SO_4 process step.

elimination of the catalyst and significant lowering of the temperature requirements for blanket operation. Electric power from the tandem-mirror-reactor direct convertor plus some power which is generated from blanket heat is used to resistively heat the SO_3–SO_2 process gases as they feed to the decomposer; thus, as indicated in Figure 9, the blanket temperature (i.e. the source of thermal heat) can be lowered to about 600°C (900 K), greatly alleviating materials problems, the level of technology required, safety problems, and cost. Figure 10 shows the progress that has been made in lowering blanket temperature and increasing the effective decomposer temperatures. The blanket temperature has been decreased by approximately 450°C while decomposer reacting temperatures have been increased by approximately 200°C.

Figure 11 illustrates an individual blanket module using heat pipes as the primary heat-transfer mechanism. The blanket is designed (1) to produce the blanket temperatures and heat required for parts of the thermochemical cycle and for driving a thermal power cycle for electricity production and (2) to breed sufficient quantities of tritium to sustain the D–T fuel cycle. The heat-pipe working fluid is either sodium or potassium.

9.4 The Fusion–Thermochemical Reactor Energy Balance – Thermal and Electrical

For the thermochemical cycles there is a requirement for both electrical energy and process heat. Two regions of the tandem-mirror fusion reactor can provide process heat: (1) the blanket and (2) the thermal part of the direct convertor. Electrical energy from the direct convertor, over and above that required to drive the reactor, could be used to resistively heat the SO_3 decomposer.

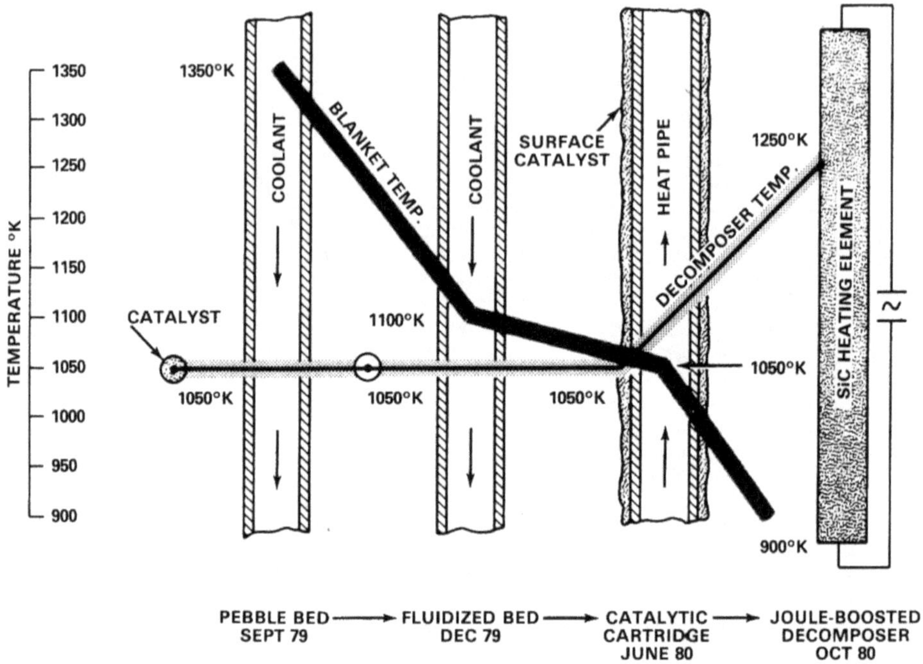

FIGURE 10 Progress in lowering the blanket temperature and raising the decomposer temperature for synfuel applications.

The percentage of the reactor output available as direct current for process chemistry tends to a limit of about 8% as the fusion energy multiplication factor Q become large (i.e. approaches 40). (Here, reactor output refers to the total fusion energy released, and Q represents the total fusion energy divided by the energy supplied to the plasma for heating.) The remainder of the electrical energy required is supplied from conventional turbogenerators operating on thermal energy.

9.5 Fusion–Thermochemical Materials Considerations

The key areas of materials needs for the design of the cauldron + heat pipe + thermochemical reactor are (1) the containment wall for liquid Li–Pb and (2) the thermochemical process system.

The selection of a material for the containment wall in contact with liquid Li–Pb presents a difficult problem since there are insufficient experimental data on corrosion of conventional alloys at 500–600°C to allow a choice to be made. It is known that in general liquid lead and Li–Pb solutions are more corrosive to iron-based alloys than are the liquid alkali metals but the mechanism and extent of liquid-lead corrosion are not well known.

SiC heating elements should function quite well in the SO_3 "Joule-Boosted" decomposer. However, the compatibility of SiC with SO_3–H_2O gases needs to be experimentally verified. Corrosive attack of the protective SiO_2 scale on SiC by steam is possible.

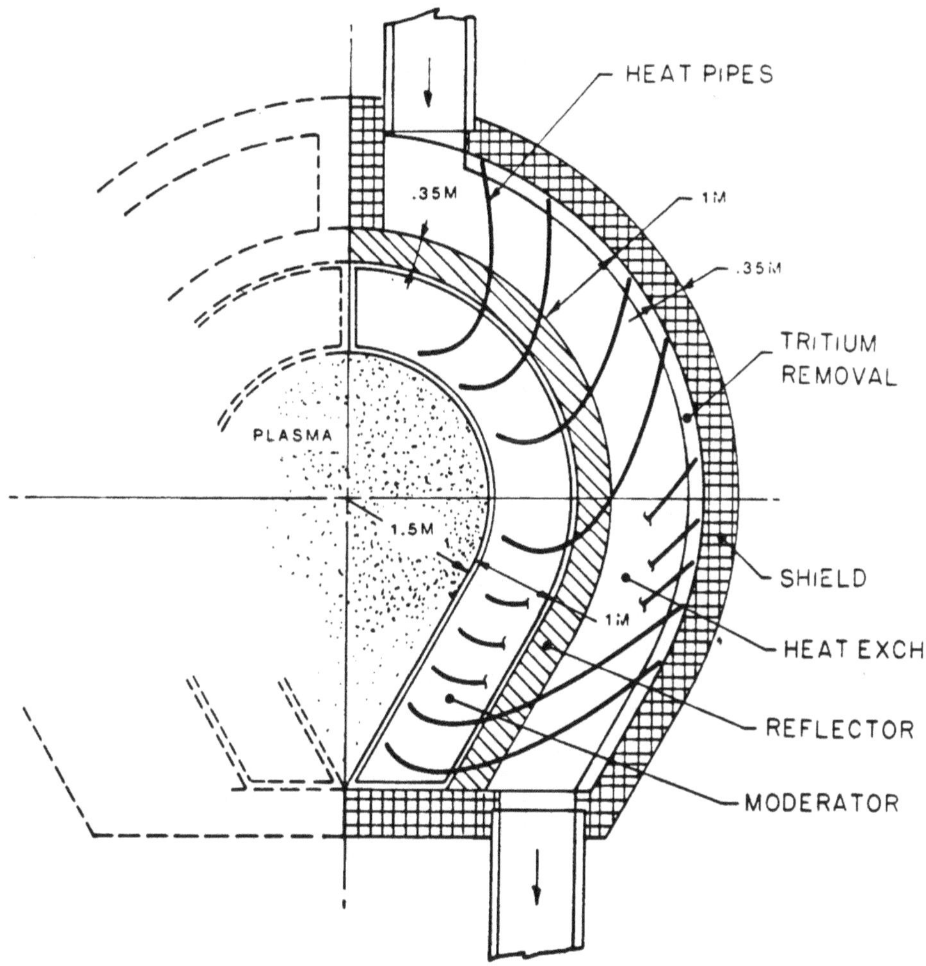

FIGURE 11 A heat-pipe blanket module.

It is expected that SiC could be used as the heat exchanger for the H_2SO_4 boiler which operates at a temperature of 400°C. Siliconized SiC (containing 10% free silicon) in the form of heat-exchanger tubes would be used for this application. Such heat-exchanger units have been built on a developmental basis but need further development for this use.

9.6 Requirements for Future Work on Thermochemical Processes

Over the near term the main technical issues for thermochemical hydrogen production based on a tandem-mirror reactor driver are centered on (a) refinement of the "Joule-Boosted" decomposer approach, (b) conceptual design of an Li–Pb blanket suitable for use at 600°C, (c) experiments to test key concepts (e.g. heat-pipe designs, tritium recovery in heat pipes, SiC–H_2SO_4(vapor) compatibility, containment of Li–Pb), and (d) systems studies aimed at obtaining a rough optimization of system efficiency and cost.

Over the longer term it will be important to conduct qualifying tests on simulated small-scale blanket modules linked to the main interface components of a thermochemical hydrogen cycle (e.g. the SO_3 decomposer and the H_2SO_4 boiler).

10 SUMMARY AND CONCLUSIONS

Gaseous and liquid fuels presently supply the major portion of world energy needs and are likely to continue in this role for the foreseeable future. Natural gas and oil supplies are dwindling and within the next few decades will have to be supplemented by synthetic fuels. These could be derived from coal or other fossil sources (oil shale, tar sands, etc.) or from any or all of the inexhaustible energy sources (fission, fusion and solar).

Fusion is a promising energy source for synthetic-fuels production. The very high temperatures that can be generated relatively easily in a suitably designed fusion blanket should result in more efficient processes for the generation of hydrogen from water decomposition. Hydrogen can be directly used as a fuel or can be combined with carbon to produce synthetic liquid or gaseous fuels. High-temperature process heat from a fusion reactor could also be used to gasify coal directly to synthetic gas.

In the US Magnetic Fusion Energy Program, scoping design studies in which a fusion reactor is coupled to a thermochemical cycle or a high-temperature electrolyzer indicate that such systems are feasible in principle. Both HTE cells and tabletop demonstrations of thermochemical cycles have been operated at high temperatures for long periods. Overall efficiencies (fusion energy to hydrogen chemical energy) are projected to be in the range 45–60%, depending on the process conditions and type of power cycle. Commercial HTE and thermochemical technology can probably be developed by the time the first commercial fusion reactors are operating in the early 21st century.

In a long-range perspective one may consider an ideal world in which fusion, burning deuterium from sea water, produces hydrogen from water; subsequently the hydrogen would then be burned back to water vapor. That should leave the biosphere secure for all time in all places.

MODERN ELECTROLYTIC PROCEDURES FOR THE PRODUCTION OF HYDROGEN BY SPLITTING WATER

H.W. Nürnberg, J. Divisek, and B.D. Struck
Chemistry Department, Institute of Applied Physical Chemistry, Kernforschungsanlage Jülich GmbH, Jülich (FRG)

ABSTRACT

Hydrogen has a number of advantages as an energy carrier. Its production by the splitting of water under technically and economically feasible conditions requires either electrolytic procedures or thermochemical hybrid cycles with an electrochemical step. These procedures for hydrogen production should in future be based on an advanced primary energy source such as nuclear or solar energy. After a review of the general principles of the electrolytic approach, the paper considers the essential features of the various types of electrolytic procedure in the low-temperature range (conventional and advanced water electrolysis), in the elevated temperature range (water vapor electrolysis in a salt melt), and at high temperatures (water vapor electrolysis in a solid electrolyte). Technical and economic parameters and requirements are critically evaluated. Advanced types of electrolysis are compared with the advanced version of the sulfuric-acid hybrid cycle. It is concluded that suitable advanced electrolysis procedures have already reached the stage where they can be scaled up to commercial processes. In contrast, the basically very attractive hybrid cycle still needs substantial research and development work on both the electrochemical step and the thermochemical part of the process (particularly the materials problems involved).

1 INTRODUCTION

Hydrogen is a valuable and attractive energy carrier. It will probably, therefore, play a significant role in any future energy economy (Figure 1) designed for a finite world (Häfele, 1981) in which fossil fuels are substituted to a large extent and remaining fossil-fuel contributions to energy consumption are utilized in an improved, more prudent, and cleaner manner by allothermic conversion of coal or heavy crude oil into more valuable synthetic gaseous and liquid fuels. Hydrogen has a number of advantages of fundamental significance (Bockris and Justi, 1980; Caspers, 1978; Häfele, 1981). It offers attractive possibilities for energy storage. This aspect is of particular importance in connection with the storage of electricity during off-peak periods. Hydrogen permits a very economical and efficient mode of energy transfer over long distances via pipelines, and it can be used readily as a fuel for heating, industrial processes, and vehicular traffic.

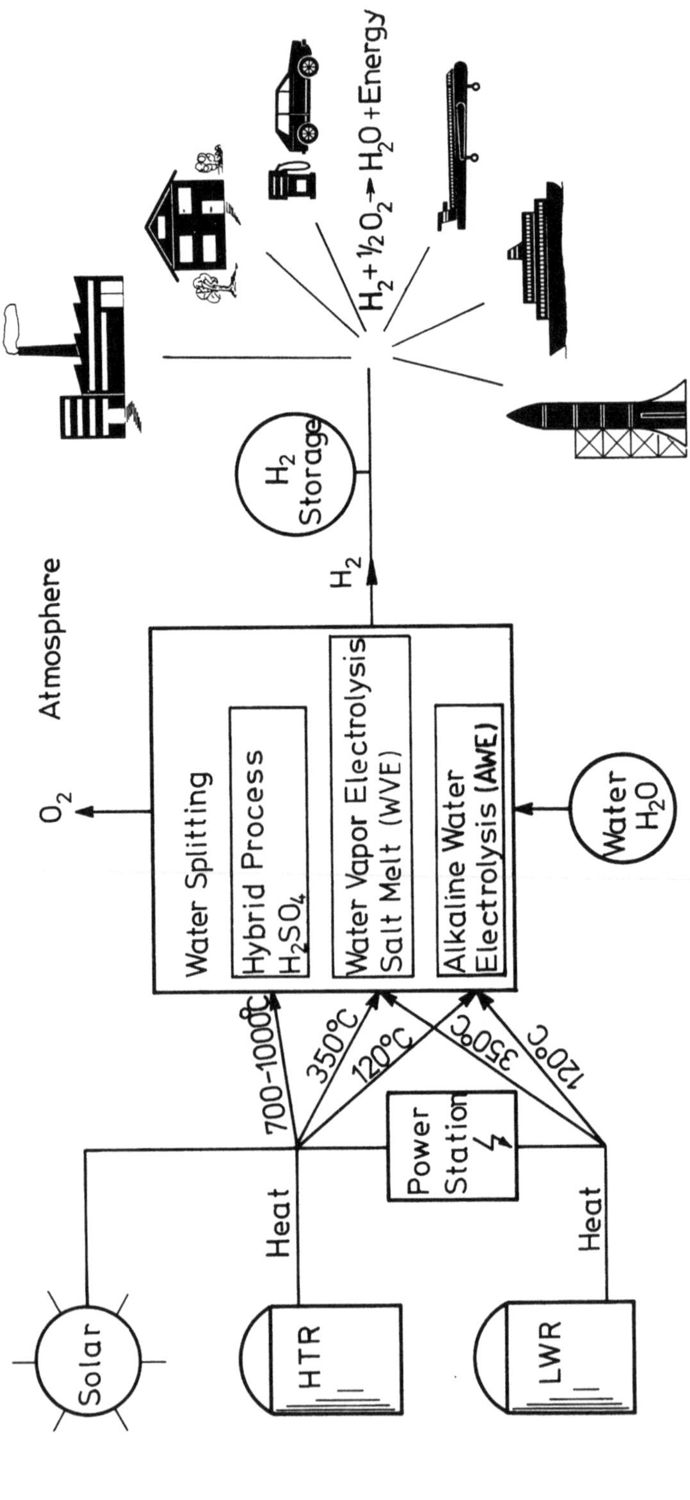

FIGURE 1 A schematic representation of a hydrogen-based energy economy.

One very important feature is that hydrogen constitutes a clean fuel that causes no fundamental or severe pollution problems in the environment since the combustion product of hydrogen is simply water. The most important, but probably "solvable", environmental problem with hydrogen combustion is connected with the production of some NO_x pollution at the parts-per-million level. Besides its use as fuel there is a substantial market potential for hydrogen as a reducing chemical in metal-ore processing and in a wide variety of chemical industries, not to mention its future large-scale use in allothermic fossil-fuel conversion which has already been referred to. In all these areas of application the use of hydrogen would substantially reduce or eliminate present burdens on the environment.

Taken alone, these very favorable features with respect to its use in a finite world in which environmental requirements are particularly significant would make hydrogen a favorite candidate for any future energy economy. But in addition, hydrogen has the unique feature that it can be obtained from water, an essentially inexhaustible resource. These advantages remain relevant if the water splitting is performed with the aid of primary energies which also meet the two fundamental conditions of being virtually inexhaustible and of producing no harmful impacts on the environment. The first choice as a primary energy source for a hydrogen economy is obviously nuclear energy (Häfele, 1981). Other primary energy sources that should become progressively more significant in the future at suitable geographical locations are solar energy and, to a certain extent, even wind energy.

In principle there are three possibilities for water splitting: (1) purely thermochemical processes, (2) hybrid cycles with both thermochemical and electrochemical steps, and (3) direct electrolysis.

Purely thermochemical processes leading either directly or indirectly, via included intermediate chemical reactions, to direct thermal water splitting suffer at present from unsatisfactorily low efficiencies and, furthermore, such enormous investment requirements that they cannot be regarded as a realistic alternative for hydrogen production (Deneuve and Roncato, 1981), at least not unless in the distant future substantially improved solar-energy devices can provide high temperatures (3000° C) for large-scale direct thermolysis of water. Thus, in the more feasible concepts of water splitting that are already partially realizable, electrolysis is inevitably involved, either exclusively as in the direct electrolytic procedures or to a certain degree in conjunction with a thermochemical step in hybrid processes. In this paper we deal with the essentials of these two applications of electrolysis for hydrogen production by water splitting and in particular we feature the research on this subject at KFA Jülich. In this context it is emphasized that the cost factors connected with the required materials and the complexity of the various technical devices can become decisive parameters for the production costs of hydrogen in addition to the overall energy efficiency and the electricity costs.

Among other parameters, the features of the primary energy source also determine the possible hydrogen-production procedures (see Figure 1). Hybrid processes always require elevated temperatures in their thermochemical step. They therefore require a high-temperature primary energy source such as a gas-cooled nuclear reactor (HTR), an advanced fusion plant, or a high-temperature solar-energy device (Struck, 1978). Direct electrolytic procedures such as Water Vapor Electrolysis (WVE) in a salt melt or advanced versions of Alkaline Water Electrolysis (AWE) can be performed (Divisek et al., 1980b, c) based on all types of nuclear reactors or on solar and wind energy. There is, however, a direct electrolysis mode at very high temperatures (900° C) in a solid electrolyte of ZrO_2 doped with Y_2O_3, the so-called "HOT ELLY", for which the HTR provides particular advantages (Dönitz, 1981).

2 ADVANCED WATER ELECTROLYSIS

2.1 General Aspects

Direct electrolysis of water, according to the process

$$H_2O \xrightarrow{+2e^-} H_2 + \tfrac{1}{2}O_2 \quad (1)$$

is an approach for the production of hydrogen which could be introduced right now as an advanced technology utilizing Light Water nuclear Reactors (LWRs) as well as fast breeders or solar-energy installations for the primary energy supply. It must however be emphasized that cost and efficiency requirements for the production of hydrogen as a secondary energy carrier dictate that advanced modes of water electrolysis must be used rather than the electrolysis procedures that are at present commercially available.

As Figure 2 shows, the conventional techniques for water electrolysis that are at present commercially available require cell voltages that are much too high even at moderate current densities and consequently yield rather low hydrogen production rates per unit electrode area. Nevertheless, water electrolysis, including the new advanced versions to be presented below, represents technology that is in principle well established and is the simplest route for the production of hydrogen.

According to basic thermodynamics, the minimum thermoneutral enthalpy ΔH required for water splitting at a pressure of 1 bar is 3.54 kWh $(m^3\ H_2)^{-1}$. This ΔH value consists of two components (Figure 3). One component is the virtually thermoneutral ΔH_v value of 3.08 kWh $(m^3\ H_2)^{-1}$ for water splitting in the vapor state. The other is the vaporization enthalpy ΔL_v of water which amounts to 0.46 kWh $(m^3\ H_2)^{-1}$. If the total ΔH of 3.54 kWh $(m^3\ H_2)^{-1}$ were introduced as electricity, a cell voltage of 1.48 V would be required. In the various practical water-splitting procedures, however, only a certain fraction of the required total enthalpy is introduced as electricity while the rest is provided as heat. In all types of low-temperature (80–150° C) water electrolyses from the liquid state the overwhelming proportion of ΔH has to be supplied as electricity. However, in high-temperature water electrolysis a fraction of ΔH which increases with increasing temperature can be introduced as heat (see Figure 3).

It should, however, be noted that the preceding, purely thermodynamic treatment does not take into account the fact that in reality, for kinetic reasons, the cathodic and anodic electrode processes inevitably require a certain overvoltage which depends on the performance of the particular electrolytic procedure. In addition, ohmic voltage losses always occur in the electrolyte, at the diaphragm, and in the conductors between the cell units of any electrolysis device. These extra voltage losses raise the effective total enthalpy ΔH above its ideal value of 3.54 kWh $(m^3\ H_2)^{-1}$. Moreover, these losses have to be compensated for in the form of electricity. Only in the purely theoretical thermodynamic

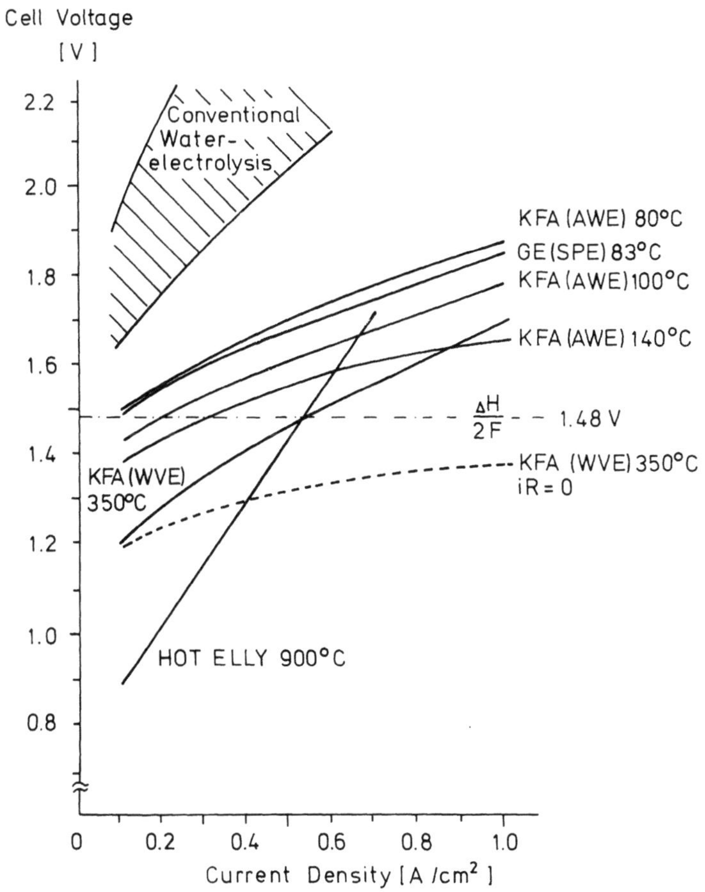

FIGURE 2 Current density versus cell voltage characteristics for various electrolytic procedures for water splitting.

case (i.e. case III in Table 2) does the electricity demand equal the free enthalpy ΔG of electrolytic water splitting (see Figure 3).

From Figure 3 it follows that, for the fraction of ΔH that can be introduced as heat Q to be significant, a prime requirement is that the process can operate at a cell voltage below 1.48 V (corresponding to 3.54 kWh $(m^3\ H_2)^{-1}$). The ideal lower limit for the fraction of ΔH that can be introduced as heat (in the absence of the losses discussed above) is given by the ΔG line in Figure 3.

Conventional water electrolysis using presently available commercial devices at ambient pressure and 80–90° C requires cell voltages between 1.8 and 2.2 V, corresponding to an overall enthalpy of between 4.3 and 5.3 kWh $(m^3\ H_2)^{-1}$. By far the largest proportion of this enthalpy has to be supplied as electricity. The difference between either 4.3 or 5.3 kWh $(m^3\ H_2)^{-1}$ and 3.54 kWh $(m^3\ H_2)^{-1}$ is dissipated as useless heat which has to be removed. Usually the cells are designed for the relatively modest current

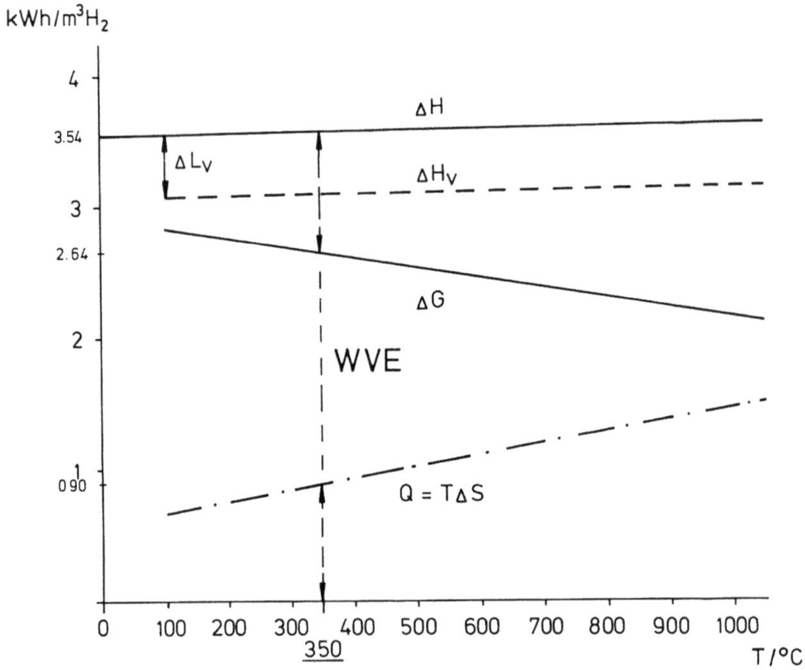

FIGURE 3 Thermodynamic functions of water electrolysis. $\Delta H = \Delta G + T\Delta S$, where ΔH is the thermoneutral enthalpy, ΔG is the free enthalpy supplied as electricity (in the absence of overvoltages and ohmic losses), and $Q = T\Delta S$ is the enthalpy increment contributed as heat. The position (corresponding to case III in Table 2) for the WVE operated at 350° C is indicated.

density of 0.2 A cm^{-2}. The substantial cell voltages that are required arise from marked overvoltages in the cathodic and anodic electrode processes and ohmic voltage losses. The effective ΔH value for water splitting is accordingly higher than the minimum of 3.54 kWh $(m^3\ H_2)^{-1}$.

The investment costs for these conventional technical cells are quoted as 1500–2500 $h $(m^3\ H_2)^{-1}$* (see for example Dönitz, 1981). Even the somewhat more advanced commercial pressure-electrolysis procedure (100° C, 30–40 atm, 0.2 A cm^{-2}) requires at least 4.3 kWh $(m^3\ H_2)^{-1}$ or 1.8 V. Note that the increased pressure has two opposing effects. It increases the theoretical cell voltage by about 0.08 V (to 1.56 V) while at the same time it reduces the ohmic losses, thus leading in practice to a somewhat reduced cell voltage at a given current density.

The overall efficiency of an electrolysis plant can be expressed in a simplified manner as follows:

$$\eta_{tot} = \eta_{el}\eta_{th} \tag{2}$$

* In other words, $1500–2000 for each installed cubic meter of H_2 hourly *capacity*. (It is assumed here that $1 ≡ DM2.)

or

$$\eta_{tot}\ (\%) = [1.48/(\text{cell voltage})]\,\eta_{th}\ (\%) \tag{3}$$

Referring to the lower limit of the shaded area in Figure 2 and assuming the present best value of 40% for η_{th}, one finds that for conventional electrolysis η_{tot} is as low as 34% (η_{el} = 85%) at 0.2 A cm^{-2} and even lower at the higher current densities which are of course desirable for large-scale hydrogen production. The situation is even more unfavorable in practice because actual voltage requirements fall well inside the shaded area rather than on its lower boundary.

The primary goal obviously has to be to increase η_{el} and consequently η_{tot} by lowering the required cell voltage if electrolysis is to approach a cost range that is tolerable for large-scale water splitting. This is also evident in the data in Table 1.

TABLE 1 Production costs of hydrogen from water splitting by advanced electrolysis[a].

Cell voltage (V)	Required electrical energy[b] (kWh)	Electricity costs[b] ($)		Total costs[b] ($)		Electricity costs as a percentage of total costs	
		I[c]	II[d]	I	II	I	II
2.09	5.0	0.125	0.325	0.195	0.395	64	82
1.86	4.5	0.113	0.293	0.183	0.363	62	81
1.65	4.0	0.100	0.260	0.170	0.330	59	79
1.45	3.5	0.088	0.228	0.158	0.298	56	77
1.25	3.0	0.075	0.195	0.145	0.265	52	74

[a] The originally quoted costs in deutschemarks have been converted to dollars assuming a rate of DM2 ≡ $1.
[b] Per cubic meter of H_2.
[c] Cost of electricity, $0.025 kWh^{-1} (see Imarisio, 1981).
[d] Cost of electricity, $0.065 kWh^{-1} (see Gassert, 1981), value extrapolated for 1990.

To facilitate comparison with Figure 2, cell voltages and corresponding amounts of electricity needed for the production of 1 m^3 of H_2 are included in Table 1. For the electricity costs two figures are quoted. The lower figures (case I) are based on the highest electricity price quoted in a recent publication (Imarisio, 1981). The more pessimistic higher figures (case II) are based on a much higher electricity price very recently quoted by the nuclear industry (Gassert, 1981) for electricity production with an LWR in 1990. The difference between the total cost and the cost of electricity alone is $0.07 $(m^3\ H_2)^{-1}$ in all cases. It has been estimated assuming an annual operating time of 7000 h and an annuity of 20% to cover the interest on the investment of 1750 $h $(m^3\ H_2)^{-1}$ as well as the operating and maintenance costs of the electrolysis plant (see for example Dönitz, 1981).

If the electricity were produced by the combustion of coal, the costs would become considerably higher. According to the most recent data DM0.233 or $0.117 are required in 1990 per kilowatt-hour of electricity. These are higher by a factor of 1.8 than the cost of electricity from an LWR (Gassert, 1981).

The figures emphasize the importance of lowering considerably the amounts of electricity required at technically reasonable current densities (0.2–0.6 A cm^{-2}) by the introduction of advanced electrolysis procedures (see Figure 2). According to a recent study (Imarisio, 1981), current densities of at least 0.5–0.7 A cm^{-2} seem to be optimal from the viewpoint of the total H_2-production costs. Recent cost analyses for our AWE and WVE procedures have shown that their cost optimum is close to 0.5 A cm^{-2}.

The primary development target in advanced water electrolysis has to be the achievement of low cell voltages, typically less than 1.5 V in the current-density range of 0.5 A cm^{-2} or more.

In this context we should also bear in mind that there is an alternative for hydrogen production which is rather favorable with respect to costs; namely, steam reforming with coal. For 1979 a total cost figure of \$0.16 $(m^3\ H_2)^{-1}$ has been quoted based on autothermic steam reforming with coal (Dönitz, 1981) which will probably have to be increased by about 20%. Nevertheless, electrolytic hydrogen production based on nuclear energy as the primary energy source will remain for the time being somewhat more expensive than steam reforming, although the discrepancies will not be too dramatic provided that the required electrical energy can be kept below 4.0 kWh $(m^3\ H_2)^{-1}$ (see Table 1). On the other hand, it must be emphasized that the somewhat more expensive electrolytic production of hydrogen provides a substantially cleaner technology with respect to environmental pollution and also saves coal resources. Strategies for future energy supply in a finite world (Häfele, 1981) cannot be based purely on production costs but will have to provide a reasonable balance between economic factors and mandatory requirements for the restriction of environmental pollution as well as resource-saving aspects which will have to be included in future decision-making procedures.

The current density versus cell voltage relationships in Figure 2 reflect the present status of various advanced versions of water electrolysis.

The main points to be improved and problems to be solved are as follows (see also Figure 5): (1) to decrease the overvoltage at the electrodes by the use of cathodes and anodes that are highly active, based on the selection of suitable electrode materials in optimal geometrical configurations, i.e. as porous electrodes with large active areas; (2) to increase the operating temperature to speed up the kinetics of the electrode process and to decrease the ohmic resistance of the electrolyte; (3) to improve the diaphragm by lowering its ohmic resistance and increasing its resistance to corrosion; (4) to improve the cell design, particularly through minimizing the distance between the electrodes and eliminating further ohmic losses by optimal assembly of the single cells to form larger electrolysis devices.

In practice, one can distinguish three main groups of electrolysis procedures, and we will now examine each in turn.

2.2 Alkaline Water Electrolysis (AWE)

In AWE the electrolyte is typically a concentrated (about 10 M) solution of potassium hydroxide (KOH) as in conventional water electrolysis. However, the electrodes are much more active and the aim of our research was to increase the operating temperature to 140 or 150° C.

As the curve in Figure 2 for our cell at 140° C reflects, our development work has reduced the cell voltage required by a considerable amount (Divisek et al., 1980 a, b; Divisek and Schmitz, 1981). The electricity required is between 3.5 and 3.8 kWh $(m^3\ H_2)^{-1}$ only for current densities from 0.2 to 0.5 A cm^{-2}. Note that the temperature of 140° C makes operation at elevated pressures essential. Our procedure is performed at 8 bar. For thermodynamic reasons, the elevated pressure will cause, as already mentioned, a slight increase of about 0.05 or 0.06 V in the cell voltage. In practice, however, this is more than compensated by the reduction of the ohmic loss. Thus the net effect is a gain rather than a loss in terms of the electricity required. The improvements are due to the following provisions. Highly active Raney-nickel electrodes have been produced by cathodic deposition of an activated Ni–Zn alloy onto a specially prepared nickel gauze and subsequent dissolution of the zinc in KOH. A new diaphragm has been obtained by a simple low-cost method whereby a porous nickel coating is deposited onto a nickel gauze. Subsequently it is oxidized in air at 900–1000° C to give NiO. This nickel oxide diaphragm has a substantially lower ohmic resistivity (below 0.1 $\Omega\ cm^2$), decreasing the voltage loss to 40–100 mV within the current density range 0.2–0.5 A cm^{-2} (Figure 4). This diaphragm has also shown excellent resistance to corrosion during a test period of over 3500 h at 110° C. Finally, it was possible to mount the electrodes immediately on the diaphragm in the sandwich arrangement that is favored for minimizing voltage losses in a bipolar cell. In this manner the electrolysis efficiency η_{el} becomes 100% at 0.3 A cm^{-2} and even exceeds 100% at lower current densities, because a fraction of the required energy is provided by ohmic heat dissipated in the electrolyte. A further decrease of the cell voltage by about 0.05 V seems to be feasible with the application of pressure and certain improvements leading to a lower overvoltage at the anode. A 30-A bipolar cell has already been operated satisfactorily for over 3500 h without any corrosion problems (Divisek and Schmitz, 1981).

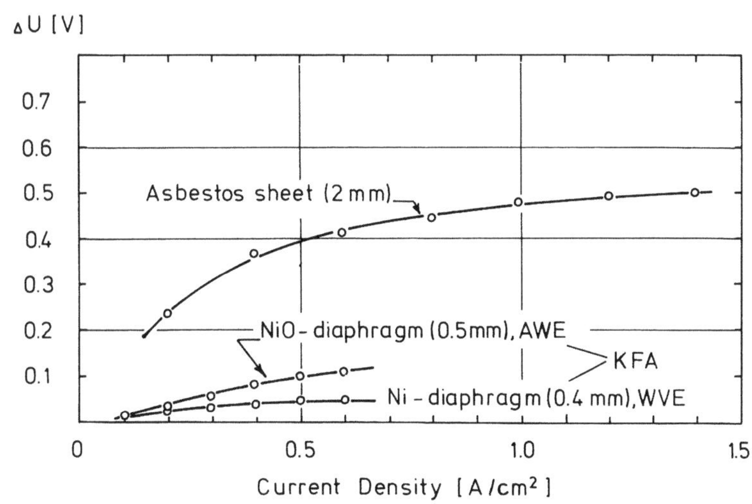

FIGURE 4 Ohmic voltage losses at diaphragms as a function of current density.

For comparison, the current density versus voltage characteristics of the Solid Polymer Electrolysis (SPE) procedure developed by General Electric (Nuttal and Russel, 1980) have been included in Figure 2. This process takes place in a device where a solid plastic sheet of a polymeric perfluorinated hydrocarbon substituted with SO_3H groups at regular molecular distances serves as the electrolyte. This "NAFION" membrane is coated on one side with a thin layer of platinum black acting as the cathode and on the other side with a thin layer of a ruthenium-based alloy catalyst constituting the anode (Sedlak et al., 1981). SPE has been regarded as one of the most advanced concepts of water electrolysis with respect to the required cell voltage. Now, however, our AWE provides similar current-density versus voltage characteristics (see Figure 2) in the relevant temperature range (80–140° C). The reason why the SPE characteristics are shown for 83° C only is that, from the material published to date, no other relevant current-density versus voltage curve is available. While from the viewpoint of the electricity demand SPE and our AWE procedure are now at comparable levels, SPE requires expensive materials for the electrodes and the NAFION membrane. The resulting high investment costs make it almost essential to operate the device at unfavorably high current densities of around 1 A cm^{-2}.

2.3 Water Vapor Electrolysis (WVE) at Elevated Temperature

A further substantial decrease in the electrical energy has been achieved in the WVE which we have developed over the last few years (Divisek et al., 1980 b, c). An NaOH–LiOH melt kept at 350° C and 1 bar serves as the electrolyte. The LiOH removes interfering peroxide components formed in side reactions. The melt contains 1–2% water which is electrolyzed at highly active nickel electrodes. The diaphragm is cathodically-protected sintered nickel which has shown satisfactory resistance to corrosion during a test period lasting more than 1200 h.

The temperature increase to 350° C has considerable kinetic and thermodynamic advantages, resulting in the lowering of the cell voltage shown in Figure 2. The water is present in the vapor state and about 0.9 kWh $(m^3\ H_2)^{-1}$ of the required enthalpy ΔH is provided by the entropic heat $T\Delta S$ according to the Gibbs–Helmholtz equation $\Delta H = \Delta G + T\Delta S$. In the present stage of development, at a current density of 0.53 A cm^{-2} a cell voltage of only 1.48 V is required, yielding an η_{el} value of 100% and an η_{tot} value of 40%. At lower current densities η_{el} considerably exceeds 100% owing to favorable kinetic and thermodynamic factors. The broken curve in Figure 2 refers to an ideal WVE with zero ohmic resistance. Although this limiting case can never be reached completely in practice, the broken curve indicates the still-considerable range for future improvements by further reduction of the ohmic losses. The investment costs for an electrolysis device are fairly moderate for the WVE described. Estimates show that, provided a reasonable credit may be assumed for the produced oxygen, the costs for hydrogen production depend predominantly on the annual depreciation for the primary energy source (i.e. the nuclear reactor, which could be an LWR or HTR) while the contribution of the annual depreciation for the electrolyzer remains marginal.

2.4 High-Temperature Water Electrolysis (HOT ELLY)

An alternative approach to water electrolysis in the vapor state is provided by the high-temperature electrolysis procedure known as HOT ELLY (Dönitz et al., 1980; Dönitz, 1981). It is operated at 900° C and therefore a solid electrolyte, ZrO_2 (which is resistant to high temperatures) stabilized by doping with Y_2O_3, is used. H_2 evolves at a porous nickel cathode and O^{2-} ions are formed, according to the scheme

$$H_2O + 2e^- \rightarrow H_2 + O^{2-}$$

The O^{2-} ions migrate through the ZrO_2, which is a purely ionic conductor for these ions, to the anode made from semiconducting mixed oxides of the perovskite type, where oxygen is evolved. As the current-density versus voltage relationship in Figure 2 shows, this electrolysis procedure clearly requires the lowest cell voltage for current densities below 0.5 A cm^{-2}. Above this current density, however, it rapidly falls behind WVE in a salt melt, and above 0.6 A cm^{-2} also behind the AWE procedure as described above. The favorable performance with regard to cell voltage and the therefore limited amount of electricity required in the lower current-density range are due to the fact that, per cubic meter of H_2, about 1.3 kWh of the total energy required can be introduced as heat, as outlined earlier for WVE (see Table 2). The 900° C operating temperature required for HOT ELLY is substantially higher than the 350° C needed for the WVE. Thus HOT ELLY requires a cheap and environmentally clean high-temperature heat source. Also technically feasible but much less economically favorable would be adiabatic operation using dissipated ohmic heat (Dönitz, 1981). However, this would require high current densities which HOT ELLY can hardly afford owing to its steep current-density versus voltage curve (see Figure 2). Thus in practice endothermic operation with heat supplied from an external high-temperature heat source remains the only realistic alternative for the low current-density range where HOT ELLY has a favorably low cell voltage. It is therefore particularly suitable for operation in conjunction with a high temperature gas-cooled nuclear reactor (HTR). In comparing HOT ELLY with WVE in the medium current-density range of 0.4–0.5 A cm^{-2} one should also bear in mind that the installation costs of HOT ELLY will be substantially higher due to the need for expensive materials that are sufficiently resistant to the higher operating temperature. Only in the current-density range below 0.4 A cm^{-2}, which is less attractive with respect to total H_2-production costs, will the gain in electricity for HOT ELLY progressively balance its higher installation costs.

3 THERMOELECTROCHEMICAL HYBRID CYCLES

3.1 General Aspects

The concept of providing a significant fraction of the total energy required for hydrogen production from water splitting in the form of heat, which is already applied to a certain extent in WVE and HOT ELLY, is developed to a greater degree in the so-called hybrid cycles. Basically these always consist of an endothermic thermochemical step and a coupled electrochemical step, and involve a chemical system X/Y which is cycled according to the general scheme:

$$Y \rightarrow X + \tfrac{1}{2}O_2 \quad \text{high-temperature endothermic step}$$
$$X + H_2O \rightarrow H_2 + Y \quad \text{electrochemical step} \qquad (4)$$

The net reaction is water splitting:

$$H_2O \rightarrow H_2 + \tfrac{1}{2}O_2 \qquad (5)$$

3.2 The Sulfuric Acid Cycle

Among the various types of hybrid cycles discussed in the literature, the sulfuric acid cycle has gained primary importance. The basic process was introduced by Westinghouse using platinum-activated graphite electrodes in the electrochemical step (Brecher and Wu, 1975). The cycle consists of the following steps:

$$H_2SO_4 \rightarrow SO_2 + H_2O + \tfrac{1}{2}O_2 \quad \text{thermochemical step at 700–1000° C}$$
$$SO_2 + 2H_2O \rightarrow H_2SO_4 + H_2 \quad \text{electrochemical step with a cell voltage of 0.6 V (Fasbman, 1979) to 0.75 V (Lu et al., 1981) at 0.2 A cm}^{-2} \text{ and 50° C} \qquad (6)$$

As can be seen, one complete cycle involves two molecules of water, of which one is split. The components of the SO_2/H_2SO_4 system are not consumed but only cycled.

3.3 The Advanced Sulfuric Acid Cycle

The electrochemical step, which consists of hydrogen evolution at the cathode and oxidation of SO_2 at the anode, has been substantially improved by us (Struck et al., 1978; 1980 a, b; 1981; Struck and Junginger, 1980). The introduction of small amounts of the system HI/I_2 as a homogeneous catalyst permitted the substitution of the platinum-activated anode by a graphite felt flow-through anode and at the same time lowered the cell voltage required. The electrochemical step is now believed to take place as follows:

$$2HI \rightarrow I_2 + 2H^+ + 2e^- \quad \text{anodic electrode process}$$
$$I_2 + SO_2 + 2H_2O \rightarrow H_2SO_4 + 2HI \quad \text{homogenous chemical } SO_2 \text{ oxidation} \qquad (7)$$
$$2H^+ + 2e^- \rightarrow H_2 \quad \text{cathodic electrode process}$$

The sulfuric acid is then concentrated from 58 to 70 wt.%, separated from very small residual amounts of HI by distillation, and subsequently decomposed as in the first step of eqn. (6). The net reaction of the hybrid cycle is again the splitting of one molecule of water.

The introduction of the homogeneous catalyst system HI/I_2 has lowered the cell voltage to 0.6 V at 0.2 A cm^{-2} and 80° C (Figure 5). There is still scope for further improvement of the cell voltage. While the anode is a graphite felt the present cathode is a cation-exchange membrane (NEOSEPTA or NAFION) doped with a small amount of platinum. This cathode type prevents the formation of hydrogen bubbles in front of the cathode and the overvoltage contributions connected with them. Furthermore, no liquid catholyte is necessary. At present the substitution of this cathode by a tungsten carbide electrode which is more favorable economically is being tested (Struck and Junginger, 1980). It is very important to prevent the migration of SO_2 from the anode to the cathode compartment where its reduction to sulfur would cause poisoning of the platinum cathode. This has been achieved satisfactorily by a three-compartment cell where the anode and cathode compartments are separated by cation-exchange membranes (NEOSEPTA or NAFION) from a third central compartment through which flows sulfuric acid of 35 wt.% (the H_2SO_4 concentration which exhibits maximum conductivity). In this manner perturbing electrolyte intermixing is also prevented. This cell with the homogenous HI/I_2 catalyst and the graphite felt anode is now being introduced in a scaled-up pilot plant at JCR Ispra (Italy).

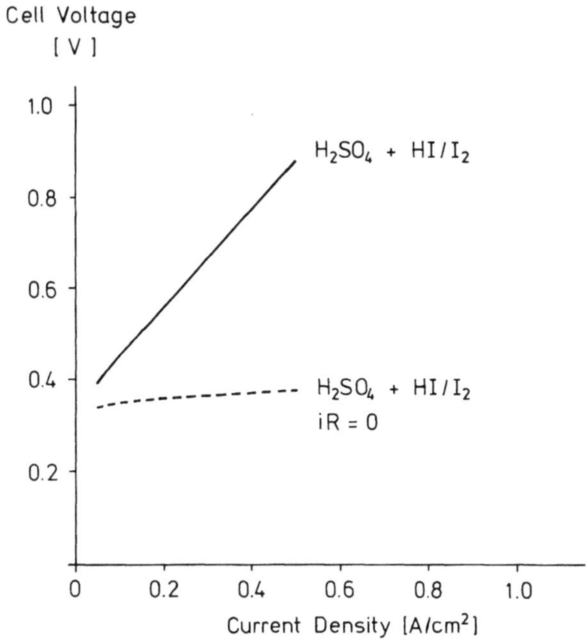

FIGURE 5 The current density versus cell voltage characteristics of the electrolytic step of the advanced sulfuric acid hybrid cycle. The broken curve corresponds to the measured performance without diaphragm ($iR = 0$).

Although considerable progress has been made during the past few years, as shown above, there still remains a large area for further improvements. This can be seen from Figure 5 where the broken curve shows the possible performance in the absence of ohmic losses in the diaphragm. There are also substantial materials problems to be solved before large-scale commercial plants can be built. These are connected mainly with severe corrosion problems, particularly if pressures of 40 bar have to be applied in conjunction with an HTR as the high-temperature heat source. Technically feasible new materials (e.g. INCOLOY 800) have recently been proposed but they still need extensive testing.

Yet a hybrid cycle, such as the sulfuric acid process with the system HI/I_2 as an additional homogenous catalyst discussed above, remains ultimately a very attractive approach for water splitting because it permits the introduction of a fairly significant fraction of the required total thermoneutral enthalpy in the form of heat. The data in Table 2 make this immediately evident.

4 GENERAL CONCLUSIONS

Table 2 compares the contributions of electricity and heat to the total enthalpy required for the production of 1 m^3 of H_2 at two typical current densities for two advanced direct electrolytic procedures with significant heat contributions and for the advanced sulfuric acid hybrid cycle. Three cases are considered: (I) the actual achievements to date with respect to cell voltage and corresponding electricity demand (see Figures 2 and 5); (II) the hypothetical case of zero ohmic resistance, which can never be fully reached but which can be approached by future development; and (III) the thermodynamic theoretical case corresponding to zero ohmic resistance and zero overvoltage for the cathodic and anodic electrode processes (see also Figure 3), which it is important to consider because it provides insights into the basic, purely thermodynamic requirements. The slight pressure dependence of ΔH has been neglected in the data reported for HOT ELLY. For a given process the fraction of the total enthalpy ΔH introduced as heat is essentially constant for all three cases. The percentages show clearly the basic attractiveness of the advanced H_2SO_4 hybrid cycle once it becomes technically utilizable, particularly in the higher current-density range around 0.5 A cm^{-2} which is desirable for reasons of capital investment.

A decisive further proviso for the utilization of the basic advantage of the process, namely introducing a substantial fraction of the required total enthalpy ΔH for water splitting as heat, is the availability of a cheap heat source which, in addition, must not cause severe environmental pollution problems. The HTR constitutes a heat source of this kind.

The correlation between cell voltage and electricity in kilowatt-hours required for the production of 1 m^3 of H_2 is depicted in Figure 6. This correlation has been used in the calculation of the required electricity demands in Tables 1 and 2. In addition Figure 6 indicates (for a current density of 0.53 A cm^{-2}) the present status of the different procedures, conventional commercial electrolysis (best value), advanced electrolysis in the low-, elevated-, and high-temperature ranges, and the electrochemical step of the advanced H_2SO_4 hybrid cycle with the homogenous catalyst HI/I_2 in the electrochemical step. Also the hypothetical limits corresponding to zero ohmic resistance are given

TABLE 2 The contributions of electricity and heat in various processes to the total enthalpy ΔH for water splitting at current densities of 0.2 and 0.53 A cm^{-2} for the production of 1 m^3 of H_2.

Process	Case I (practical)			Case II ($iR = 0$)			Case III (theoretical)			Heat	Heat percentage of ΔH		
	Cell voltage (V)	Electricity demand (kWh)	ΔH (kWh)	Cell voltage (V)	Electricity demand (kWh)	ΔH (kWh)	Cell voltage (V)	Electricity demand ΔG (kWh)	ΔH (kWh)	introduced[a] Q (kWh)	I	II	III
Current density = 0.2 A cm^{-2}													
WVE[b]	1.28	3.1	4.00	1.23	2.96	3.86	1.1	2.64	3.54	0.90	22.5	23	25
HOT ELLY[c]	1.02	2.4	3.71	–	–	–	0.94	2.23	3.54	1.31	35	–	37
H_2SO_4 hybrid cycle[d]	0.60	1.45	4.41	0.385	0.90	3.86	≈0.25	0.58	3.54	2.96	72	77	84
Current density = 0.53 A cm^{-2}													
WVE[b]	1.48	3.54	4.44	1.32	3.2	4.1	1.1	2.64	3.54	0.90	20	22	25
HOT ELLY[c]	1.48	3.54	4.85	–	–	–	0.94	2.23	3.54	1.31	27	–	37
H_2SO_4 hybrid cycle[d]	0.90	2.3	5.26	0.40	0.95	3.91	≈0.25	0.58	3.54	2.96	56	76	84

[a] Including ΔL_V.
[b] 350° C, 1 bar.
[c] 900° C, 20 bar.
[d] With HI/I_2 and at 700–1000°C in the thermochemical step; calculated for 1 bar.

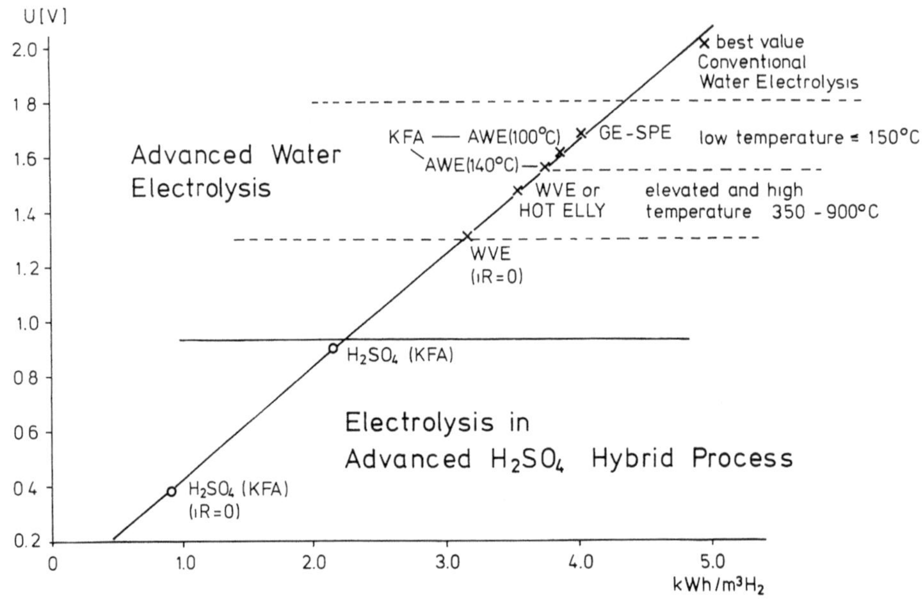

FIGURE 6 Relationship between required electricity in kilowatt-hours per cubic meter of H_2 and cell voltage U. The points inserted for various electrolytic procedures refer to a current density of 0.53 A cm^{-2}.

for certain procedures which still have further development potential (see also Figure 2). This is also true for the AWE at 140° C for a lowering of the cell voltage by another 0.05 V, particularly by further improving the anode. It can be seen that the margin for future development is significantly higher in the electrochemical step of the advanced H_2SO_4 hybrid cycle with the HI/I_2 system than in the direct electrolysis procedures which have already reached a state fairly close to their theoretical optimal performance.

In conclusion we may state that the advanced electrolysis procedures AWE and WVE have reached a stage which makes their scaling up to technical dimensions feasible within this decade. In the next decade, then, corresponding hydrogen-production plants based on the LWR as primary energy source could go into use. To a more limited extent this also applies to HOT ELLY in conjunction with an HTR, albeit under the restrictions mentioned at the end of Section 2.

The thermoelectrochemical hybrid cycles, which are very attractive in principle, still have a significantly longer way to go in their development, even in the case of the most developed advanced H_2SO_4 process with the catalytic HI/I_2 system. Substantial research efforts should be invested, however, in view of the very desirable goal of producing hydrogen by such a hybrid cycle with an HTR as primary energy source. At any rate, based on the present state of knowledge, commercial usage of the hybrid-cycle approach is not expected much before the year 2000.

REFERENCES

Bockris, J.O'M. and Justi, E.W. (1980). Wasserstoff – Die Energie für alle Zeiten. U. Pfriemer, Munich.

Brecher, L.E. and Wu, Ch.K. (1975). Electrolytic Decomposition of Water. US Patent 3,888,750.

Caspers, M.S. (Editor) (1978). Hydrogen Manufacture by Electrolysis, Thermal Decomposition and Unusual Techniques. Noyes Data Corp., Park Ridge, Illinois.

Deneuve, F. and Roncato, J.P. (1981). Thermochemical or hybrid cycles of hydrogen production – technico-economical comparison with water electrolysis. International Journal of Hydrogen Energy, 6:9.

Divisek, J. and Schmitz, H. (1982). Bipolar cell for advanced alkaline water electrolysis. International Journal of Hydrogen Energy, in press.

Divisek, J., Schmitz, H., and Mergel, J. (1980a). Neuartige Diaphragmen und Elektrodenkonstruktionen für die Wasser- und Chloralkalielektrolyse. Chemie-Ingenieur-Technik, 52:465.

Divisek, J., Mergel, J., and Schmitz, H. (1980b). Improvements of water electrolysis in alkaline media at intermediate temperatures. In T.N. Veziroğlu, K. Fueki, and T. Ohta (Editors), Hydrogen Energy Progress, Vol. 1. Pergamon Press, Oxford, pp. 209–219.

Divisek, J., Mergel, J., and Niessen, H.F. (1980c). Production of hydrogen by the electrolytic decomposition of water in fused sodium hydroxide. International Journal of Hydrogen Energy, 5:151.

Dönitz, W. (1981). Die Wasserelektrolyse zur Wasserstofferzeugung. Bundesministerium für Forschung und Technologie, Bonn (Editor), Elektrochemische Energietechnik, pp. 281–298.

Dönitz, W., Schmidberger, R., Steinheil, E., and Streicher, R. (1980). Hydrogen production by high temperature electrolysis of water vapour. International Journal of Hydrogen Energy, 5:55.

Fasbman, G.H. (1979). The Westinghouse sulphur cycle production process. In T.N. Veziroğlu and W. Seifritz (Editors), Hydrogen Energy Systems, Vol. 5. Pergamon Press, Oxford, p. 2485.

Gassert, H. (1981). Status and Perspectiven des HTR für Strom und Prozeßdampferzeugung. Paper presented at HTR-Statusseminar, September 21, 1981, Jülich.

Häfele, W. (1981). Energy in a Finite World. Ballinger, Cambridge, Massachusetts.

Imarisio, G. (1981). Progress in water electrolysis at the conclusion of the First Hydrogen Programme of the European Communities. International Journal of Hydrogen Energy, 6:153.

Lu, P.W.T., Garcia, E.R., and Ammon, R.L. (1981). Recent developments in the technology of sulphur dioxide depolarized electrolysis. Journal of Applied Electrochemistry, 11:347.

Nuttal, L.J. and Russel, J.H. (1980). Solid polymer electrolyte water electrolysis – development status. International Journal of Hydrogen Energy, 5:75.

Sedlak, J.M., Lawrence, R.J., and Enos, J.F. (1981). Advances in oxygen evolution catalysis in solid polymer electrolyte water electrolysis. International Journal of Hydrogen Energy, 6:159.

Struck, B.D. (1978). Wasserspaltung unter Einsatz von HTR-Wärme. Jahresbericht 1977, Kernforschungsanlage Jülich, pp. 9–14.

Struck, B.D. and Junginger, R. (1980). Development of an electrolytic cell for the anodic oxidation of sulfur dioxide and the cathodic production of hydrogen within the sulfuric acid hybrid thermochemical cycle. In A.A. Strub and G. Imarisio (Editors), Hydrogen as an Energy Vector. Reidel, Dordrecht, pp. 80–100.

Struck, B.D., Junginger, R., Boltersdorf, D., and Gehrmann, J. (1978). Problems Concerning the Electrochemical Step of the Sulfuric Acid Hybrid Cycle in Hydrogen as an Energy Vector. EUR 6085. Commission of the European Communities, Brussels, pp. 109–123.

Struck, B.D., Junginger, R., Boltersdorf, D., and Gehrmann, J. (1980a). The anodic oxidation of sulfur oxide in the sulfuric acid hybrid cycle. International Journal of Hydrogen Energy, 5:487.

Struck, B.D., Junginger, R., Boltersdorf, D., and Neumeister, H. (1980b). A three compartment electrolytic cell for anodic oxidation of sulfur dioxide and cathodic production of hydrogen. In T.N. Veziroğlu, K. Fueki, and T. Ohta (Editors), Hydrogen Energy Progress, Vol. 4. Pergamon Press, Oxford, pp. 2315–2322.

Struck, B.D., Junginger, R., Neumeister, H., and Dujka, B. (1982). A three compartment electrolytic cell for anodic oxidation of sulfur dioxide and cathodic production of hydrogen. International Journal of Hydrogen Energy, 7:43.

THE DENSE PLASMA FOCUS AS A SOURCE OF NEUTRONS

H. Conrads
Association EURATOM–KFA, Kernforschungsanlage Jülich GmbH, Jülich (FRG)

ABSTRACT

The neutron economy of a fusion–fission hybrid system with a plasma focus acting as a source of fusion neutrons is considered. After an outline of the essential features of the plasma focus device, the potential scale of neutron yield is examined and the electrotechnical, heat-transfer, and tritium-containment systems of a "zero-breeding assembly" are discussed.

1 THE PLASMA FOCUS AS A NEUTRON SOURCE IN A FISSION-REACTOR ECONOMY

We will take as our starting point the work of Harms and Heindler (1978). The following breeding chains based on transmutation will be considered:

$$n + {}^{238}U \longrightarrow {}^{239}U \xrightarrow[24\ \text{min}]{\beta^-} {}^{239}Np \xrightarrow[2.4\ \text{d}]{\beta^-} {}^{239}Pu \tag{1}$$

$$n + {}^{232}Th \longrightarrow {}^{233}Th \xrightarrow[22\ \text{min}]{\beta^-} {}^{233}Pa \xrightarrow[27\ \text{d}]{\beta^-} {}^{233}U \tag{2}$$

Here β^- represents beta decay, and the related half-lives are stated below the arrows. The fertile nuclei have to be contained in a structure cooled by a heat-exchanger fluid and any consequent degradation in neutron economy has to be made up by neutron multipliers (n, xn reactions). Neutron multipliers can also be used to lift the number c_p of fissile atoms bred per fusion neutron above 1. In the literature, c_p values between 0.8 and 5.75 have been reported. The neutron multiplication for each 14-MeV source neutron in an infinite pure medium for some materials discussed for breeders is as follows: ^{238}U, 4.2; ^{232}Th, 2.5; Be, 2.7; ^{7}Li, 1.8; Pb, 1.7; D, 1.4.

For the case of ^{7}Li, the tritium bred in the reaction

$$n + {}^{7}Li \longrightarrow T + {}^{4}He + n \tag{3}$$

is considered to represent one of the total of 1.8 neutrons, since the fuel for controlled thermonuclear reactions (CTR) has to be bred anyway, usually from ^{6}Li in a blanket under neutron absorption.

The most probable reactions in a plasma focus as described later are

$$D + T \longrightarrow {}^{4}He + n \tag{4}$$

and

$$D + {}^{7}Li \longrightarrow {}^{8}Be + n \tag{5}$$

Equations (4) and (5) represent two neutron-producing channels which are statistically independent from each other. In both equations the energy for each neutron is larger than 10 MeV. Breeding devices with $c_p < 1$ and $c_p > 1.5$ are called low-gain and high-gain breeders, respectively.

For pure breeders, i.e. devices which are not also used simultaneously for highly efficient electrical or thermal power production, it might be advantageous to use $c_p <$ 1, which is preferable due to smaller heat generation and smaller after-heat (continuing decay heat from radioactive products after shutdown) and thus eases the technical difficulties associated with the breeder. A fission-suppressed breeder using the thorium cycle according to eqn. (2) is described by Jakeman (1979). The fission-suppressed breeder might be composed of a ^{7}Li shield for multiplying neutrons when source neutrons enter the shield and a structure filled with ^{6}Li and ^{232}Th for breeding ^{3}H and ^{233}U, respectively. This latter structure would be immediately behind the ^{7}Li shield.

Harms and Heindler (1978) and Heindler and Lang (1978) give the required neutron yield of the plasma focus versus the thermal power of the supported fission reactor and versus the effective conversion ratio of a fission reactor of 1.5 GW thermal power for $0.5 < c_p < 2.5$. Their results show that for $c_p < 1$ a yield of about 5×10^{19} neutrons per second is required from the plasma focus for a reactor having a thermal power larger than 1 GW and an effective conversion ratio smaller than 0.8. The required neutron yield decreases sharply although c_p remains smaller than unity for advanced converter reactors with a conversion ratio larger than 0.8 if reactors with a thermal output of less than 1 GWth are desired. The required neutron yield has been evaluated to provide a "self-sufficient fission converter reactor", which is self-sufficient as far as its lifetime fissile fuel needs are concerned. Such systems would aid in controlling fuel inventories, would cut down dependence on fuel sources abroad, and would generally stretch the supply of fissile fuel for fission reactors. What are the prospects of the plasma focus serving such a purpose?

2 THE NEUTRON YIELD OF THE PLASMA FOCUS

The term "plasma focus" is used for both a special formation of plasma, small in configuration space but large in phase space, and for the device producing such a plasma. The device consists of a power supply, a power amplifier and power compressor, and a vessel with tritium-containment and heat-exchanger systems, as well as peripheral systems such as pumps, a tritium-storage and tritium-cleaning system, and control equipment.

The power supplies include rectifiers fed by a high-voltage transformer connected to the mains and capacitor banks or inductances for intermediate storage and the initial power amplification.

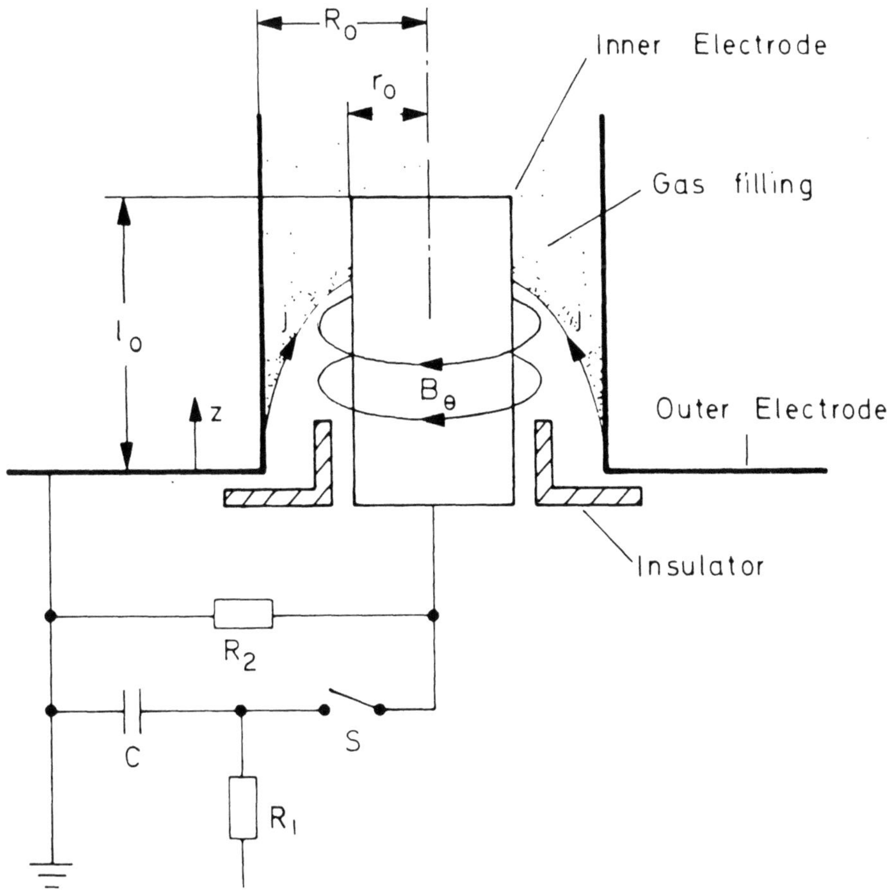

FIGURE 1 The start-up of a plasma focus discharge (schematic).

The final power amplification is achieved by a set of coaxial conductors (Figure 1) that are filled with gas at low pressure and which act on the circuit for most of the time as a variable inductance and for a short period as a variable resistance as well. As soon as the switch S is closed, the capacitor bank C – charged via the resistor R_1 – is discharged into the power amplifier by an electrical breakdown in the vicinity of the insulator of the power amplifier. A radial current starts to flow. Under the effect of induced magnetic fields the current sheath moves up in the *z*-direction and finally collapses at the open end of the coaxial line where the current also reaches its maximum (Figure 2). While the inductance of the circuit increases as the current sheath moves up the coaxial conductors,

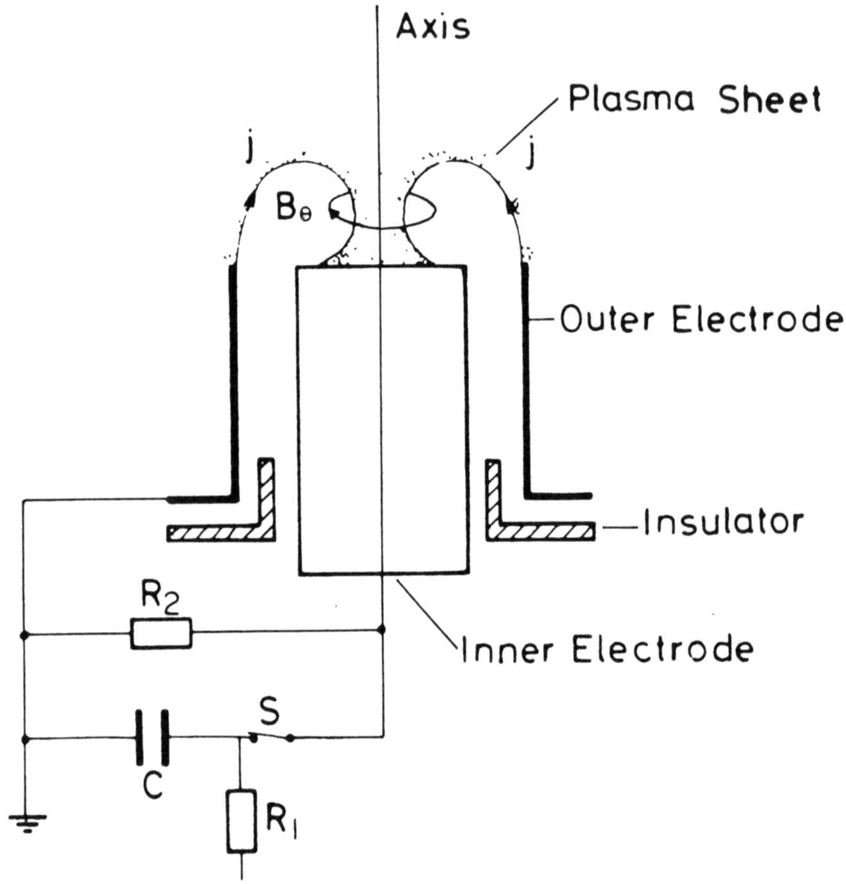

FIGURE 2 The final phase of a plasma focus discharge (schematic).

the resistance of the sheath remains almost unchanged during this time but increases by more than a factor of 10 after the current is squeezed towards the axis. The power $P = i^2R$ and power density D reach their maxima at the moment of maximum compression of the current sheath.

This power amplifier is contained in a vessel. The electrical discharge in the power amplifier of the plasma-focus device produces a plasma from the gas filling the vessel. In the instant of maximum power compression (e.g. 10^{12} W cm^{-2}; Bernard et al., 1979) the plasma focus is formed. Figure 3 shows interferograms of the formation of the plasma focus (Decker and Herold, 1980); the exposure time is 1 ns. The pictures were taken in a single sequence. The numbers above each picture mark the time in nanoseconds after maximum compression of the plasma, which is not necessarily the instant of maximum current density. The electrons, having quite a different mobility from ions because of their different charge-to-mass ratio, are locally no longer space charge compensated by the ions. The result is a quite large electric field, which could be as high as 10^{10} V cm^{-1} over a distance of several micrometers. Ions entering such field regions accelerate rapidly

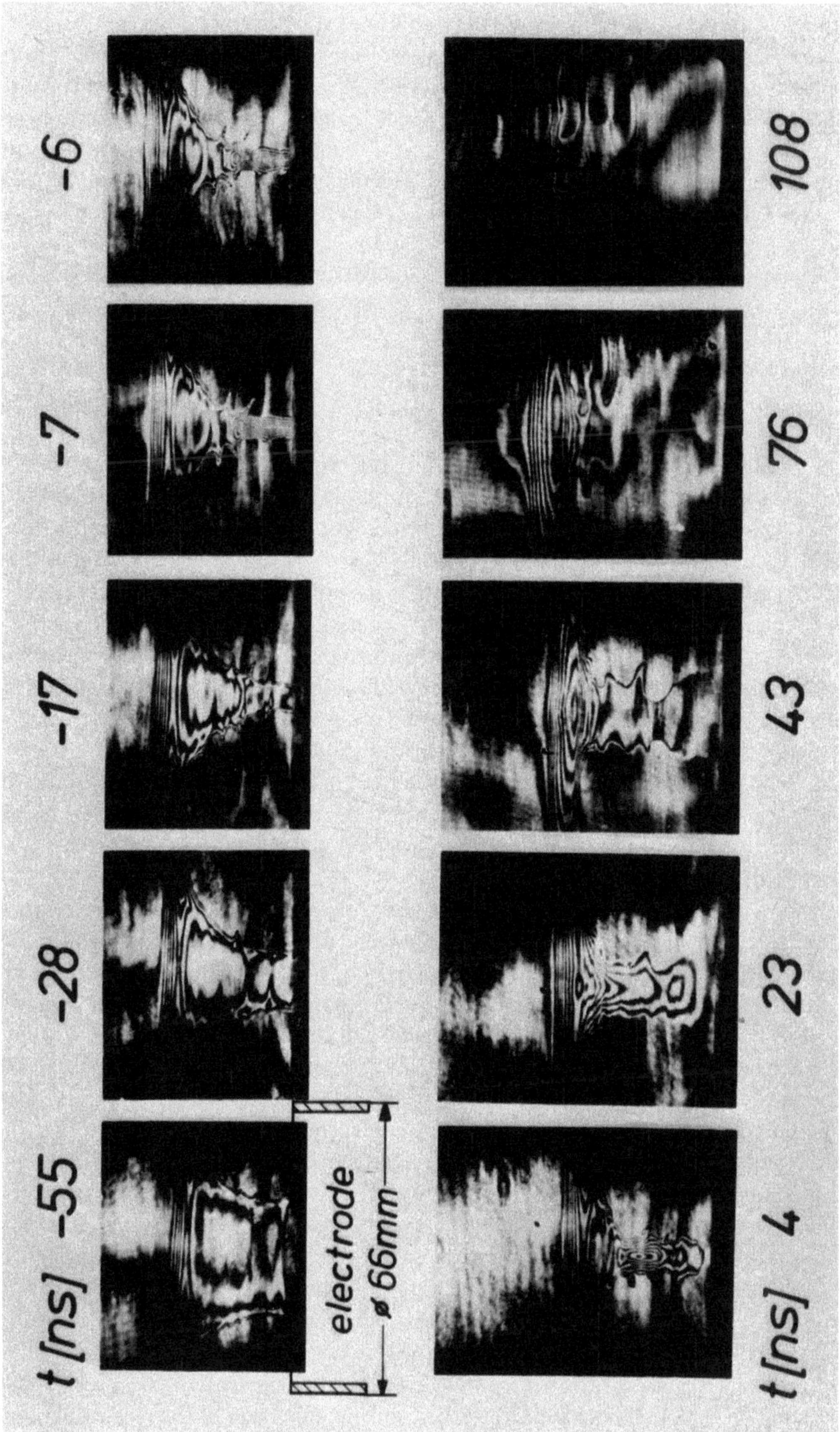

FIGURE 3 Interferograms of the formation of the plasma focus.

(Cloth and Conrads, 1977; Filipov, 1979; Bertalot et al., 1980). In contrast to accelerators using solid-state electrodes, these "plasma electrodes" have a strong tendency to vanish as a result of collective density fluctuations (Conrads et al., 1973; Post and Marshall, 1974) inside the plasma, restoring local charge neutrality. Depending on the local plasma conditions, these "plasma electrodes" appear here and there and from time to time (Salge et al., 1978) in a random way. The maximum neutron emission appears when the focus is disintegrating, i.e. 46 ns after maximum plasma compression in Figure 3.

The neutron yield Y of the fusion reactions described in eqns. (4) and (5) is equal to the reaction rate:

$$Y = N_1 N_2 \langle \sigma \cdot u \rangle$$

where N_1 and N_2 are the densities of the reacting species, u their relative velocity, σ the cross-section of the related fusion reaction, and $\langle \sigma u \rangle$ the average of σu. For eqn. (4) $\langle \sigma u \rangle$ peaks for $u_0 \simeq 2 \times 10^8$ cm s^{-1} and for eqn. (5) it peaks for $u_0 \simeq 8 \times 10^8$ cm s^{-1}. If u shows not a monoenergetic but rather a wider (e.g. Maxwellian) distribution, $\langle \sigma u \rangle$ peaks at about $u_0 = 10^8$ cm s^{-1} and $u_0 = 4 \times 10^8$ cm s^{-1}, respectively. Elastic Coulomb collisions are more frequent by several orders of magnitude than fusion collisions, even if u is in the vicinity of u_0; i.e. only relatively few particles react with partners in binary fusion collisions. If fast particles enter a cold environment they lose their energy rapidly (free-expansion model) and the total neutron yield is low. If the same fast particles are trapped in a hot environment (high-temperature plasma with $u = u_0$) the whole assembly gains energy (trapped-ion model) and the neutron yield is high.

Due to the high power density, the large electric fields, and the erratic formation of the "plasma electrodes", external aids to raise the neutron yield are not very promising. An increase in plasma current, however, has been proved to be very effective in increasing the neutron yield markedly (Figure 4).

During the 1970s, plasma-focus devices have already produced 2×10^{12} neutrons per discharge from ^{2}H–^{2}H reactions. It has also been demonstrated during this time, but in smaller devices, that the neutron yield from a ^{2}H–^{3}H gas could be 50–100 times larger than that from pure ^{2}H gas. Since about five years ago, (1–2) $\times$ 10^{14} neutrons per discharge dissipating about 0.3 MJ of stored energy can be considered as the state of the art.

For several years the plasma focus was the strongest plasma neutron source. What are the future prospects for this source? As soon as the velocity u of the majority of the accelerated particles approaches u_0, an increase in current can only produce a linear increase in neutron yield as long as the current does not increase the density in the focus. Interferograms such as those in Figure 3 show that there is little if any hope for such an effect. The magnetic fields increase with current and are able to trap the accelerated ions in a region of high plasma temperature only if the field strength is high enough and if the field configuration is favorable. The latter condition has so far proved elusive. No strategy has yet been proposed that produces a field geometry suitable for plasma confinement in the focus with increasing current. Such confinement, if attained, would allow a further strong increase of the neutron yield with plasma current.

Based on the observations made earlier, some comparison of the free-expansion versus the trapped-ion models may be obtained from measurements of the total neutron

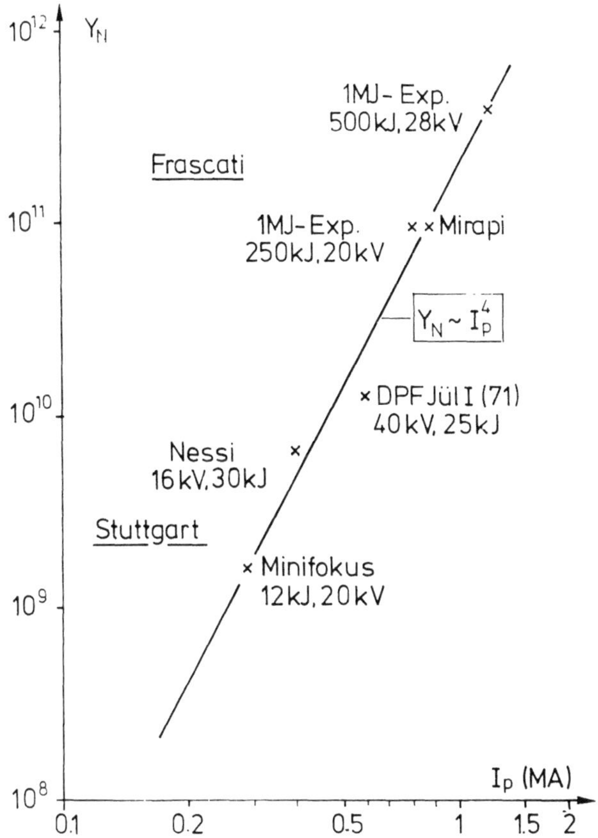

FIGURE 4 ^2H–^2H neutron yield as a function of plasma current.

yield from primary and secondary nuclear reactions. For example, the reaction given by eqn. (4) can occur as a secondary reaction to the process

$$D + D \longrightarrow T + n \tag{6}$$

The triton released in this reaction has an energy of about 1 MeV.

For a free-expansion situation, the ratio of neutron yields $Y_{D,T}/Y_{D,D}$ is 5×10^{-6}. Measured data (Hübner et al., 1981) give $Y_{D,T}/Y_{D,D} = 7 \times 10^{-3}$, i.e. the measured ratio is more than a factor of 10^3 larger. It is not unrealistic to assume that the fast tritium ions have a curved flight path of 150 cm in a volume of only a few cubic centimeters. The slower deuterium ions would be even more efficiently trapped. The more ions are trapped, the better is the isotropy of the emitted neutrons. The neutrons emitted when unidirectional beams enter a target have a strong directional anisotropy. The isotropy of neutrons emitted from a plasma focus tends to improve as the plasma current increases. Thus, although not yet accessible to a priori design considerations, there is hope for an active trapping effect and for an increase of neutron production more than proportional to the plasma current even if the velocity of most of the particles has reached u_0.

3 A ZERO-BREEDING PLASMA FOCUS

In a commercial breeder facility the energy necessary to produce a specific number of neutrons has to be balanced against the energy gained from fission of the bred materials. Taking into account a fission-reactor conversion ratio of 0.8 and the known efficiencies in producing electricity from fission power, a minimum output of 10^{10}–10^{11} neutrons per joule of energy fed to the focus has to be achieved (Heindler and Lang, 1978). If this condition were only met exactly, there would be no net energy gain from the breeding facility, although it would be possible to supply enough fuel to a nuclear power station to cover its own electricity consumption. The facility is therefore called a "zero breeding" device. As mentioned in Section 2, the state of the art is 6×10^8 neutrons per joule.

According to the scaling of neutron yield versus plasma current given in Figure 4, a plasma current of about 15 MA will be required for a zero-breeding plasma focus based on a D–T mixture in the vessel. In such a device the plasma current would have to be increased by a factor of 3–4 beyond the present state of the art.

3.1 The Power Supplies

The consideration of layout has to start with the intermediate storage and the power amplifier and power compressor, which are connected in series with each other and which form a damped LC circuit. $L = L_{\text{tot}}$ is the total inductance and C is the total capacitance of this circuit. Ideally, the circuit would (1) be aperiodically damped and (2) have the smallest possible energy content in the intermediate storage section. The largest ohmic resistance in such a circuit is presented by the plasma during its axial compression. This has been measured in different devices to be as high as 500 mΩ (see Bernard et al., 1979). If the circuit does not oscillate, most of the stored energy is fed to the plasma during focus formation and therefore does not wear out the power compressor without producing neutrons.

A capacitative intermediate storage of 1 MJ is feasible as far as present knowledge of the technical problems is concerned. The upper part of Figure 5 shows a schematic circuit for calculating voltage and current during focus formation (as shown in the lower part of the figure) (Braunsberger and Salge, 1981). The intermediate storage consists of capacitors, switches between these capacitors and the power compressor, and buswork. The capacitance is given primarily by the capacitor bank (32 μF), and the ohmic resistance primarily by the switches (1–2 mΩ during focus formation); the capacitors, switches, and buswork all contribute significantly to the total inductance ($L_{\text{device}} = 1$ nH). R_{device} and L_{device} can be kept small because many elements can be switched in parallel. For the power compressor a value of $L_{\text{focus}} = 6$ nH has been assumed; this is difficult to achieve but is within the scope of present experience. The discharge starts at a voltage level of 250 kV. The peak current is 14.7 MA and the peak voltage at the focus is 1.6 MV, as shown in Figure 5. The reaction according to eqn. (5) could multiply by a factor of two all the numbers quoted for neutron yield because its cross-section is close to that related to eqn. (4) for such high voltages. The assumption here is that the trapping times in the magnetic field for deuterium, tritium, and lithium are all similar.

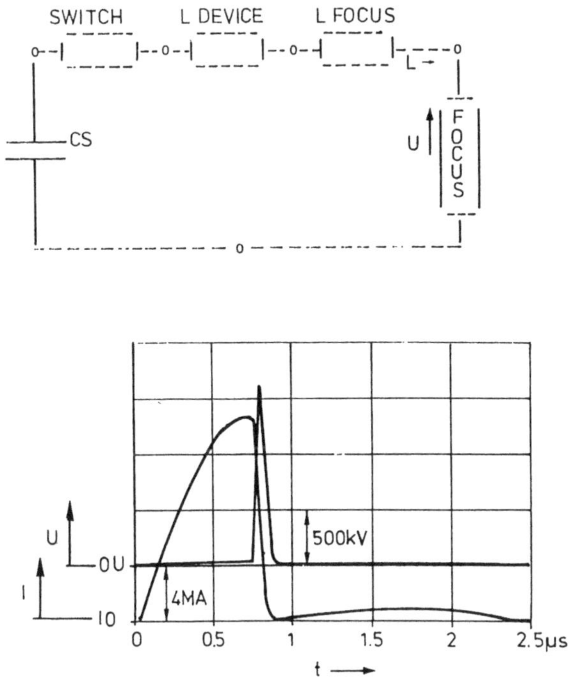

FIGURE 5 (Top) Auxiliary circuit of a plasma focus device. (Bottom) Voltage across and corresponding current in the power-amplifier compressor: $W = 1$ MJ; $C = 32$ μF; $L_{tot} = 7$ nH; $U = 250$ kV.

An intermediate storage of 1 nH inductance such as that envisaged here might, for example, be composed of 160 subunits. Each subunit would have a switch connecting it together with others in parallel. It would further contain two capacitors connected by a switch in series and rated for 125 kV. This latter switch would be activated simultaneously with the first one and the switching would take place once every second. Similarly, the charging unit could consist of a primary controlled transformer directly connected to the grid and a rectifier. The required power for charging the intermediate storage would be about 3 MW if the storage has to be charged within 0.7 seconds.

A fission reactor plant of 1 GW thermal power would require about (3–5) $\times$ 10^{19} neutrons per second for breeding with $c_p < 1$. If the scaling of the neutron yield with plasma current were substantially better than $Y \propto i$ but not necessarily as good as $Y \propto i^4$, an intermediate storage facility requiring 100 MW electric power could be feasible. The handling of such power levels is not a crucial electrical engineering problem for the fusion devices presently under construction. Also, the level of circulating electrical power of 25% could be quite acceptable.

3.2 The Heat-Transfer and Tritium-Containment System

In contrast to the electrical power engineering, the handling of the thermal power in the plasma-focus device is a problem and requires further consideration.

Thermal wall loadings of several kilowatts per square centimeter are under discussion in today's plasma physics experiments. The cooling systems have thin walls (of the order of 1 mm thick) and require high pressure to avoid film boiling of the coolant. For a device such as the plasma focus a thermal level of 200 W cm^{-2} would be more adequate because it is unavoidable that once in a while the plasma touches the bare wall. In what follows we will again refer to the zero breeding facility, which might be a first step in the development process.

In order to transfer the heat of the plasma-focus device to larger heat-exchanger surfaces, the vessel is designed as a heat pipe using, for example, gaseous lithium as the transfer medium. This is accomplished by designing the walls of the vessel in such a way that they are "sweating" liquid lithium, as explained below. The concept for such a vessel was described in detail by Cloth et al. (1975) and is shown in Figure 6. This concept is for a purely experimental facility. Conceivably the areas designated as sample stations and radiation shield, and also the space outside the vessel, could be used for placing a breeder blanket. However, no detailed design for a commercial plasma-focus breeding device has been developed.

From the position of the plasma focus in the device, it is obvious that about 50% of the total of 1 MW will have to be dissipated by the inner and 50% by the outer structure. This means that a surface of about 2500 cm^2 would be required for heat transfer to both the inner electrode and the vacuum vessel. The inner electrode is hollow and contains a pool of liquid lithium, part of which is vaporized by the plasma jet of the focus discharge. The pool is kept (over a time average) at a constant level. In this way the acute heat load on the solid wall of the inner electrode might easily be reduced by more than a factor of ten, so the critical area could shrink by a factor of three in its linear dimension – to about 15 cm in diameter. The increase of the heat-exchanger surface of the vacuum wall to 1 m^2 should not be a problem. Also, it should be easy to supply a film of liquid lithium on these walls for instantaneous evaporation at the moment of maximum heat load. Lithium, and more particularly lithium hydride, has the capability of wetting the surfaces of specific metals remarkably thoroughly. It would be sufficient to have a pattern of tiny holes in the wall through which lithium or lithium hydride could enter the chamber. Either liquid would wet the whole wall perfectly. The similarity to a sweating skin is the reason for calling such an idea the "sweating-wall concept". The sweating wall would encounter no difficulty with gravity. Every part which is supposed to be wet can be kept wet easily. Deuterium and tritium are both very soluble in lithium and hence their escape from the vessel can be prevented.

Such a system has to be kept above the melting points of both lithium and lithium hydride, i.e. around 600° C. The required pressure of the ^2H–^3H mixture of about 10 mbar can be achieved by controlling the temperature of the wall only. No fuel input (i.e. tritium circulation after each discharge) would be necessary. The tritium is captured almost exclusively in the lithium (Cloth et al., 1975). The tritium inventory for such a zero-breeder device could be kept at a level equivalent to 50–100 kCi, values not very different from those required in tokamaks and other fusion devices. The values considered and compared here are the necessary tritium inventories for the containing vessel. Tritium breeders are not treated here and so neither are their tritium inventories.

Walls at 600° C are subject to enhanced interstitial diffusion of hydrogen. Tritium, although mainly caught in the wetting lithium, can still penetrate the wall. This tritium

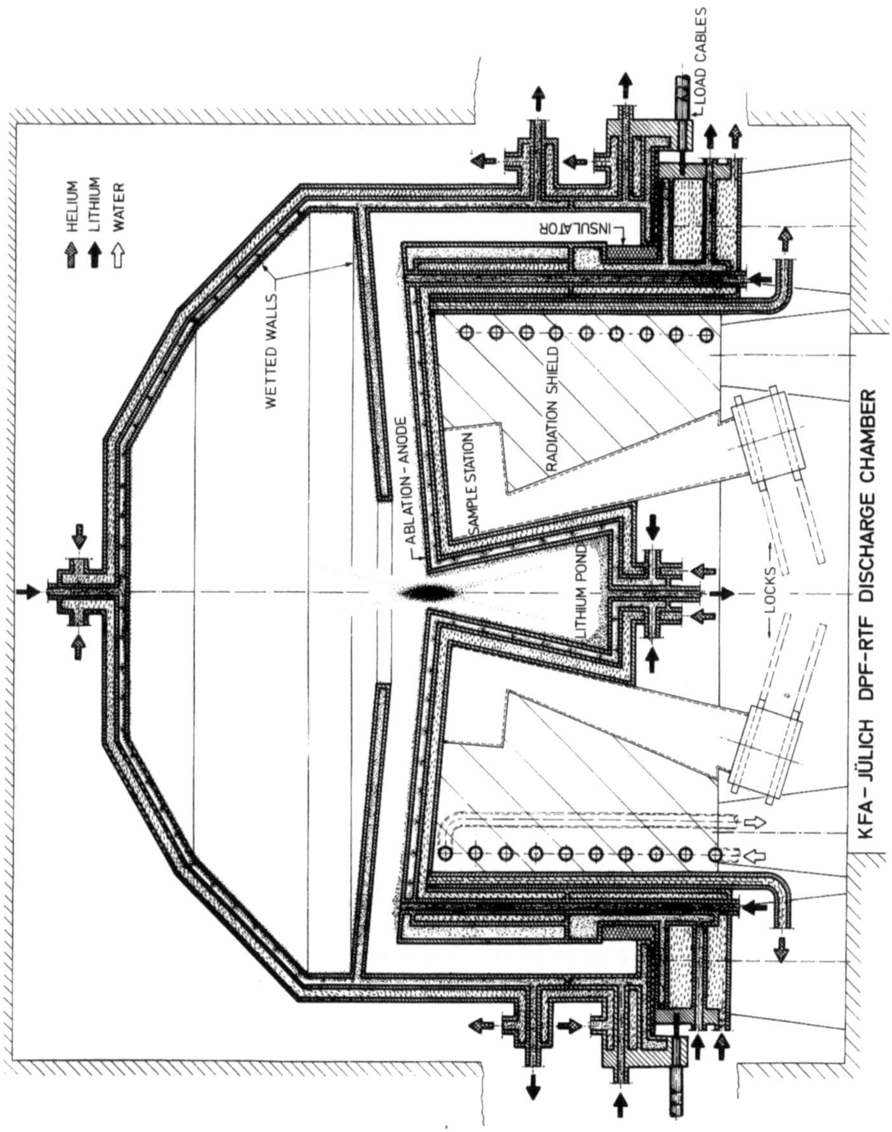

FIGURE 6 Conceptual design of a plasma focus as a neutron source.

could be contained by systems of lithium and of helium-cooled aluminum-covered heat exchangers as shown in Figure 6 and described in more detail in Cloth et al. (1975).

4 THE NEXT STAGES

In addition to the many experiments that will be necessary to study the increase of neutron yield with plasma current, one of the most urgent problems is to demonstrate whether the heat-transfer and tritium-containment problem in a device like a zero breeder can be mastered. The "sweating wall" is one possible solution. Such a scheme would require the following: (1) experimental demonstration of a sweating wall; (2) measurement of heat transfer in a chamber with sweating surfaces; (3) construction and testing of a power compressor with a sweating wall and an effective insulating gap between the electrodes; (4) single focus discharges in a sweating-wall chamber at a neutron level of 10^{10} neutrons per discharge; (5) design of breeders for tritium and ^{233}U; (6) development of remote-handling tools; (7) repetitive discharge at a level of 10^{14} neutrons per discharge; (8) repetitive discharge at a level of 10^{16} neutrons per discharge; (9) test runs at full rated parameters; (10) construction and testing of a breeder for tritium and ^{233}U to demonstrate feasibility. The intermediate storage section would primarily require the development of long-lasting switches.

ACKNOWLEDGMENTS

The author would like to thank U. Braunsberger, K. Hübner, J.P. Rager, J. Salge, and K. Steinmetz for their contributions to this paper and H. Herold for supplying an original interferogram.

REFERENCES

Bernard, A., Garconnet, J.P., Jolas, A., Breton, J.P., and De Mascureau, I. (1979). IAEA-CN-37/U-3-4. International Atomic Energy Agency, Vienna. Also, Nuclear Fusion Supplement, 1979, pp. 159–172.

Bertalot, L., Herold, H., Jäger, U., Mozer, A., Oppenländer, T., Sadowski, M., and Schmidt, H. (1980). Physics Letters, 79A(5, 6):389–392.

Braunsberger, U. and Salge, J. (1981). Über die Bemessung eines Stromkreises zur Speisung eines Plasma Fokus. Untersuchungsbericht 165 des Instituts für Hochspannungstechnik der Technischen Universität Braunschweig, Braunschweig.

Cloth, P. and Conrads, H. (1977). Nuclear Science and Engineering, 62:591–600.

Cloth, P., Conrads, H., Ihle, H.R., Gourlan, C., Maissonier, Ch., and Robouch, B.V. (1975). Proceedings of the International Conference on Radiation Test Facilities for the CTR Surface and Materials Program. ANL/CTR-75-4. Argonne National Laboratory, Argonne, Illinois, pp. 498–526.

Conrads, H., Gollwitzer, D., and Schmidt, H. (1973). In the Proceedings of the European Conference on Controlled Fusion and Plasma Physics, 6th, Vol. 1. Joint Institute for Nuclear Research, Moscow.

Decker, G. and Herold, H. (1980). Physikalische Blätter, 11:328–333.

Filipov, N.V. (1979). Zhurnal Eksperimental'noi i Teoreticheskoi Fiziki, 76(May):1547–1550.
Harms, A.A. and Heindler, M. (1978). Nuclear Science and Engineering, 66:1–8.
Heindler, M. and Lang, H. (1978). Atomkernenergie, 32:142–152.
Hübner, K., Rager, J.P., and Steinmetz, K. (1981). Secondary ^2H–^3H reactions in the Frascati plasma focus. Plasmabericht 2/81 der Institut für angewandte Physik II der Universität Heidelberg, Heidelberg, FRG.
Jakeman, D. (1979). Energy Multiplication and Fissile Material Production in a Fission–Fusion System. Report AEEW-R-1301. UKAEA Winfrith, UK.
Post, R.S. and Marshall, T.C. (1974). Physics of Fluids, 17:452.
Salge, J., Braunsberger, U., Fell, B., Uemo, J., and Conrads, H. (1978). Nuclear Fusion, 18:972.

POSSIBLE APPLICATIONS OF A HYBRID THERMONUCLEAR ENERGY SOURCE BASED ON A DPF DEVICE IN MODERN ENERGY COMPLEXES

V.A. Gribkov and M.G. Tyagunov
P.N. Lebedev Physical Institute of the Academy of Sciences of the USSR, Moscow Energetics Institute, Moscow (USSR)

ABSTRACT

A source of thermonuclear energy based on the dense plasma focus (DPF) device in a hybrid fusion–fission version is proposed. The device would be physically compact (chamber diameter 3 m), would require a small uranium blanket (1–4 t), and could be expected to deliver 100–300 MWe; fuel replacement would be necessary every 2–5 years and construction and maintenance should be relatively simple. In its initial operating phase such a facility would be a net energy consumer and would breed fissile material; as the fissile content in the blanket increases, the installation would become a net energy producer. Under the proposed scheme of blanket operation, up to 50% of the uranium could be burned while maintaining electrical output and without refabrication of fuel elements. If desired, operation could continue after the fuel is almost completely exhausted to burn the nuclear waste. It is thought that the new source could become both technologically and economically feasible in the near future. Smooth control should present no problem and the speed at which the device could be brought up to full load should greatly improve the flexibility of the overall electrical supply system. Decentralized use of the new sources could reduce transmission losses and associated problems, while the sources could also operate in the same way as conventional fission power stations for heat production; moreover, the new hybrids could be used to replace the boiler sections of outworn fossil-fuel stations whilst still utilizing the existing turbines, generators, etc. Finally, due to its long fuel cycle the proposed source is essentially proliferation safe.

1 INTRODUCTION

Investigations of the development of energy systems to ensure the survival of mankind have to be founded on system-analytical principles, taking into consideration not only how to produce the energy (i.e. which technology to use) and possible uses of the energy by industrial and private consumers but also the probable ecological, social and political consequences of the application of these systems. Of course, this does not mean that such a complete analysis has to be performed by every specialist working in this field. On the contrary, we think that it is impossible for specialists with much more narrow scientific interests to estimate competently all the possible issues of the application of various methods of energy production.

With this in mind, and taking into account the existence of special institutes working on the problems of systems analysis, we shall examine one possible means of ensuring energy supplies for human needs, taking into consideration the development tendencies of the actual energy systems and its present and future problems. We shall base our examination on a proposal for a special type of hybrid thermonuclear installation.

It is clear that in considering tomorrow's energy supply options one must directly take into account today's problems because, in such an inertial system as the energy industry, technological succession – the successive and coherent conversion from one energy technology to another – must be smooth. This means that (1) new energy installations must conform with modern technological systems of industry and energy consumption, and with their probable development in the immediate future; (2) new energy systems must take into account evolving sociopolitical relations; (3) systems for the conversion and transportation of the energy generated by a source of a given type must be technologically realizable in the immediate future; and (4) new energy systems must comply with the modern goals of biological protection of the environment and of the people involved in energy-production processes.

We shall look in particular at questions connected with the production of electricity, whose share in the world energy economy is steadily increasing and will, according to forecasts, reach 50% by the year 2000.

2 GENERAL PROBLEMS OF ENERGY SUPPLY SYSTEMS

International investigations in the field of electricity production show that systems for ensuring energy supplies will be based on the creation and development of large-scale energy facilities, combining electric power stations, energy-transportation lines, and centers of energy distribution (Kochs, 1980; Wolfberg, 1981). This is also reflected in the reports and conclusions of major international conferences, such as the 1979 Summer Meeting of the IEEE Power Engineering Society (IEEE, 1980). The development of energy production systems presupposes the steady and proportional development of existing technologies (thermal, nuclear, and hydro-electric power stations) together with the deployment of alternative sources of electric energy for use in the future. However, any estimation of this development must not be restricted to static parameters which are usually applicable for this purpose, such as predicted maximum energy consumption, resource constraints, and the installed power of the various types of generating stations. The dynamic parameters of these systems must also be taken into account. These include the daily, weekly, and seasonal variations in energy consumption, the system load factor (as measured by the ratio of the maximum load to the minimum loading), the frequency of load changes, the quality of the electric power (stability of voltage and frequency, admixture of higher harmonics), the distribution of reserves between the various types of stations, the change of the generating mix and, connected with this, criteria of optimal development and control of the energy systems.

Clearly, many of these problems are very closely interconnected. They touch upon important issues such as the question of whether the energy sources are to be centralized or decentralized, the complex use of energy resources, environmental protection, operability, etc. This determines a range of problems which have to be resolved by properly using the respective energy sources within the overall structure of the developing energy system. As mentioned already, we will not examine the entire energy problem, but will concentrate on those aspects which are directly connected with the type of energy source we want to propose here. Moreover, it should be stressed that we will not attempt to make comparative estimates of efficiencies (on whatever criterion) for our own proposal or other types of thermonuclear energy source. On the contrary, we shall only examine the possibilities of contributing to the solution of part of the energy problem by using a special type of thermonuclear facility, namely sources based on Controlled Thermonuclear Reaction (CTR), in particular the Dense Plasma Focus (DPF) device in a hybrid fusion–fission version.

We will consider the following major problems:

(1) the technological realizability of the proposed energy source taking into account its size, its energy-resource requirements, and the policy for its use;
(2) the structure of an industrial version of the proposed energy facility;
(3) the controllability of the facility to ensure the reliability of energy system operation by means of effective switching between normal and peak power operation;
(4) the possible utilization of the facility as an energy accumulator to permit the extension of its own control range and also to accommodate the necessary operating conditions for less flexible electricity generating stations;
(5) the use of the device in both centralized and decentralized energy supply systems;
(6) the maintenance of the quality of the generated electricity;
(7) the use of the device for the production of heat;
(8) the use of the device for "burning" nuclear wastes (transmutation of radioactive elements);
(9) the possibility of energy transportation over long distances;
(10) the possibility of the reconstruction of present power stations so that they can use the new source of energy;
(11) the utilization of the device in autonomous energy systems;
(12) the maintenance of safety in the facility;
(13) the environmental effects caused by the facility.

Before addressing these points, we should mention a number of other problems which might be solved by the use of this type of installation and which are connected with the social and political aspects of ensuring mankind's continued existence.

3 SOCIOPOLITICAL ASPECTS OF THE USE OF A NEW TYPE OF ENERGY SOURCE

The main sociopolitical aspects of the introduction of a new type of energy installation include the following. First, these devices must permit conversion to the new energy technologies over as short a period of time as possible to prevent an energy crisis when present-day natural energy resources become exhausted. (For oil and natural gas this period is 20–40 yr; for resources required in nuclear fission stations – excluding breeders – it is of the order of 100 yr.) This requirement could even have an adverse effect on those types of energy sources (e.g. breeders or solar energy), that otherwise appear very favorable.

Second, for any kind of nuclear power station there are very important questions connected with disposal of radioactive wastes and the possibilities of accidents (from the point of view of a station's location) as well as ecological considerations related to the station's use of space and its effects on the local environment.

No less serious are the many political problems arising from natural resources. For example, the consequences for the political stability of the world of the fact that oil appears at present to be concentrated in a relatively small number of developing countries are very well known. Currently there are indications of possible negative consequences for political stability arising from the projected transition to coal-based energy systems. These consequences are connected with the fact that the main resources of this raw material are concentrated in the United States, the Soviet Union, and China. (This is in addition to the very serious *ecological* consequences of coal energetics.)

The next highly important question is the proliferation of nuclear arms. In principle, within the framework of the world energy economy, it should be possible to organize a process for the delivery of nuclear fuel in such a way that refabrication plants are concentrated in a limited number of countries and a new batch of fuel for use in the reactors is delivered only in exchange for the waste fuel. Nevertheless, it is clear that it would be even more preferable to have facilities in which it is possible to burn a substantial fraction of the uranium fuel (instead of the very low 0.5–1% for present-day hot-water reactors) without intermediate regeneration and refabrication of fuel so that no high-grade fissile material can be diverted for arms production.

4 OPERATING PRINCIPLES AND OPERATIONAL REGIMES FOR THE PROPOSED ENERGY DEVICE

One of the possibilities in nuclear energy development that is being widely discussed at the moment is a hybrid thermonuclear installation consisting of a thermonuclear neutron source and a surrounding fission blanket (Velikhov et al., 1977; Bethe, 1979). One possibility is the use of pulsed thermonuclear sources of neutrons of the laser-type CTR or DPF–CTR types (Feoktistov et al., 1978; Harms, 1978; Gribkov, 1980).

We shall discuss here the aforementioned questions with reference to DPF-type installations. As shown by Gribkov (1980), the present level of understanding of the physical processes taking place in these devices, together with the accumulated experimental data allow us to predict in the very near future a thermonuclear output which is about 10% under the stored feeding energy. This efficiency demands a stored feeding energy that is an order of magnitude higher than that achieved at present. Estimates on the basis of scaling laws for a DPF and data on blanket calculations by Feoktistov et al. (1978) lead to the following possible parameters for such a device: dimension of chamber, 2.5–3 m; input energy per cycle of system operation, 2–10 MJ; mean frequency of system operation, 10 Hz; mean thermonuclear power, 10 MW; weight of uranium blanket, 0.8–4 t; thermal power in the blanket, 500–1000 MW; mean available electrical power on the bus-bars, 100–300 MW; periods between fuel replacements (i.e. the time taken to burn 50% of the uranium) 2–5 yr; time to reach break-even between energy consumption and generated electric energy, 1–2 months.

The chamber of the device is cylindrical with a diameter of 2.5–3 m, i.e. it is very compact. It will require a special cooling system, which can, however, be constructed at the level of present technology. The pulsed power supply requirements can be met by discharge of capacitor banks or by an inductive storage system, and hence, probably, directly from the bus-bars of an electric system. The main advantages of a thermonuclear neutron source of this type are its very low cost, the absence of complicated technology, the easy access to any chamber elements as a result of its simple structure, the small fuel inventory (i.e. the small blanket mass), and the possibility of a long period of operation without replacement of the working gas within the chamber.

The major peculiarity of this scheme of blanket operation (Feoktistov et al., 1978) is the fact that up to 50% of the uranium is burned, with net electricity production throughout and without refabrication of the fuel elements. The fuel elements can, in principle, be extracted at the moment of maximum plutonium concentration (9%), i.e. at the peak of the installation's power. Alternatively, they can be left in for more complete burning of radioactive waste produced, i.e. extending the operation into the last stage shown in Figure 1, characterized by energy *consumption* by the system.

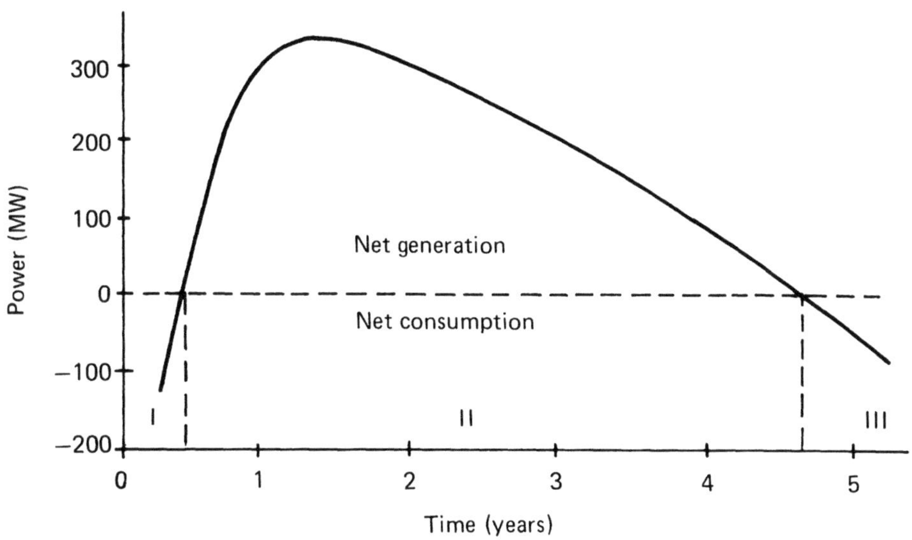

FIGURE 1 The temporal cycle of system operation for the proposed hybrid: I, initial stage; II, working stage; III, final stage.

5 WHAT MAKES THE INSTALLATION USEFUL?

Before examining the ways in which this thermonuclear hybrid system can be used, we would like to comment on the determination of the effectiveness of such an installation.

There is a widespread opinion that the effectiveness of an energy installation is determined by its efficiency based on present-day (single-purpose!) application. This view would of course be correct if the installation's only function is, for example, electricity generation. For such a single-purpose-oriented view,

$$\eta = E_{\text{useful}}/E_{\text{consumed}} > 1$$

really appears to be one of the most important indexes of effectiveness or, more generally, usefulness. However, if an energy installation is to meet several goals, some of which have more social rather than economic aspects, the approach used to determine its usefulness should be different. Because at present it does not appear to be possible to calculate the general (overall) figure of merit of such a "multipurpose" installation, the most productive approach to determining its usefulness is the examination of how closely it meets the various demands. From this position we shall try to show the usefulness of the proposed energy installation for providing solutions to the complex problems formulated in Sections 2 and 3.

5.1 The Technological Realizability of the Installation

One advantage of this type of energy source is its compactness: the thermonuclear source has a small diameter and there is no great quantity of uranium within the blanket. This makes the source suitable for industrial purposes for economic reasons as well as from the point of view of resource conservation. It also allows the device to be utilized at a level of power much lower than that usually considered for thermonuclear systems (several gigawatts thermal), namely in the region of 100–300 MWth, making it possible to implement the thermonuclear technology in the immediate future, without having to solve the engineering problems which inevitably arise in the deployment of larger-size industrial energy systems.

The technological realizability of the energy plant and the possibility of its industrial exploitation are also favored by the short period (an estimated 2–3 months) required to reach a coefficient of power amplification higher than 1 (according to calculations (Feoktistov et al., 1978; Gribkov, 1980) it will reach a coefficient of 2–3).

Apart from the characteristics of this energy installation pointed out here and in Section 4, its technological realizability in the immediate future is also based on the moderate power requirements for its thermonuclear part and the fact that the capacitor banks and inductive storage required are already feasible using present-day technology. Also important are the moderate calculated neutron flux through the blanket and first wall and the possibility of easy replacement of the blanket and first wall (Gribkov, 1980).

These factors allow us to speak about such a device as a power plant of the relatively near future.

5.2 The Structure of the Industrial Energy System

Future energy-supply complexes based on commercial thermonuclear energy installations, according to expert predictions, will involve combined (successive) use of the thermonuclear energy in elements based on the MHD (magnetohydrodynamic) generator, thermoionic, and classical methods for transforming energy into electricity (Miskolczy and Huffman, 1977). The application of such multistep use of the initial energy within the electricity-generating installation will allow its overall efficiency to be increased. However, even based on a single-step heating cycle, analogous to the cycle used in present-day thermal power stations (using either nuclear or organic fuels), the efficiency of a commercial hybrid facility would be sufficient for its introduction into the market. Calculations of various specific economic indicators for such hybrid electric stations show that, depending on circumstances, they will be between 20% and 300% more economic than fission power plants with the same parameters (Bethe, 1979).

Hence in the following we shall examine only electric power stations based on a heating cycle.

5.3 The Controllability of the Power Plant

To ensure a high level of reliability of energy supply, in a reasonably economic way (as determined mainly by economic use of energy resources), the overall energy system has to have various subsystems which can be switched into the system as demand dictates. Characteristic times to bring these subsystems from the completely cold state up to full operational load range from 30 seconds to 4 hours. The subsystems are considered to be the more effective the shorter their switching time.

Special demands are made on the reserve energy plants to ensure not only their reliability but also the quality of the energy they supply. Thus one can specify three broad classes of flexibility associated with the switching in of several reserve steps: "instant", over a period of less than 30 s; "fast", in less than 15 min; "slow", within 4 h (Falch et al., 1980). Such reserve systems, with minor variations, provide back-up generating power in the majority of developed countries.

Present experience with both nuclear and conventional thermal power stations indicates that they are much better suited to base-load operation and that they have little flexibility as reserve power generators. This is because of technical difficulties in regulating the reactors and steam boilers and the generally poor economics of running the stations in this role.

The hybrid installations proposed do not have these sort of disadvantages which are based on "system inertia". Switching out the source or putting it into operation involves only the raising or lowering of the voltage on the device; in other words, the duration of the source switching is determined only by the time constant of the communication equipment and the control electronics. The "inertia" of the whole electricity generating system will thus be determined by the time constant of its thermal component. Thus the thermonuclear energy systems can meet peak power demands and can also act as "slow" reserves for the overall energy system. While it is possible to provide flexible operational reserves by combining the thermonuclear generating plants with small-capacity hydropower storage installations in united energy complexes, the use of the proposed energy installation for smooth regulation of the load is also promising. It is known, for example, that the neutron yield of this thermonuclear installation is proportional to the fourth power of the device's current, to the square of the plasma density, and to the frequency of the feeding voltage. These parameters, separately or in any combination, can be smoothly regulated, which makes it possible, by decreasing the neutron yield within the source, to control the energy system power very smoothly without decreasing its effectiveness; i.e. it is possible not to control the "output" energy, as in fission reactors, but rather the "input" energy, without exhausting the uranium of the blanket. If such a control mode is economically expedient for the installation, it can also operate as a fast or almost instant reserve for the energy complex.

If the installation is operated with a stepped use of the output energy (MHD, thermoionic, and steam cycles), the aforementioned properties can be further improved.

5.4 Use for Energy Accumulation

As mentioned before, a number of energy-generating elements in use at present are not suitable for operation in the variable regime. On the other hand, energy demand can fluctuate considerably, with the ratio of peak to normal power between 1.2 and 2.0 (see for instance Falch et al., 1980). This creates the well-known problem of the schedule "gap" during which (depending on the structure of the electricity generating system) base-load plants, even those unsuited to such intermittent operation, have to be switched in and out of the system. At present, to create a "comfortable" operating regime for the base-load plants, hydraulic energy storage is used, and in future it is proposed to use also pneumatic and various electric energy storage devices.

The problem of smoothing the operating regimes of the base-load electric power stations can be solved by the use of hybrid thermonuclear energy devices of the type examined here, operating with their multiplication coefficient smaller than 1, i.e. during the first 2–3 months of their exploitation (see Figure 1, zone I). In other words, during its initial breeding phase this hybrid installation can be treated as an energy-accumulating facility, building up an energy resource without requiring extra equipment. Later exploitation of this resource does not require any change in the way in which the installation operates.

5.5 Use in Centralized and Decentralized Energy Systems

In the last few years decentralized energy systems have aroused considerable interest. This interest can be explained first of all by the fact that large energy systems consisting of powerful electricity generating stations and high-capacity transmission lines create disturbances in the surroundings (heat pollution, dust pollution, changes of local climate, etc.) which are proportional to the power of the energy-production plants and which create a number of potentially negative economic and social consequences that are difficult to predict (Seifritz, 1980). Besides this there are a number of more practical objections against centralized systems, for example, that large energy generating blocks demand reserves of considerable unit size to ensure their steady operation, that losses in long electricity-transmission lines sharply increase as their length is increased, that the power-transmission lines have a large metal inventory, etc.

However, decentralized systems also meet with some objections: the difficulties of skilled servicing, the required increase in total reserve power since every installation must have a separate reserve, the weakening of the control of industrial safety measures, the required increase in manufacturing capacity, the need to ensure protection of the surroundings, etc. In this context we have to distinguish between "pure decentralized" systems (i.e. systems of the "generator–load" type which are completely disconnected from other energy systems), the disadvantages of which are those just given, and systems with "distributed para-units", i.e. systems consisting of local "generator–load" energy systems but linked in the aggregate network with shared tasks for the maintenance of energy quality.

The compactness and simplicity of the examined hybrid installation and the possibility of making a facility of small unit power which is easy to adjust all favor its use in a "pure decentralized" energy system. In other words, the proposed energy installation can compete with the other energy sources which have been proposed for use in a pure decentralized variant. At the same time, in our view, the more promising structure for the energy economy is the one with the distributed units, which will allow the creation of an energy system without the disadvantages of either the centralized or the pure decentralized energy systems. In a distributed-units energy system the thermonuclear hybrid installations can be used in the same way as in a pure decentralized system, but their power can be larger and they can benefit from the advantages of centralized service and the ensuring of industrial safety measures. Also, such energy systems make it possible to decrease the system's total power reserve while at the same time increasing the reliability of each local mini energy system.

In summary, the hybrid energy installations proposed can be used in a straightforward way either within large centralized energy systems (where they would function at least in part as energy-accumulating installations) or in the more decentralized complexes just described.

5.6 Ensuring the Quality of the Electric Energy Supplied

Given the operational characteristics of a thermonuclear energy source of the kind examined, and in particular its pulsed regime of working, it is necessary to provide technical and organization-control measures for reducing the influence of high harmonics within the power-supply network. This problem can be examined from two points of view.

First, the use of the pulsed source of energy in local (decentralized) systems does not need to be accompanied by special technical measures for the filtering out of high harmonics if the consumers of the energy are not too sensitive to this phenomenon. This category of consumers will include some industrial and agricultural as well as private and public energy consumers. In connection with this, the thermonuclear energy installations can of course be used in local energy supply systems of the decentralized kind if they are located in regions where the required supply quality is not so high.

Second, various filter designs are known which can decrease the number of harmonics within a network. Such filters have already been extensively developed for large consumers who make high demands on the energy quality. Thus, the use of energy installations with pulsed thermonuclear sources would not require the introduction of completely new devices into the energy system. Moreover, the operation of hybrid thermonuclear installations within powerful centralized energy systems will perhaps not have much influence on the quality of the electricity if the dumping capacity of the overall energy system is taken into account. In cases where this is insufficient, other energy installations can be used for dumping by combining them with the thermonuclear hybrid devices in united energy complexes, as shown above.

5.7 Application for Heat Production

This type of use of the proposed energy systems appears to be no less feasible than the analogous application of fission power stations in district heating plants. However, in contrast with fission power stations, the compactness of the hybrid devices means that they can also be used within decentralized systems of heat supply. In this variant they could simultaneously serve as electric power regulators located very near the final consumers of the energy. The combination of these properties of the hybrid devices with their capability for regulation and the possibilities for their use in a power-accumulating mode make the application of the installations for heat production look very promising.

5.8 "Burning" of Nuclear Waste

One of the main problems of modern nuclear energy cycles is the regeneration or disposal of used uranium fuel with highly radioactive contaminants present. The problem of burial of radioactive wastes is at present at a crucial point, especially with regard to the location of the waste depositories. Is it possible to "burn out" the long-lived components to leave the residual wastes in a safer condition?

The use of the hybrid installations under the operational mode described earlier, with the power multiplication coefficent below 1, looks quite attractive for "burning out" the nuclear fuel without any fundamental change in the technology or design of the installation. In this mode, the hybrid functions as a regulated consumer of energy, allowing it to partially smooth out the fluctuations in demand; i.e. it accomplishes two functions at the same time.

5.9 Energy Transmission

The most material- and capital-intensive part of the energy system is the network of long-distance power-transmission lines. Besides the difficulties of building the lines, there is also the problem of reducing the energy loss in them, which increases with increasing line length. Moreover, it is sometimes necessary to deliver power via transmission lines to users more than a thousand kilometers away from the generating station. To reduce energy losses, methods of transmitting energy at ultrahigh voltages have been developed which, in turn, lead to a whole range of new engineering and ecological problems.

The use of the proposed hybrid energy installations allows delivery over long distances of, not the electricity itself, but the basic energy resource (i.e. bred nuclear fuel) as is usually the case with organic fuels if the installation is only used within zone I (see Figure 1). Such energy transportation decreases the losses and ensures fuller use of the scarce energy in a simple enough way. Furthermore, in comparison with organic-fuel transportation, the delivery of partially enriched fission fuel and tritium has the advantage that the energy-equivalent mass of these fuels is many times lower. This advantage, in principle, makes the transportation of the "accumulated energy" economically possible even though expensive special safety measures have to be taken. In addition, if similar hybrid energy installations are used in the consumer's region as well, there is no need for the regeneration and refabrication of the nuclear fuel.

5.10 Reconstruction and Adaptation of Existing Generating Stations

One important aspect of present-day energy systems is the reconstruction of old power stations by the installation of new, more reliable and efficient equipment. In this way the operation of the reconstructed power plants can be made more economical.

As has been shown, there is little difference between the structures of an electric power station based on the hybrid thermonuclear energy system and one using the conventional thermoelectric system. It is therefore possible to reconstruct stations which at present work on organic fuel by substituting only the boiler (reactor) section, while preserving not only the associated industrial buildings but also a considerable part of their equipment (the steam turbine, the generator, the electrical equipment, etc.). Thus a further reduction in the cost of creating such energy systems can be achieved as a result of the savings in building time and in the expenditure of materials.

5.11 Safety Factors

The proposed installation has all the characteristics inherent to hybrid thermonuclear devices of any kind. However, it differs from classical devices for the generation of powerful relativistic particle beams (to which class it belongs) in three main ways: its feeding voltage is 10 times lower than the energy of the particles generated, it is very compact, and the amount of radioactive materials it contains and the level of X-ray radiation generated are both relatively low due to the anomalous mechanism of electron beam absorption inside the plasma (Gribkov and Krokhin, 1981; Gribkov et al., 1979).

5.12 Protection of the Surroundings

In this respect, the basic problems are the same as for any comparable installation. However, the small size of the proposed system and the relatively low power (and consequent small amount of fission and fusion fuels involved) again give it an obvious advantage.

6 PROBLEMS AND PERSPECTIVES

To conclude this qualitative analysis, we note that first of all it will be necessary to conduct intensive investigations and engineering studies on the questions of construction and use of the proposed system. These studies should include the following: (1) detailed calculations of blanket and first-wall loading; (2) precise economic calculations; (3) investigation of the possibilities of applying the stationary sources for feeding this type of installation; (4) detailed calculations of the harmonics of current and voltage which appear under the influence of the consumed power pulses; (5) investigation of possible ways of operating the installation within centralized and decentralized energy systems in order to determine its optimal operating parameters and economic characteristics.

Experience with installations of megajoule scale as regards stored feeding energy that have already been constructed in various countries will allow us to judge the technological realizability of the installation. Also, it seems possible to increase the thermonuclear efficiency of the installations by injecting hot plasma streams to improve plasma heating and discharge ignition (Gribkov and Krokhin, 1981). More detailed investigations of this operational regime should be carried out. Success in this direction – even if not very great – for example, a threefold increase in the neutron yield (which has already been demonstrated; see Kaliski et al., 1975, and Gribkov, 1980) brings either the possibility of feeding a greater number of heating fission stations with plutonium from an installation of the same power (benefits of economy!) or the possibility of decreasing the system size (benefits of compactness!).

One of the main peculiarities of this system is that inside the installation a plasma with a very high mean energy per particle is produced. Recent investigations indicate that this energy tends to increase almost linearly as the feeding current of the installation is increased. This makes this type of device very interesting from the point of view of using promising future thermonuclear fuels such as pure deuterium and, possibly, boron-11 to take advantage of the increased feeding-energy capacity.

REFERENCES

Bethe, H. (1979). Fusion hybrids. Physics Today, 32(5): 44.

Falch, C., Larsson, S., Mantinen, R., and Svoen, Y. (1980). Coordination des reserves communes d'exploitation dans le réseau norfique. Electra, No. 71: 71–82.

Feoktistov, L.P., Avrorin, E.N., Varganova, L.F., Gadzhiev, A.D., Lykov, V.A., Neehai, V.Z., and Shibarshov, L.I. (1978). Kvantovaya Electronica, 5(2):349–358.

Gribkov, V.A. (1980). Atomkernenergie, 36:167.

Gribkov, V.A. and Krokhin, O.N. (1981). Powerful neutron source based on the laser micropinch with the combined plasma heating. Preprint of the Lebedev Physics Institute, No. 57.

Gribkov, V.A., Krokhin, O.N., Isakov, A.I., Nikulin, V.Ya., and Semenov, O.G. (1979). Paper presented at the International Conference on High Power Electron and Ion Beams, 3rd, Novosibirsk.

Harms, A.A. (1978). Atomkernenergie, 32(1):3.

IEEE (1980). Proceedings of the Summer Meeting of the IEEE Power Engineering Society, Vancouver, British Columbia, 1979. MIREK, Munich.

Kaliski, S., et al., (1975). Soviet Physics – Technical Physics, 16:187.

Kochs, H.-D. (1980). Internationaler Stand und Trend in der elektrischen Energietechnik. Elektrotechnische Zeitschrift, 101(3):160–163.

Miskolczy, G. and Huffman, F. (1977). Evaluation of MHD–thermionic–steam cycles. Thermo-Electron, Waltham, Massachusetts; also in Mechanical Engineering, 100(5):94.
Seifritz, W. (1980). Sanfte Energietechnologie-Hoffnung oder Utopie? Karl Thiemig, Munich.
Velikhov, E.P., Kadomtsev, B.B., and Orlov, V.V. (1977). Teploenergetica (Thermal Engineering Review), 24(11):22–28.
Wolfberg, D.B., Ershevitch, V.V., Kotler, V.R., Lebedev, B.P., and Shlimovitch, V.D. (1981). Obzor energitiki mira. Energokhoziaystvo za rubejom, 1:1–39.

Section Four

Accelerator Breeding

ACCELERATOR SPALLATION REACTORS FOR BREEDING FISSILE FUEL AND TRANSMUTING FISSION PRODUCTS

Meyer Steinberg
Department of Nuclear Energy, Brookhaven National Laboratory, Upton, New York 11973 (USA)

ABSTRACT

A summary of the status of the development of accelerator spallation reactors for breeding fissile fuel and for transmuting fission products is presented. The development of linear accelerators for generating protons in the GeV range and the subsequent spallation-reaction production of neutrons are reviewed. Target–reactor designs for production of fissile fuel and the advantages and disadvantages of the various applications of the spallation reactors are given. The linear accelerator fuel enricher–generator (LAFER) can supply significant amounts of fissile fuel to conventional light water reactors (LWRs) in an economical manner. A fuel cycle is designed which recycles the transuranics and long-lived fission products for transmutation and waste management.

1 INTRODUCTION

This report constitutes a summary review of the status and prospects of the development of accelerator spallation reactors for breeding fissile fuel and transmuting fission products.

The concept of utilizing a high-energy particle accelerator in conjunction with a target to generate neutrons for use in converting U-238 to Pu-239 dates back to 1947 when E.O. Lawrence recommended this method for the US weapons program (Kouts and Steinberg, 1977). A project called MTA was established and studies were made on producing Pu-239 with the use of a 500-MeV deuteron beam and a 320-mA current on a primary beryllium target and a secondary uranium target surrounded by a uranium lattice blanket. Liquid NaK was conceived as the coolant for the primary and secondary targets and water for the uranium blanket. The incentive for the project was the lack of natural uranium resources (U-235) for the nuclear weapons program. On discovery of ample supplies of uranium in the United States and development of reactor technology, the MTA project was terminated in 1953.

However, the technical feasibility of the idea was established and it is interesting that even at that early stage of development the design efficiency of the linear accelerator was as high as 50% (ratio of power into the machine to beam power out).

The next development was a detailed design study by Wilson et al. of Atomic Energy of Canada Limited (AECL) for an Intense Neutron Generator (ING) in the early 1960s (Lewis et al., 1965; Bartholomew and Tunnicliffe, 1966). The design was based on measurements of spallation neutrons performed at the Brookhaven National Laboratory (BNL) Cosmotron (Fraser et al., 1965) in a joint project between AECL and Oak Ridge National Laboratory (ORNL). The ING was a 65-MW linear accelerator (65-mA, 1000-MeV beam) machine with a Pb–Bi liquid metal target and coolant to generate high-energy neutrons for basic research purposes. However, Wilson always had in mind that this could also be of value as a suitable machine for fissile-materials production.

In 1967 workers at BNL first proposed the use of an accelerator spallation reactor for transmuting the long-lived Cs-137 and Sr-90 fission-product waste (Gregory and Steinberg, 1967). This was proposed because it was determined that thermal fission reactor cores could not produce the high neutron flux necessary to shorten effectively the half-lives of these long-lived radioactive products by neutron transmutation. A study was again undertaken in 1976 using a ThO_2 target with helium gas cooling (Steinberg et al., 1976a). It was shortly thereafter that studies were undertaken at BNL on a linear-accelerator breeder for producing nuclear fissile fuel material from fertile material as a possible backup to the Liquid-Metal Fast Breeder Reactor (LMFBR) (Steinberg et al., 1976b; Steinberg, 1977). The concept of the Linear-Accelerator-Driven Reactor (LADR) and the Linear-Accelerator Fuel Producer (LAFP) was developed at that time. ORNL, Los Alamos Scientific Laboratory (LASL), and the Lawrence Livermore Laboratory (LLL) then undertook studies on their own. A conference on accelerator breeding was convened at BNL in 1977, where a number of concepts concerned with the production of fissile material with spallation-induced neutrons were reviewed (Kouts and Steinberg, 1977).

In April 1977 the Carter administration in the United States established a nonproliferation policy which forbade the reprocessing of nuclear fuel because of the risk of nuclear weapons proliferation. This policy for all practical purposes tended to retard the progress towards commercialization of the fast breeder reactor in the United States. The US Department of Energy set up studies for developing a Nonproliferation Alternative Systems Assessment Program. A number of alternative candidates for nuclear reactors and systems were proposed and investigated (US Department of Energy, 1979). Several linear-accelerator fissile-fuel production concepts were considered and one particular concept, the Linear-Accelerator Fuel Enricher–Regenerator (LAFER), was studied in depth by BNL (Grand et al., 1978).

The LAFER study was presented at the First International Meeting on Advanced Nuclear Reactor Concepts at Graz, Austria, in April 1978. The accelerator spallation breeder was included in the International Nuclear Fuel Cycle Evaluation (INFCE) studies only as an advanced concept because the system was not considered to be a demonstrated technology.

In addition to the foregoing, several waste-management concepts (Powell et al., 1980; Takahashi and Mizzo, 1980; Steinberg, 1980) based on transmutation by accelerator spallation neutrons and decay were studied by BNL and were presented at the Second International Meeting on Advanced Nuclear Reactor Concepts at Lucerne, Switzerland in April 1980 and at an International Meeting on Waste Transmutation at Austin, Texas, in July 1980 (Davidson, 1980; Takahaski and Mizzo, 1980).

2 NEUTRON SPALLATION–EVAPORATION CHARACTERISTICS AND YIELDS

The basis for an accelerator spallation reactor is the neutron release from nuclei obtained when a high-energy particle having an energy greater than approximately 50 MeV interacts with a target atom. The principal types of interactions with atoms are electron ionization–excitation and inelastic collisions. These latter interactions are called spallation–evaporation or intranuclear cascade evaporation processes. Some experimental measurements of the yield of neutrons produced in the spallation–evaporation process by energetic protons have been made. Figure 1 shows the results of experiments conducted with the 3-GeV Cosmotron at BNL. It should be noted that the neutron yield for U-238 is twice that obtained for lead. The difference is partly due to fast high-energy neutron (> 2 MeV) fission occurring in U-238.

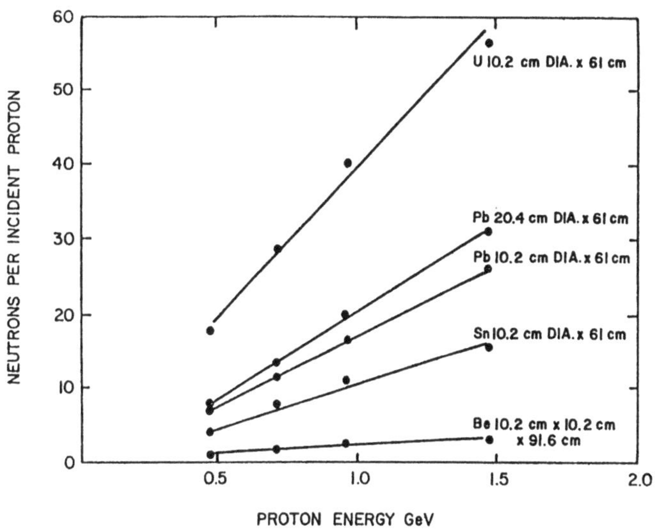

FIGURE 1 The experimental yield of neutrons obtained by bombardment of a heavy-metal target with high-energy protons.

Figure 2 shows the calculated energy spectra of neutrons produced by 800-MeV protons incident on solid targets of lead and U-238. Here again the difference introduced by fission neutrons is clearly indicated. The yields and energy spectrum resulting from the intranuclear cascade alone (> 5 MeV) are about the same for both materials.

The neutron yields have been calculated from models using the Monte Carlo method with the Nuclear Meson Transport Code (NMTC). Table 1 indicates the calculated neutron yield when 1-GeV protons are injected into the center of an infinite medium of different types of material of interest for a target assembly. It can be seen that, even for H_2O–UO_2 lattices of widely varying H_2O–UO_2 volume ratios, the neutron yield does not vary greatly. This calculation only includes reactions induced by neutrons

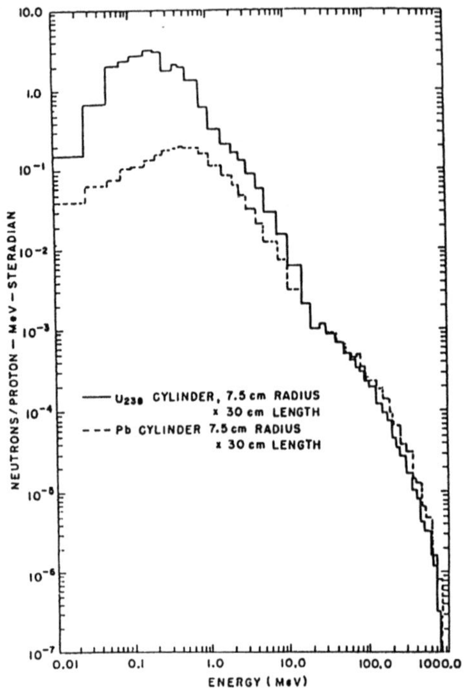

FIGURE 2 Neutron energy spectra produced by 800-MeV protons incident on lead and U-238 targets.

TABLE 1 Monte Carlo calculations for lead and UO_2: neutrons produced by reactions of greater than or equal to 15 MeV excitation energy (neutrons per 1-GeV proton)[a].

Target material	Average yield[b]
Pb	35
UO_2	35
H_2O–UO_2 (H_2O/UO_2 effective volume ratio, 1.4)	28
H_2O–UO_2 (H_2O/UO_2 effective volume ratio, 0.7)	29
H_2O–UO_2 (H_2O/UO_2 effective volume ratio, 0.17)	30

[a] Infinite medium, 1-GeV proton injected into center.
[b] Does not include fission neutrons of energy less than 15 MeV.

of more than 15 MeV. The interactions for neutrons of less than 15 MeV are included in the target/blanket calculations using neutron transport codes with neutron cross-section libraries.

Experimental studies have indicated that for the interaction of 660-MeV protons with a metallic uranium lattice the total neutron yield amounts to about 50 neutrons per proton (Vasilkov et al., 1978). Calculations by Boroshenkov and Toneev (1973) indicate

TABLE 2 Initial fissile-fuel production rates for a Pb–Bi target and fuel-element blanket (0.3-A, 1.5-GeV proton accelerator)[a].

Design no.	Fertile material	Coolant	Density of coolant ($g\ cm^{-3}$)	Initial neutron yield Y_n (includes fission reaction)	Initial production rate of fuel material ($kg\ year^{-1}$)
1	UO_2 (natural U)	D_2O	0.7	53.8	1010 (Pu)
2	ThO_2	D_2O	0.7	46.6	850 (^{233}U)
3	UO_2 (natural U)	H_2O	0.7	74.1	1000 (Pu)
4	UO_2 (natural U)	H_2O	0.35	65.8	1050 (Pu)
5	UO_2 (natural U)	H_2O	0.175	64.9	1070 (Pu)
6	ThO_2	H_2O	0.7	46.5	850 (^{233}U)
7	ThO_2	H_2O	0.35	48.6	890 (^{233}U)
8	ThO_2	H_2O	0.175	49.0	900 (^{233}U)

[a] Moderator/fuel volume ratio, 0.8; efficiency (ratio of beam power output to electric power input), 50%; plant factor, 80%.

TABLE 3 Initial and final blanket thermal power (in megawatts) for a 1-year irradiation period starting with different enrichment values (Pb–Bi target; fuel-element blanket; 0.3-A, 1.5-GeV proton accelerator)[a].

Case	Value	Initial enrichment			
		0%	0.7%	2%	3%
1	Initial	–	300	430	570
	Final		460	620	780
2	Initial	55	–	420	850
	Final	220		610	1040
3	Initial	–	380	530	620
	Final		550	660	740
4	Initial	–	380	530	650
	Final		540	680	810
5	Initial	–	450	640	810
	Final		620	830	1030
6	Initial	50	–	750	1030
	Final	500		1000	1260
7	Initial	60	–	550	770
	Final	310		710	940
8	Initial	80	–	560	820
	Final	260		730	1040

[a] Moderator/fuel volume ratio, 0.8; efficiency (ratio of beam power output to electric power output), 50%; plant factor, 80%. 0% and 0.7% pertain to the natural content of fissile material in Th and U, respectively. Numbers of cases correspond to those in Table 1.

a yield of 100 neutrons per proton for 1-GeV protons. Further neutron-yield experiments are needed to confirm these results which include the fast fissioning of U-238.

It is also of interest to note that in the 1-GeV region deuterons give a 20% increase in neutron yield compared to protons. Although this indicates that accelerating deuterons might be advantageous, the higher cost and design constraints of a deuteron accelerator tend to negate the advantage of a higher neutron yield.

Calculations have been made for the production of fissile material and the generation of thermal power in various target/blanket assemblies. Table 2 shows the production rate of fissile material for a number of arrays based on a 1.5-GeV proton beam with a 0.3-A current. The target is lead surrounded by various blankets of fertile material plus coolant. Table 3 gives the initial and final (after irradiation for 1 year) calculated blanket thermal power for different values of fissile-fuel enrichment in the blanket. This thermal power and the direct target thermal power deposited by the proton beam are recoverable for the generation of electrical power for use by the accelerator in making the system self-sufficient in power. The neutron-yield and thermal-power calculations are considered as conservative. A prime experimental effort for developing the accelerator spallation reactor should be the actual measurement of the neutron yield and the heat released in a real lattice structure.

3 ACCELERATOR DEVELOPMENT

The reason that spallation reactors can be seriously considered as a production tool today is that the linear accelerator has been developed and engineered to a high degree over the past 30 years for the purposes of basic research in high-energy physics. Linear-accelerator technology is at the point where a Continuous-Wave (CW) high-capacity machine can be designed and built with a great deal of confidence that it can be made to operate with a high degree of reliability.

Protons and heavier particles can be accelerated to energies above 1 GeV most economically in synchrotrons. These accelerators are limited, however, to pulsed operation and very low average intensity. A continuously operating high-current beam accelerator is required for fissile-fuel production. The linear accelerator is the only type of accelerator available today that is capable of accelerating a steady continuous beam of protons at a current of over 100 mA. With linear accelerators, peak proton currents of 300 mA have been accelerated, and the possibility of accelerating even higher currents is available if necessary. Table 4 lists a number of relevant accelerators (most of them already operating) together with their characteristics. Storage rings are listed to show that very large currents (large for accelerators) can be circulated and contained for very long periods essentially without beam loss.

In addition to its unique capability of accelerating high currents, the linear accelerator has the advantages of (a) providing the most efficient means of energy conversion (electrical energy to beam energy) and (b) having the lowest beam loss factor (particles lost during the acceleration process).

Minimization of beam loss is an important criterion for this application. In all circular accelerators a certain amount of beam loss is inevitable at injection and extraction, which makes all equipment located in these areas highly radioactive. For the very high beam-power application considered here, even the lowest fractional beam loss that has been achieved is still several orders of magnitude higher than that which is acceptable. With a linac there is no extraction loss and, since the injection energy is typically less than 1 MeV, injection losses are manageable. In addition, because the beam travels in a straight

TABLE 4 Parameters of existing relevant accelerators.

Laboratory	Accelerator	Type	Ion type	E_{max} (MeV)	I_{max} (A)	Duty factor (%)	Status
LASL	LAMPF	Linac	Proton	800	0.020	12.0	Operating
LLL	MTA	Linac	Deuteron	500	0.320	100.0	Proposed
BNL	AGS	Linac	Proton	200	0.2	0.5	Operating
FNAL		Linac	Proton	200	0.3	0.2	Operating
HEDL	HFNS	Linac	Deuteron	35	0.1	100.0	Proposed
CERN	ISR	Storage rings	Proton	3×10^4	40.0	–	Operating
BNL	ISA	Storage accelerator	Proton	4×10^5	6.0	–	Proposed

line the loss during the acceleration is much lower than in all other types of accelerators. These simple considerations lead directly to the conclusion that the linac is the optimum choice for a production-type facility.

The charged particles are produced in an ion source. They are first accelerated to an energy which is typically about 750 keV in a Direct-Current (DC) electrostatic accelerator. The 750-keV beam is then injected into the linac and is accelerated to the final energy by the Radio-Frequency (RF) field in the cavities. New developments in linear accelerators indicate that the injection energy required for high beam currents might be relaxed. In particular, work carried out in the Soviet Union shows that 100-keV injection might be possible. Many different linac cavity designs are available, each having its own merits for special applications. The choice of the specific structure and rate of acceleration per unit length is based on cost, beam-loading efficiency, and other construction and operating features.

Above the lower limit of beam energy set by ionization loss the neutron yield in the target is roughly proportional to the beam power, which is in turn proportional to the RF power for a given beam loading. A major part of the cost of the linac (roughly about 70% of the total cost of the accelerator) is in the RF supply. This cost is proportional to the beam power and hence the fuel yield. Thus for a given fuel yield the total accelerator cost is primarily affected by the cost of the RF system. The remaining costs (of the cavity structure, the control and monitoring system, the linac housing, etc.) amount to about 30% of the total. The costs of these components are generally proportional to the length, and hence the energy, of the linac. Thus for a given neutron yield the total cost of the linac is somewhat lower for lower energy and higher current.

A beam current of 300 mA has been obtained at FNAL, though only in a pulsed mode. However, during the pulse a steady state is reached insofar as the transfer of power from the RF supply to the beam is concerned. The pulse length is limited only by the size of the RF supply. Hence, a 300-mA CW beam current is considered to be a realistic goal attainable with high confidence. At LASL a particle energy of 0.8 GeV has been achieved in a linac, and further acceleration will require only iteration of stages that are now in use. This demonstrates the total practicality of attaining energies above 1 GeV. One reference design is based on an accelerator that will provide a beam of 1.5-GeV protons at 300 mA to produce, at a conservative estimate, more than 1 ton of fissile fuel per year.

This is enough fissile material to support three LWRs. These parameters are not optimized, and it is likely that optimization will result in somewhat different design characteristics.

Higher beam energies are well within the expectations of linac technology. A higher energy, with the same beam current, will yield a correspondingly higher fuel-production rate with little or no reduction in probability of success. However, the unit cost of the fuel produced will not be significantly lower.

It should be noted here that linear accelerators at BNL, the Centre Européen pour la Recherche Nucléaire (CERN), FNAL, and LASL have impressive records for operational reliability (more than 90% of scheduled time with an overall plant factor of better than 80%). It is expected that the linac part of the fuel-regenerator facility will at least match this performance.

A discussion of the linear-accelerator design and development needs can be found in Grand et al. (1978). The major parts of the accelerator can be categorized as follows: (1) the injection system, (2) the low β-linac section, (3) the high β-linac section, (4) the RF systems, and (5) the control and diagnostic system. Improvement in efficiency and cost must necessarily come from development in the RF supply. If a Soviet development of an RF system called the Gyrocon materializes, this could increase the overall conversion efficiency of DC to RF power to over 90% and would greatly reduce the cost of the linac.

It is generally agreed that a 50% efficiency of electrical power input to beam power output can be readily achieved with state-of-the-art linac technology. Improvements could ultimately bring efficiencies to 70% or more.

4 TARGET–REACTOR DEVELOPMENT

The target–reactor design generally depends on a combination of the following four factors: (1) the primary-target material; (2) the coolant of the primary target; (3) the blanket configuration; (4) the coolant of the blanket.

If the primary target is a fertile material such as uranium or thorium, the neutron yield will be considerably higher than if the target were lead. However, much greater thermal power and radiation damage will occur in a solid uranium or thorium target than in a liquid lead target. The primary-target and coolant properties will determine whether a window or barrier between the electron machine and the primary target is necessary. If a gas coolant (e.g. helium or steam) or a liquid coolant (e.g. light or heavy water) is used, a window or thin-walled barrier between the vacuum of the linac beam tube and the primary target becomes necessary. A beam window becomes a design problem of balancing beam energy losses with thermal and structural stability. In addition to materials with high strength-to-weight ratios such as steel, aluminum, and titanium, graphite has been suggested as a window material. Liquid coolants such as molten lead, sodium, or salts do not normally require a window because of their low vapor pressure.

The bred fuel can be produced in the primary target or in a surrounding blanket in which fertile material is placed. The blanket is subject mainly to a neutron flux. A contained or uncontained coolant surrounding the blanket fertile material may be used.

An example of a workable target–blanket design developed at BNL is given in Figure 3. The assembly is designed to generate fuel in situ in a fuel assembly bundle of the Pressurized-Water Reactor (PWR) type. This configuration is also useful for production

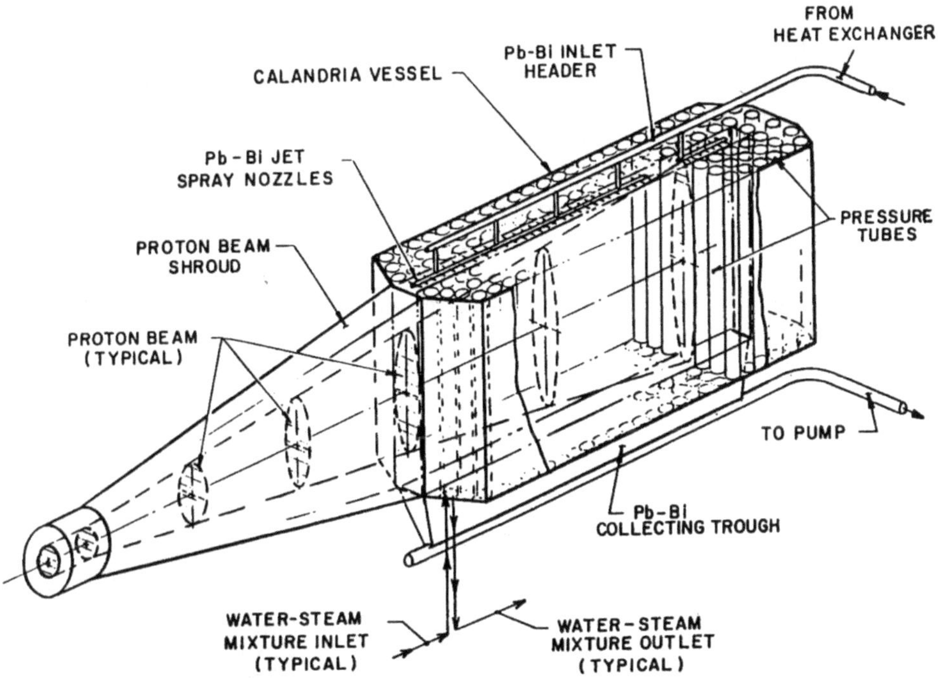

FIGURE 3 A PWR target assembly.

of fissile fuel from fertile uranium or thorium. The bundles are placed into pressure tubes which surround a central slot through which jets of liquid Pb–Bi metal are projected down along the length of the calandria assembly of pressure tubes. The proton beam spreads elliptically from the beam-transport tube entering the side of the calandria vessel with no interfering walls and interacts with the jets or columns of falling liquid lead. Neutrons are spallated and evaporated isotropically from the lead by the high-energy protons and the high-energy spallation-and-evaporation neutrons enter through the walls of the pressure tubes and are absorbed by the PWR-type fuel-element rods (Zircaloy-clad UO_2) forming fissile Pu-239 and U-238 in situ within the element. The purpose of the jets is to disperse the dense Pb–Bi target so as to allow a longer penetration path in the target for the purpose of distributing the neutron flux to a level which will be tolerable from a power-density point of view. The Pb–Bi has a vapor pressure of 10^{-4} torr at the expected temperature of 300° C, thus maintaining an adequate low-vacuum condition for the proton beam. The Pb–Bi is collected at the bottom of the vessel and is cooled in a separate circuit. In order to prevent loss of neutrons the outside area of the calandria vessel is covered with a neutron reflector (i.e. graphite) including top and bottom sections. The assembly should be long enough to accommodate a full-length PWR bundle (8 ft long) (Figure 3).

Steam is used for cooling purposes because it is necessary to provide a hard spectrum; however, since a high heat flux is also expected, wet steam or two-phase evaporative cooling is used. The calandria tubes are made of Zircaloy, as is the cladding of the

PWR elements. Wet steam prevents corrosion of the Zircaloy. Liquid heavy-water coolant is also an adequate coolant which would not require two-phase cooling. The thermal hydraulics of the system have been analyzed (Grand et al., 1978). Power can be recovered from the primary lead target as well as the secondary coolant. Pumping power would be less than 5% of the power required by the accelerator.

Another target-assembly arrangement of the Heavy-Water Reactor (HWR) type is shown in Figure 4. This is a horizontal-pressure-tube type of assembly similar in concept to the CANDU reactors. The Pb–Bi jets are in the center slot with the pressure tubes surrounding. The fuel elements are inserted horizontally through the pressure tubes. Shorter elements can be used in this arrangement for fissile-fuel production with reprocessing.

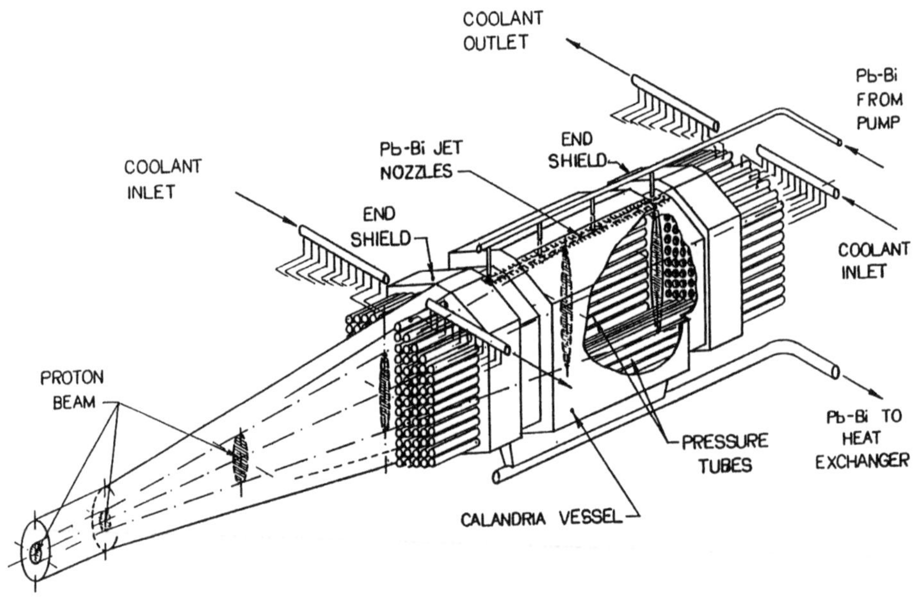

FIGURE 4 An HWR-type target assembly.

5 SYSTEM APPLICATIONS

The accelerator spallation reactor can be applied in several basic modes.

5.1 The Linear Accelerator Driven Reactor (LADR)

The LADR uses a subcritical reactor assembly in which the neutrons generated by the spallation reaction cause fission multiplication and a net production of power. This configuration is shown in Figure 5. The LADR becomes economically viable when the power produced in the target is five times that required by the accelerator.

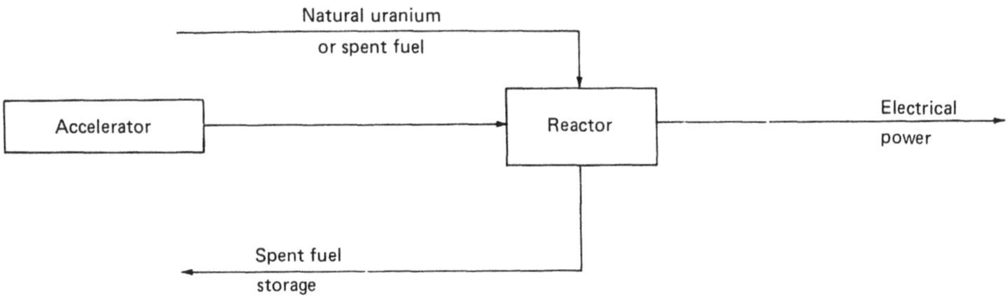

FIGURE 5 The scheme of the LADR.

The advantages of the LADR are (1) the fuel elements can be put through the reactor once without reprocessing, (2) power can be obtained from natural uranium in subcritical assemblies moderated by light or heavy water (a subcritical assembly has numerous reactor-safety advantages), (3) the uranium resource can be stretched, and (4) the assembly does not require enriched fuel.

Disadvantages are (1) each power reactor requires an accelerator, (2) construction and operation are more complex than for conventional power reactors, and (3) longer burnup may be limited by radiation damage to cladding and structural material.

5.2 The Linear Accelerator Fuel Enricher Regenerator (LAFER)

The LAFER produces fissile material in a fuel element designed for use in a conventional power reactor of the Light-Water Reactor (LWR) type. After burnup in the LWR, the fuel is regenerated in the LAFER target/blanket assembly to build back the concentration of fissile material in the element to LWR feed criticality values. A schematic diagram of the concept is given in Figure 6. A number of advantages accrue with this concept. (1) There is no departure from existing LWR power-reactor practice. The utility operator does not have to learn a new technology as would be needed with an LMFBR. (2) No reprocessing is required; reprocessing would only become necessary in the event of radiation damage to the fuel element with an increasing number of enriching–burning cycles. (3) One accelerator spallation reactor can supply several LWR power reactors. (4) The LAFER stretches the natural uranium fuel resource by a considerable amount. (5) The accelerator is kept separate from the power reactor; this makes for a higher availability of the total system. (6) Enrichment capacity can be considerably reduced; no new enrichment facilities would be required. (7) The LAFER substantially reduces reprocessing requirements when reprocessing is allowable. (8) No new scientific principles or technologies have to be developed or demonstrated.

The only disadvantage of the LAFER is the limitation of the durability of the element on long-term burnup, due either to fission-product buildup or to radiation damage to cladding.

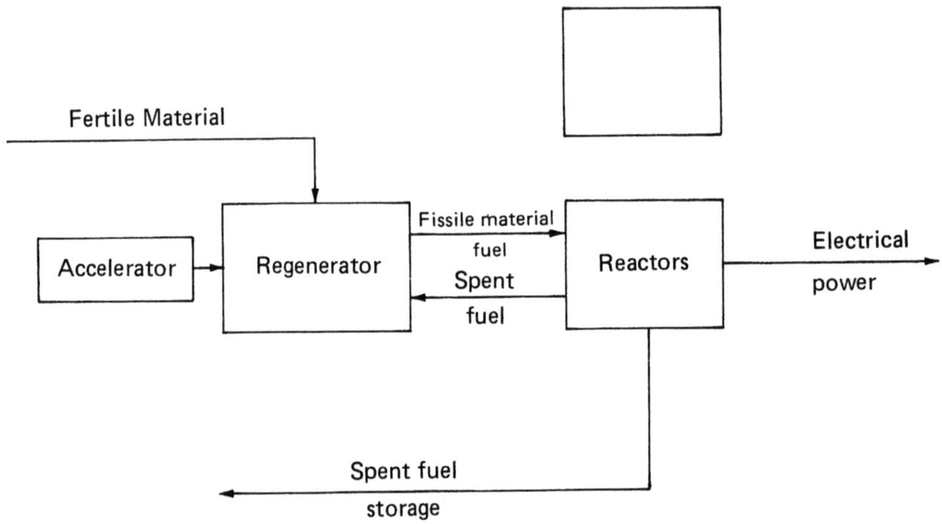

FIGURE 6 The scheme of the LAFER.

5.3 The Linear Accelerator Fuel Producer (LAFP)

The third concept is the LAFP. It is used to produce fissile material from fertile material in conjunction with reprocessing and fabrication operations. The fissile material is incorporated into fuel elements for use in power reactors. After burnup the spent fuel from the power reactors is returned to the reprocessing plant to recover the fissile material remaining and to remove the fission products which are sent to waste storage. A schematic of this system is shown in Figure 7. The advantages of the LAFP are as follows: (1) the fuel cycle comes closest to the conventional cycle; (2) the need for enrichment is eliminated; (3) the LAFP makes full use of the fuel resource and is therefore a true breeder; (4) one accelerator can supply fissile fuel to several power reactors; (5) the LAFP can produce U-233 from thorium for use in high convertors. The main disadvantage of the system is that it depends on reprocessing the fuel elements.

6 SYSTEM ANALYSIS FOR THE LAFER

A system analysis of the LAFER concept illustrates some of the advantages and economics of the LAFER fuel cycle.

The conventional LWR nuclear fuel cycle is shown in Figure 8. Natural uranium feed is isotopically enriched to 3.2% U-235 in a gaseous diffusion plant, fabricated into Zircaloy-clad uranium oxide fuel, and then burned in the LWR. If reprocessing is allowed, the plutonium is extracted or recycled and the separated radioactive waste fission products are sent to long-term storage. If reprocessing is prohibited, which is the present US nonproliferation policy, the spent fuel elements must be stored in pools. For a 1000-MWe reactor, over a 30-year life, 6300 tons of natural yellow cake (U_3O_8) are required to

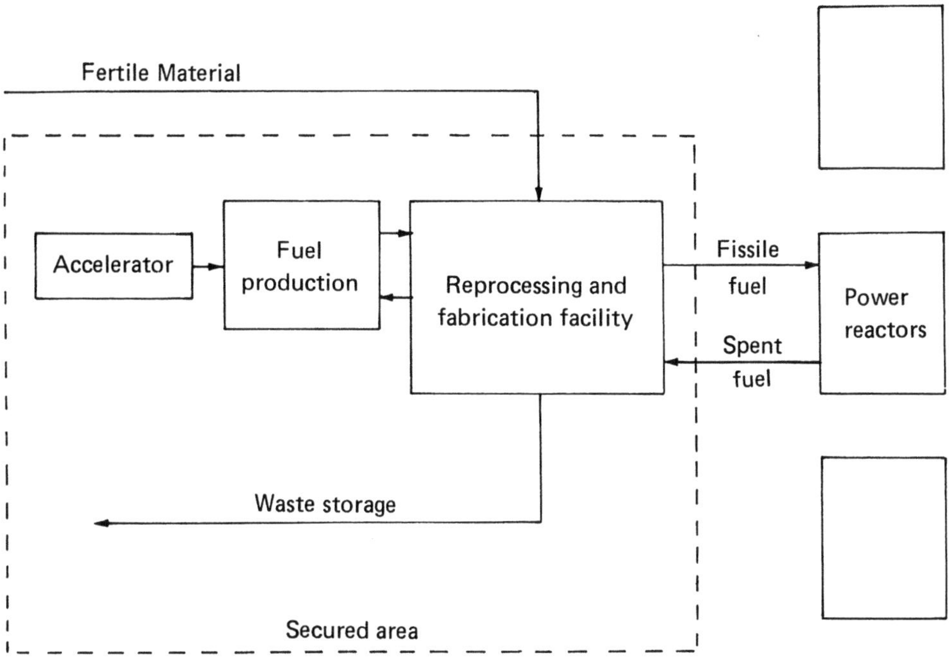

FIGURE 7 The scheme of the LAFP.

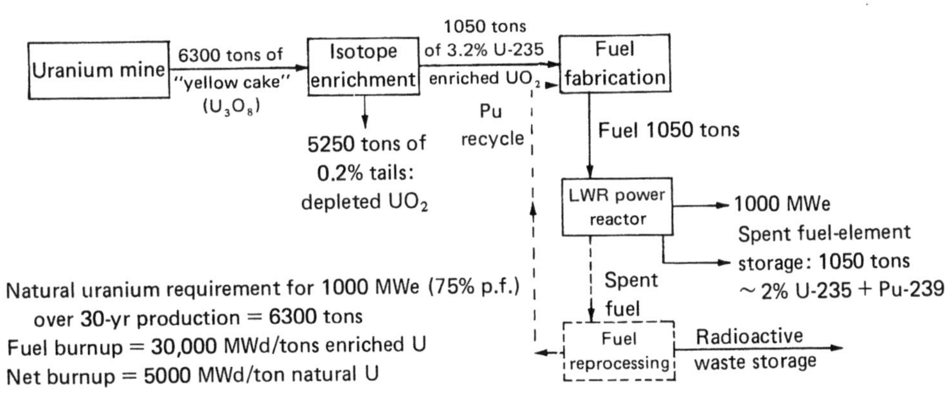

FIGURE 8 The conventional LWR nuclear fuel cycle.

produce 1050 tons of enriched fuel, which ends up in spent fuel elements containing about 1.0% Pu-239 and 1.0% U-235. The net burnup is equivalent to 5000 MWd per ton of natural uranium. By comparison, heavy-water-cooled reactors (CANDU type) exhibit a fuel utilization of about 8000 MWd ton^{-1} and require no enrichment. The isotope-enrichment plant wastes a great deal of the uranium resource, requiring 6 tons of raw natural uranium material to produce 1 ton of fuel; 5 tons are waste tailings.

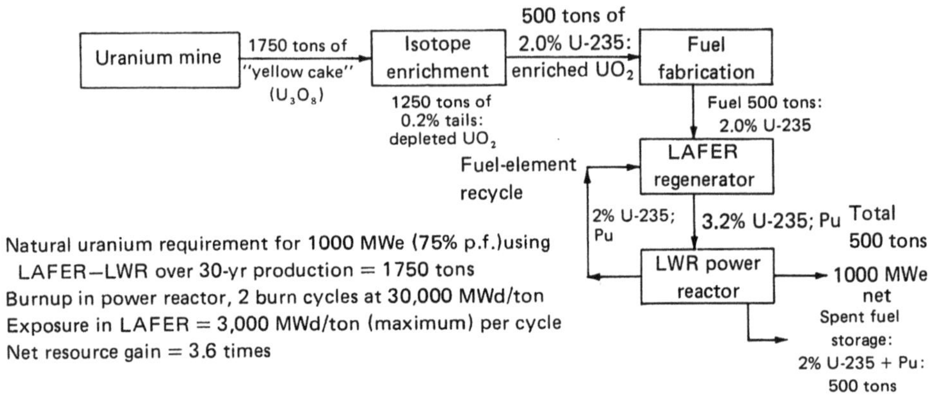

FIGURE 9 A LAFER fuel cycle with no reprocessing.

The proposed LAFER cycle starts off with either no enrichment or a lower U-235 enrichment and builds into the fuel the required Pu-239 reactivity. One illustrative cycle is shown in Figure 9. Yellow cake is isotopically enriched to 2.0% U-235, fabricated into Zircaloy-clad oxide fuel, and placed in the LAFER target where the fissile content is brought up to 3.2% fissile material by breeding in situ the additional 1.2% Pu-239. The element is then ready for generating power in an LWR. After one burn cycle of 30,000 MWd ton^{-1} the fissile material is reduced back to 2% and the fuel element is reinserted into the LAFER where fissile material is once again increased from 2% to 3.2%. Without reprocessing, the net fuel to storage is 500 tons or half that for a conventional cycle. The number of allowable fuel-regeneration cycles depends primarily on the radiation damage to the cladding. Conservatively, the number of burn cycles has been limited to two, meaning that the element experiences a burnup of 60,000 MWd ton^{-1} in the reactor with another burnup of 6000 MWd ton^{-1} in the accelerator. Figure 10 indicates the buildup of Pu-239 and the decrease of U-235 in the LAFER, followed by buildup of fission products and burnup of the fissile material in the reactor. It is interesting to note that the thermal and epithermal neutron absorption due to fission products is decreased in the LAFER irradiation cycle. This indicates that some of the fission products are transmuted from strongly absorbing isotopes to weaker neutron-absorbing species in the process.

There has been much successful operation of conventional LWR elements to burnups of 60,000 MWd ton^{-1} in LWR power reactors. Present projected burnup designs for LMFBRs call for 100,000 MWd per ton of fuel. The total amount of natural uranium (yellow cake) needed for the LAFER cycle is 1750 tons over the 3-year life of the 1000-MWe LWR. Thus the resource gain is 3.6 times that in the conventional cycle. Furthermore, no additional capacity of expensive enrichment plants is necessary. The LAFER is a positive enrichment machine because it converts fertile U-238 to fissile Pu-239. The isotope-enrichment plant is actually a depletion plant since it only utilizes the very limited natural U-235 resource and throws away the bulk of the natural fertile material.

Based on the foregoing concept, a system was devised, as shown in Figure 11, where three 1000-MWe reactors are supplied throughout their 30-year lifetime with fuel from

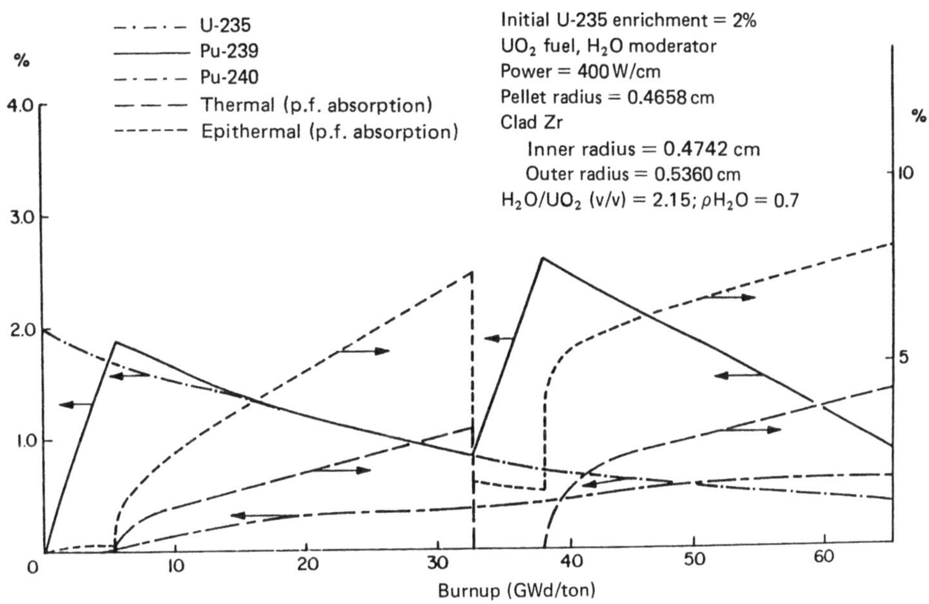

FIGURE 10 The concentration of U-235, Pu-239, and Pu-240 neutron absorption due to fission products in a LAFER.

one LAFER. The 450-MW beam from the accelerator is produced using 900 MW of electrical power, 450 MW of which is obtained by recovery of the heat developed in the target. A comparative cost estimate of the LAFER systems has been made. Table 5 shows the LAFER capital investment using 1978 dollars and escalating to 1986 using industry cost accounting. In terms of depreciation of the \$1.5-billion capital investment for the LAFER machine, the charge to electrical energy production amounts to 1.16 cents per kWhe delivered by the power reactors. Table 6 gives the entire fuel-cycle cost and compares it to the conventional LWR cost. It is interesting to note that yellowcake enrichment and fabrication costs are reduced by substantial factors. The cost of makeup linac power charged at the full production-cost rate and the amortization charges increase the fuel-cycle cost to the point where the LAFER fuel-cycle cost is

TABLE 5 LAFER fuel capital cost (1978 dollars escalated to 1986).

Linac cost (0.3 A, 1.5 GeV)	\$500 × 10^6
Target reactor cost (450 MWe × 1000 × \$700/kWe)	\$315 × 10^6
LAFER cost (1978 dollars)	\$815 × 10^6
Escalation and interest charges (7% and 9% repectively)	\$675 × 10^6
LAFER cost (1986 dollars)	\$1490 × 10^6
Amortization (15%)	\$223 × 10^6
Energy from three 1000-MWe reactors (75% plant factor)	19.2 × 10^9 kWhe $year^{-1}$
LAFER fuel capital cost	1.163 cents $kWhe^{-1}$

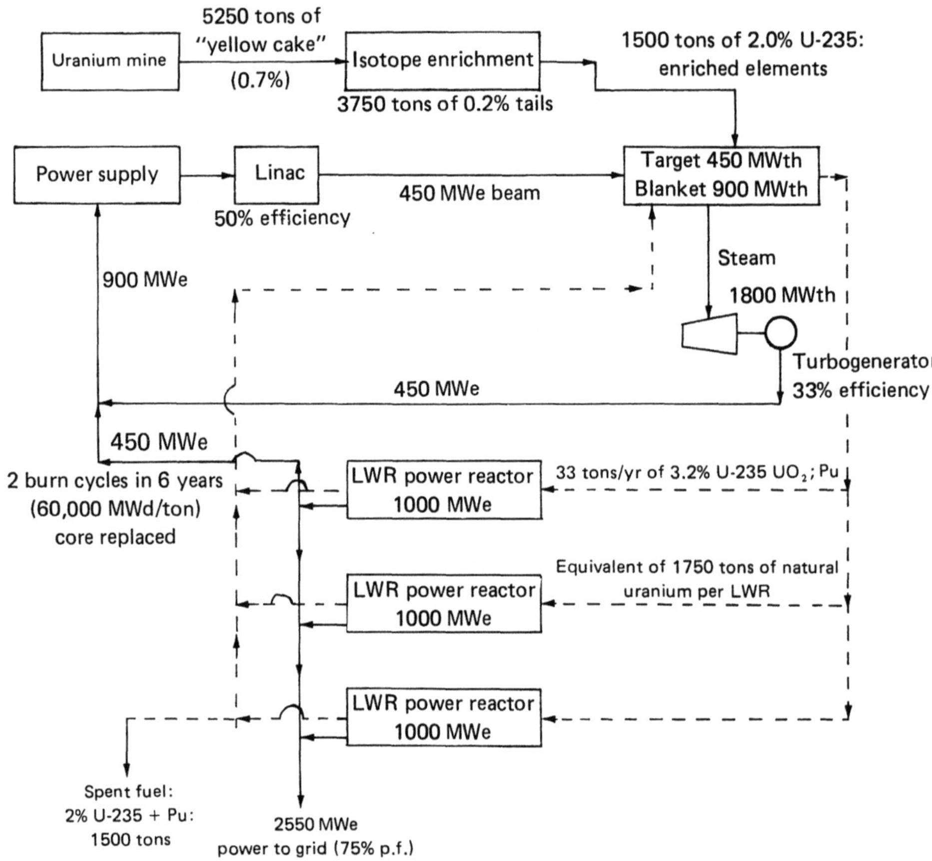

FIGURE 11 A system of three LWRs supplied by one LAFER.

twice the conventional LWR cycle cost. An optimization study should bring these values down, especially if the LAFER can be made self-sufficient in power by generating sufficient heat in the target to produce enough power to supply the linac. Even with the increased fuel-cycle cost, the total production cost for power is only 35% higher than that for the conventional fuel cycle, as shown in Table 7. This is about the same incremental estimate that is projected for the LMFBR breeder. However, this cost differential could easily disappear in the next 20 years depending on the growth of the nuclear industry and the availability of natural uranium. Moreover, this potential cost penalty has to be weighed against the enormous cost benefit of more than tripling the uranium fuel resource and reducing the amount of radioactive waste by half, all within the context of supplying the well-established and accepted conventional LWR nuclear power reactor economy.

It should be noted that the LAFER cycle is a general one in that the initial content of fissile material can vary from 0% (natural thorium) to 0.2% (depleted uranium) to 0.7% (natural uranium) to any amount of enriched fuel that makes economic sense. The amount of makeup fissile fuel needed to supply different types of convertors decreases

TABLE 6 Fuel-cycle cost in mills per kilowatt-hour of electricity generated (escalated to 1986).

Item	Unit cost[a] (dollars)	Conventional LWR[a]	Reduction factor	Two-cycle LAFER–LWR
Yellow cake	\$50 (lb U_3O_8)$^{-1}$	3.94	3.62	1.09
Conversion	\$11 kg^{-1}	0.29	3.62	0.08
Enrichment	\$100 SWU^{-1} [b]	2.90	4.33	0.67
Fabrication	\$200 kg^{-1}	2.05	2.0	1.03
Storage and carrying charge	\$400 (kg HM)$^{-1}$ [c]	3.87	2.0	1.94
Transportation	\$30 (kg HM)$^{-1}$ [c]	0.31	2.0	0.16
Amortization of LAFER (15%, 80% plant factor)		–		11.63
Electrical power (450 MWe) to LAFER at 58.3 mills kWhe^{-1}		–		9.46
Operation and maintenance		–		2.00
		13.36*		28.06

[a] Source: ANS (American Nuclear Society), US Industry Report (1976–1977).
[b] SWU, Separative Work Unit.
[c] HM, U^{238} metal.

TABLE 7 The comparative economics of the LAFER with LWRs.

Item	Conventional LWR[a]	LAFER–LWR
Capital cost		
Base capital cost (gross) (\$ kW)	600	600
Completion cost (gross) (\$ kW) (escalated for 1986 operation)	1100	1100
Resource cost		
Fuel requirement over 30 years (tons natural U)	6300	1750
Power-generation cost (average for first 10 years)		
Capital charges (15%, 70% plant factor) (mills kWhe^{-1})	26.9	26.9
Fuel (5% escalation per year) (mills kWhe^{-1})	13.4	28.1
Operation and maintenance (mills kWhe^{-1})	3.3	3.3
Total (mills kWhe^{-1})	43.6	58.3

[a] ANS, US Industry Report (1976–1977).

with increasing conversion ratios. Figure 2 indicates that the number of 1000-MWe reactors that can be supported by fuel supplied by one accelerator of 300-MW beam power increases from three for a conventional LWR with a 0.6 conversion ratio to five for an advanced reactor (0.73 conversion ratio) to 14 for a high-conversion reactor (0.9 conversion ratio).

The foregoing LAFER and LAFP are based on a conservative design using a neutron yield of 35 neutrons per 1-GeV proton with a lead target. A more recent design and analysis of an accelerator spallation reactor has been made for fuel production based on a uranium-metal target yielding approximately 100 neutrons per 1-GeV proton. The

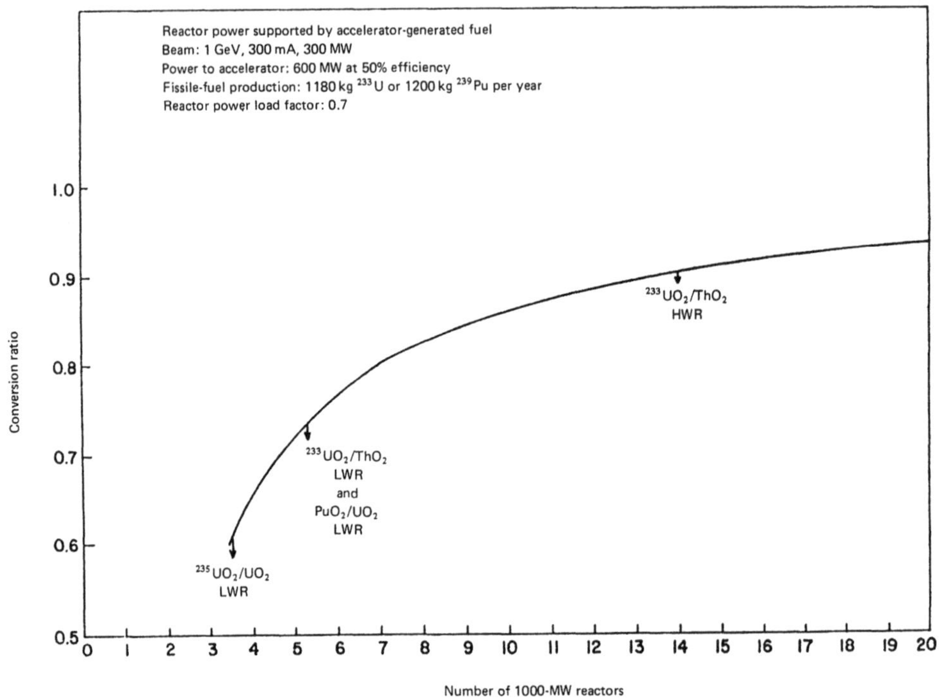

FIGURE 12 The reactor power supported by accelerator-generated fuel.

uranium-metal target is surrounded by Zircaloy-clad UO_2 fuel elements. The design characteristics are listed in Table 8. This analysis indicates that 3.7 tons of fissile material can be produced with a 2-GeV, 300-mA proton beam and that sufficient energy is developed in the target as a result of fast fission to supply the electrical power to the front end, thus making the system entirely power self-sufficient. This system could supply fuel for 12 LWRs at an economically attractive rate.

TABLE 8 Characteristics of an accelerator spallation reactor for supporting an LWR power reactor economy, for a UO_2 containing target.

Proton energy	2 GeV
Current	300 mA
Beam power	600 MW
Machine efficiency	50%
Power to accelerator	1200 MWe
Power generated in target	3600 MWth (self-sufficient)
Plant factor	80%
Pu-239 fissile-fuel production	3700 kg year^{-1}
Fuel for one 1000-MWe LWR	300 kg year^{-1}
Number of LWRs supported	12

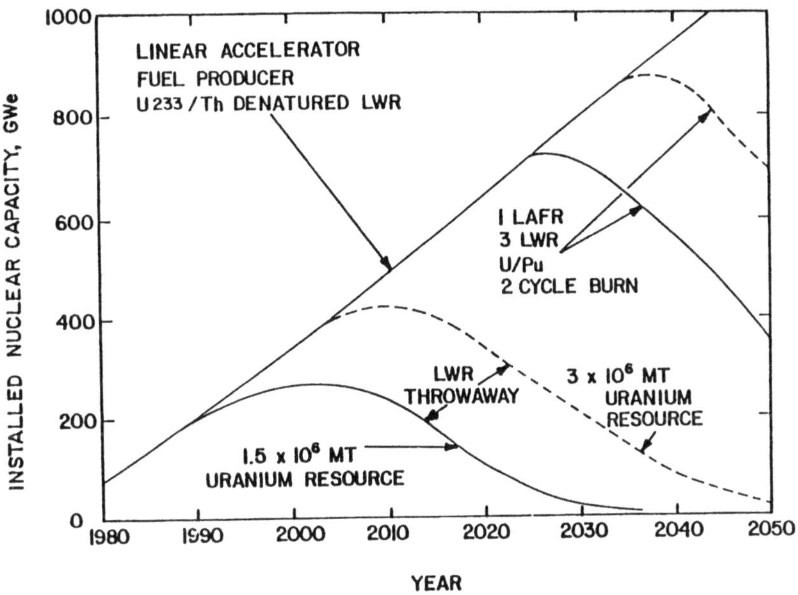

FIGURE 13 Nuclear power growth patterns: the effect of LAFER systems (small U_3O_8 supply) for a capacity growth rate of 15 GWe year^{-1}.

Figure 13 indicates how the LAFER and the LAFP significantly improve the utilization of the nuclear fuel resource while maintaining an LWR economy. The data are based on the conservative value of three LWRs per accelerator. Assuming various growth rates for nuclear power and various quantities of uranium resource, a peaking in fuel availability will occur at a higher capacity and at a later date with the accelerator fuel producer than without it. The more recent design, indicating 12 LWRs per accelerator, makes the supply picture look even better. The accelerator fuel generator essentially replaces the isotopic-enrichment plants. Allowing reprocessing of spent fuel, the accelerator fuel producer acts as a true breeder while maintaining a long-term LWR power reactor economy. The LAFP can act as a backup to the fast breeders or indeed it can be a strong competitor or even a supplier to the fast breeder. The only other near competitor is the futuristic fusion–fission hybrid which still requires scientific proof of feasibility followed by long-term technical development. Linear accelerators are already with us today in the form of research tools. They can be converted to fuel-production machines with relatively little additional development investment. Countries that desire to become self-sufficient in nuclear fuel and want to maintain the LWRs that are now universally accepted for power production need only install an accelerator-spallation-reactor fuel producer to maintain a continuous assured fuel supply. The LAFER and LAFP cycles appear to be the missing link in the long-term usage of an LWR nuclear power reactor technology.

7 TRANSMUTATION AND WASTE-MANAGEMENT CYCLES

The accelerator spallation reactor can also be used to burn up transuranic materials and to transmute fission-product wastes. Several concepts have been investigated. In the APEX concepts, after several LAFER cycles the fuel is reprocessed and partitioned. The stable low-hazard materials with short half-lives (< 2 years) are disposed of and the transuranics are returned to the nuclear fuel cycle along with the remaining fissile and fertile fuel. Selected fission products such as Sr-90 and Cs-137 can either be recycled along with nuclear fuel (APEX 1) or be transmuted separately (APEX 2). A composite schematic diagram of the APEX waste-management fuel cycles is shown in Figure 14. The Airox process indicated in the diagram involves a dry partial reprocessing by hydrogen reduction followed by oxidation of the uranium oxide fuel pellets. In this system the volatile fission products are removed and the remaining fuel can be refabricated and burned further in the LWR. In this fuel cycle the transuranics with high neutron cross sections are fissioned and the fission products with low cross sections are allowed to build up and decay within the fuel cycle so that mainly stable fission products are sent to storage. Computer calculations indicate that a fuel cycle in which an accelerator enriches fuel, burns transuranics, and transmutes fission-products can be designed so that the equilibrium concentrations of transuranics and long-lived fission products would be about the same as those anticipated in the 100,000 MWd ton^{-1} burnup cycle planned for the LMFBR fuel.

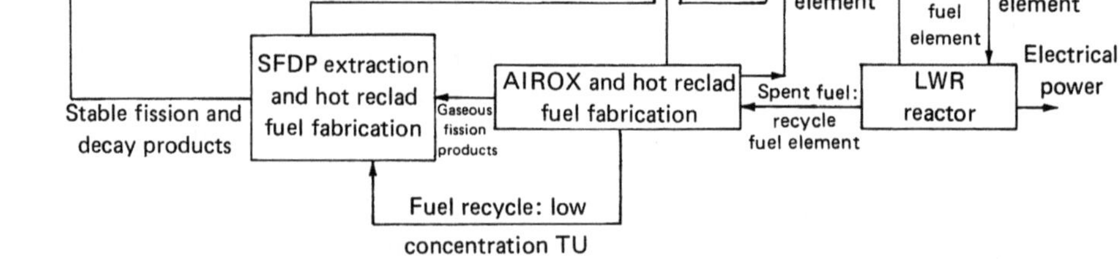

FIGURE 14 The APEX LAFER–LWR fuel cycle and waste management (spent-fuel storage): FE, fuel elements; SFDP, stable fission and decay products; TU, transuranics. Note that fuel elements can be fabricated with enriched uranium, depleted uranium, or natural uranium.

In the direct transmutation of the long-lived fission products cesium and strontium it might be necessary to separate Cs-137 isotopically to maintain a high-neutron economy. This would not be necessary for the Sr-90.

It has also been suggested that the military waste should be disposed of by incorporating the residual plutonium, other transuranics, and long-lived fission products into the growing LWR fuel cycle.

The APEX concepts depend on reprocessing the wastes with a high degree of decontamination to minimize the loss of long-lived fission products to the waste stream. Fissile material and fission products do not have to be concentrated; only losses need to be minimized. The APEX concepts then provide an alternative to the long-term geological-age disposal of long-lived radioactive nuclear waste. Additional studies are needed to evaluate thoroughly the tradeoffs and economics of each of the alternative APEX cycles to establish the optimum system.

ACKNOWLEDGMENTS

This report was prepared under Contract No. DE-AC02-76CH00016 with the US Department of Energy.

REFERENCES

Bartholomew, G.A. and Tunnicliffe, P.R. (Editors) (1966). The AECL Study for an Intense Neutron Generator. AECL 2600. Atomic Energy of Canada Limited, Chalk River, Ontario.

Boroshenkov, V.S. and Toneev, V.D. (1973). Interaction of particles and nuclei of high and ultra-high energy with nuclei. Atomnaya Energia, 35:163.

Davidson, W. (Editor) (1980). Proceedings of the International Conference on Nuclear Waste Transmutation, The University of Texas, Austin, Texas, July.

Fraser, J.S., Green, R.E., Hilborn, J.W., Milton, S.C., Gibson, W.A., Gerss, E.E., and Zucker, A. (1965). Physics in Canada, 21(2):17.

Grand, P., Kouts, H.J.C., Powell, J.R., Steinberg, M., and Takahashi, H. (1978). The conceptual design and economic analyses of a light water reactor fuel enricher–regenerator. BNL 50838. Brookhaven National Laboratory, Upton, New York.

Gregory, M.V. and Steinberg, M. (1967). A nuclear transformation system for disposal of long-lived fission product waste in an expanding nuclear power economy. BNL 11915. Brookhaven National Laboratory, Upton, New York.

Kouts, H.J.C. and Steinberg, M. (Editors) (1977). Proceedings of the Information Meeting on Accelerator Breeding. CONF 770107. Brookhaven National Laboratory, Upton, New York.

Lewis, W.B. (Editor) (1965). AEC Symposium on the Generation of Intense Neutron Fluxes, Chalk River, Ontario, April. AECL 2177. Atomic Energy of Canada Limited, Chalk River, Ontario.

Powell, J.R., Steinberg, M., Takahashi, H., and Grand, P. (1980). The APEX accelerator cycle for transmutation of long-lived fission waste. BNL 28282. Brookhaven National Laboratory, Upton, New York.

Steinberg, M. (1977). Nuclear power without nuclear bombs. New Scientist, July 7, pp. 14–16.

Steinberg, M. (1980). Keynote address at the International Conference on Nuclear Waste Transmutation. BNL 28193. Brookhaven National Laboratory, Upton, New York.

Steinberg, M., Powell, J.R., Takahashi, H., Grand, P., and Kouts, H.J.C. (1976a). The clean breeder machine: a preliminary analysis and proposal. BNL 22114. Brookhaven National Laboratory, Upton, New York.

Steinberg, M., Powell, J.R., Takahashi, H., Grand, P., and Kouts, H.J.C. (1976b). Linear accelerator breeder: a preliminary analysis and proposal. BNL 50592. Brookhaven National Laboratory, Upton, New York.

Steinberg, M., Powell, J.R., Takahashi, H., Grand, P., and Kouts, H.J.C. (1978). Electronuclear fissile fuel production. Atomkernenergie, 32(1):39–48. Also BNL 24356 (March 1978), Brookhaven National Laboratory, Upton, New York.

Steinberg, M., Powell, J.R., Takahashi, H., Grand, P., and Kouts, H.J.C. (1980). The linear accelerator fuel enricher (LAFER) and fusion product transmuter (APEX). Atomkernenergie, 36(1):42–46. Also BNL 26951 (April 1980), Brookhaven National Laboratory, Upton, New York.

Takahashi, H. and Mizzo, N. (1980). Use of linear accelerator for incinerating the fission products of ^{137}Cs and ^{90}Sr. BNL 28779. Brookhaven National Laboratory, Upton, New York.

US Department of Energy (1977). Nuclear proliferation and civilian nuclear power. Report of the Nonproliferation Alternative System Assessment Program. Report No., DOE/NE 0001. US Department of Energy, Washington, D.C.

Vasilkov, R., Goldinski, V.I., Pimonov, B.A., Chistyokov, Yu., and Chistyokov, N. (1978). Atomnaya Energia, 44(4):329–335.

ACCELERATOR BREEDER APPLICATIONS

J.S. Fraser
Accelerator Physics Branch, Atomic Energy of Canada Limited Research Company, Chalk River Nuclear Laboratories, Chalk River, Ontario K0J 1J0 (Canada)

ABSTRACT

Since the earliest proposals for the accelerator-based breeding of fissile material over thirty years ago, impressive developments in accelerator physics and technology have greatly enhanced the prospects for the technical feasibility of the accelerator breeder. Continuing investigations of the economic outlook for the accelerator breeder reaffirm that it will have an economic role to play in the thermal fission power field when the cost of uranium in constant dollars rises above the current level by a factor of about three. In spite of the low present cost of uranium, the probability of a cost rise of that magnitude is high if potential alternative sources of electrical energy such as the fast breeder reactor, fusion, and photovoltaics prove to be too costly for large-scale adoption. It would appear timely to begin a modest development program directed towards the practical demonstration of key components of the accelerator breeder. The starting point clearly has to be the injector and initial acceleration of the total beam current that is forseen. To this end, a design study for a 300-mA, 10-MeV proton accelerator has been initiated at the Chalk River Nuclear Laboratories.

1 INTRODUCTION

Intense sources of fast neutrons have been proposed for a variety of applications in the nuclear power industry. These applications range from the electronuclear breeding of fissile material for use in fission reactors (Lewis, 1952; Livermore Research Laboratory, 1954), as inventory for Fast Breeder Reactors (FBRs) (Schriber et al., 1977), and for reenrichment of used reactor fuel (Steinberg et al., 1979), to tritium production (Nucleonics Week, 1981) and even to the burning of long-lived fission products (Steinberg et al., 1979). Two types of neutron sources have been proposed: a high-energy proton accelerator producing neutrons by the spallation reaction in a heavy-element target and a fusion reactor whose neutrons are used in a surrounding blanket to breed fissile material in addition to the tritium required to sustain operation. This paper will discuss the accelerator-based neutron source in terms of the economic outlook for its use as a fissile-fuel breeder and in terms of a possible staged development leading to operating accelerator and target systems and culminating in a full-scale plant. The lead time required to develop and demonstrate an electronuclear breeding system is such that there is still time to carry out the development before it is needed.

Studies of the economic outlook for an accelerator breeder have shown that the production capacity of fissile material would have to be 2.5 kg per day or more (Tunnicliffe et al., 1976). To put the accelerator breeder in perspective, a production rate of 2.5 kg per day would require a proton beam power of the order of 300 MW and a target/blanket assembly with a total power production of the order of 1500 MWth.

2 DESCRIPTION OF THE FACILITY

The main components of an accelerator breeder as currently envisaged are shown in Figure 1. An injector consisting of an ion source and 50-kV extraction column delivers a beam to the initial accelerator structure called a "Radio-Frequency Quadrupole" (RFQ). This structure captures and accelerates 300 mA of protons with high efficiency. At an energy of 2 MeV a conventional drift-tube structure takes over to boost the beam energy to 150 MeV. From 150 MeV to 1000 MeV a coupled-cavity linear accelerator structure is used. The choice of structure at each of the four stages in the accelerator is dictated by its suitability and efficiency in the energy range concerned. Finally, the 300-MW beam is delivered to a liquid metal or molten salt target containing a high proportion of heavy element. The spallation reaction which then occurs in the target yields a large number of energetic neutrons which in turn are multiplied by the fission reaction in actinide elements or are captured to produce the desired fissile product. The power multiplication that results from the fission induced by the fast neutrons and charged particles can be as large as a factor of five or more with an enrichment factor in the blanket of 4–5%. With the addition of a turbogenerator system this energy is converted to electric power to supply approximately enough power to drive the accelerator.

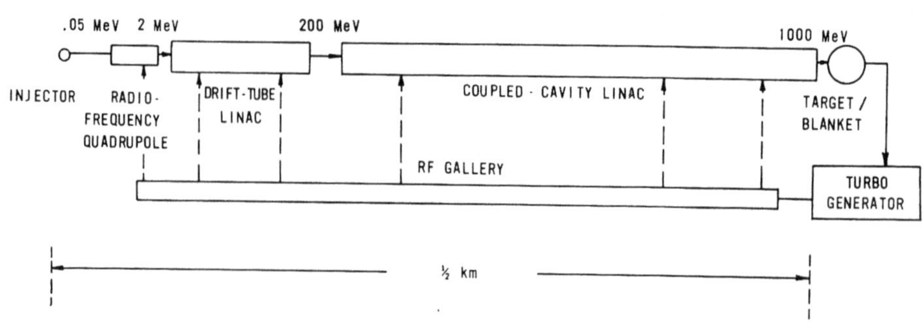

FIGURE 1 A schematic diagram of the main components of an accelerator breeder.

3 ELECTRONUCLEAR BREEDING OF FISSILE MATERIAL

The conversion or breeding of ^{233}U from ^{232}Th is technically possible in thermal reactors in which the neutron economy is excellent. Both ^{233}U and ^{239}Pu could be bred in a thermal neutron flux if the fertile material were irradiated for a time that is short compared with the growth time of the fissile product and were then withdrawn to allow the fissile isotopes to grow in. A more attractive alternative is to irradiate the fertile material in an undermoderated fast neutron flux that is at least as hard as that in the FBR. Some fast-neutrón-induced fission of the fissile product occurs, but at a much lower rate than would be the case in a thermal reactor. The D–T reaction in a fusion reactor and the spallation reaction in an accelerator target are sources that can produce in practice a harder spectrum than that which exists in an FBR. The use of either a fusion reactor or an accelerator to provide an intense flux of fast neutrons for fertile to fissile conversion is called "electronuclear breeding". The distinctively subcritical nature of electrical breeding systems allows their designers a degree of freedom not available to the fission-reactor designer.

Interest in the possible application of the spallation reaction to fissile-fuel breeding began soon after its discovery in the late 1940s. Goeckerman and Perlman (1948) observed that the fission of bismuth under bombardment with 190-MeV deuterons was preceded by the evaporation of 12 neutrons. Later the same year a similar phenomenon was found when uranium was bombarded by 380-MeV α particles (O'Connor and Seaborg, 1948). The copious accelerator-based neutron production from the spallation reaction began to be exploited in the MTA program (Livermore Research Laboratory, 1954) with the objective of producing ^{239}Pu and ^{233}U in commercially useful quantities. That project was terminated in 1952 not because it was unpromising but because of the discovery of high-grade uranium ores in Colorado. At about the same time Lewis (1952), at the Chalk River Nuclear Laboratories, recognized the significance of the spallation reaction especially for the use of ^{232}Th in the CANDU reactor. This recognition led to two series of experiments (Bell and Skarsgard, 1956; Bercovitch et al., 1960) on the neutron yield from heavy elements bombarded by intermediate-energy protons. Later, during the Chalk River Intense Neutron Generator Study (Bartholomew and Tunnicliffe, 1966; Church, 1967), experimental data on neutron yields were extended to targets thick enough to stop a 1-GeV proton beam.

Although the Intense Neutron Generator project was not funded, the nucleus of the accelerator physics and engineering group was retained at Chalk River to continue development of the technology required for the accelerator breeder. This work on the technical aspects of high-power accelerators continues and the progress realized and the prospects for future work are outlined in Section 5.

4 ECONOMIC OUTLOOK AND DEVELOPMENT STRATEGY

An assessment of the economic feasibility of the accelerator breeder must be carried out with a specific model of the nuclear industry in mind. In Canada the appropriate model is based on the use of the CANDU thermal reactor operating on a near-breeding ^{232}Th–^{233}U fuel cycle. The total unit energy cost can be mimimized in this system if the conversion ratio is reduced to about 0.9 and the fuel burnup is increased to about three times that for a self-sustaining fuel cycle (Critoph et al., 1976). The role for the accelerator breeder, then, is to provide externally produced fissile material to allow less frequent reprocessing. One way of calculating the economic feasibility of electrically breeding fissile material for the CANDU system is to estimate the capital cost allowance made possible by the use of externally supplied fissile material. Such a calculation indicates (Hoffmann, personal communication, 1981) that the allowed capital cost for an accelerator breeder is such that it should be economic to build one only if the price of ^{235}U were at least US\$150 g^{-1} in 1981 US dollars. This is about three to four times the current price.

A corresponding analysis of a fusion–fission hybrid breeder (Hoffmann, personal communication, 1981) yields a similar conclusion, namely that the allowed capital cost is currently too low by a factor of about two to three. A possibly significant difference in the two types of breeders, however, is in the minimum magnitude of the investment required for them to be economically viable. An accelerator breeder of the minimum economic size might require about half the investment needed for a viable fusion breeder. For the accelerator breeder the capital cost is approximately US\$(0.8–1.3) $\times$ 10^9 for a fissile production capacity of 2.5 kg per day whereas double this production capacity from a fusion hybrid would require a capital outlay of the order of US\$(1.7–2.5) $\times$ 10^9.

The economic outlook for electronuclear breeding of fissile material thus depends primarily on the fueling costs of thermal reactors. At the present time the nuclear power industry could absorb a large increase in the price of fissile material. Beyond that, however, further price increases engendered by the scarcity of ^{235}U might be limited only by the cost of electrical energy from competitive sources such as the FBR or photovoltaics. According to Weinberg (1981), in that eventuality the cost of fissile material might rise by an order of magnitude above the current price. Clearly, electronuclear breeding, especially with the accelerator breeder for reasons to be discussed in Section 5, will offer the means of supplying fissile material at a cost that will tend to reduce the rate of increase in the costs of electrical energy.

What, then, is an appropriate strategy for the development and demonstration of the accelerator breeder? Although the technical feasibility of the accelerator system itself now rests on a secure basis of proven technology there remain a number of crucial factors that must be demonstrated under realistic conditions. The technical feasibility and engineering practicality of the high-power target/blanket has not yet been tackled seriously, although it will very likely resemble a subcritical liquid metal or molten salt FBR core. In an accelerator breeder target the power density is more concentrated near the beam impact zone than the approximately sinusoidal distribution in a fast reactor. Moreover, the production rate of fissile material is greater near the beam impact zone, augmenting the concentration of heat production by fission as the enrichment grows. Preliminary designs that address these twin problems have been discussed in the literature (Grand, 1979; Furukawa et al., 1981).

A staged development of accelerator technology and target/blanket engineering would probably require 20 years to achieve the level of a full-scale demonstration. The necessary preliminary steps being planned at Chalk River to demonstrate the technical feasibility of launching and initially accelerating a high-current proton beam are described in the next section.

A thorough review (Dreyfuss et al., 1978) of alternative nuclear breeding methods has been published in the United States. It compared external-source breeders (accelerator-based and fusion-reactor-based breeders) with internal-source breeders (Liquid-Metal (LM) FBRs) of fissile material to be consumed largely in light-water reactors. Its conclusions were, briefly, that "a commitment to breeder development should be broadly based and should include adequate investment in external-neutron-source breeders, as well as in reactor breeders, such as the LMFBR" and that, whereas the hybrid fusion breeder technology awaits the achievement of significant D–T burning, accelerator breeder technology should be readily demonstrated with a major engineering effort.

5 TECHNICAL FEASIBILITY DEMONSTRATION

The foundations of the necessary accelerator technology have been laid in the design, construction, and successful operation of the Los Alamos Meson Physics Facility (Hagerman, 1979). That concept (800 MeV, 20 mA peak, 12% duty factor) provides the basis for designs of linear accelerators with a 100% duty factor and very high current that would be required for breeding. Although the feasibility of accelerating a peak current of over 200 mA of protons was demonstrated many years ago in the CERN 3-MeV experimental pulsed linac (Warner, 1972), the acceleration of such large currents in a continuous mode is a largely engineering project that remains to be carried out. The radio-frequency (RF) power tubes of megawatt capacity of the type that would be required have been built and efficiencies of 75% have been achieved (Lien, 1970; Giebeler, 1969).

The technical feasibility of accelerator breeding can be demonstrated in a logically staged progression with the outlay in hardware and effort increasing as the several steps are completed successfully. The first step is the formation and initial acceleration of a high-current proton beam. The first element in this step is an ion source capable of delivering a high-quality beam of protons at a current level of well over the required 300 mA, say 400 mA. At Chalk River an experimental program has been underway for a number of years with this objective. The source is a multiaperture duoPIGatron-type device which has delivered over 500 mA of unanalyzed hydrogen ions reliably. The development program continues with good prospects of achieving at least 300 mA of analyzed proton beam.

An integral part of the ion source is a low-voltage extraction column which forms and accelerates the beam to about 50 keV. A high degree of reliability has been demonstrated for Direct-Current (DC) columns in this voltage range. An injector test experiment (Figure 2) has been mounted at Chalk River in which thorough tests of the injector required for an accelerator breeder will be carried out in the near future.

As shown in Figure 1, the stage following the injector is an RFQ. This is a radically new type of accelerating structure that was invented in the Soviet Union (Kapchinskii and Teplyakov, 1970) and subsequently developed more fully in the United States (Potter et al., 1979). This device is a breakthrough in the field of high-intensity continuous-current linear accelerators because it allows us to bypass a difficult problem in high-voltage (up to 750 kV) DC injectors which are prone to electrical breakdown with ion beam currents of the magnitude required for the accelerator breeder. A typical RFQ is shown schematically in Figure 3. A detailed description of the structure can be found in

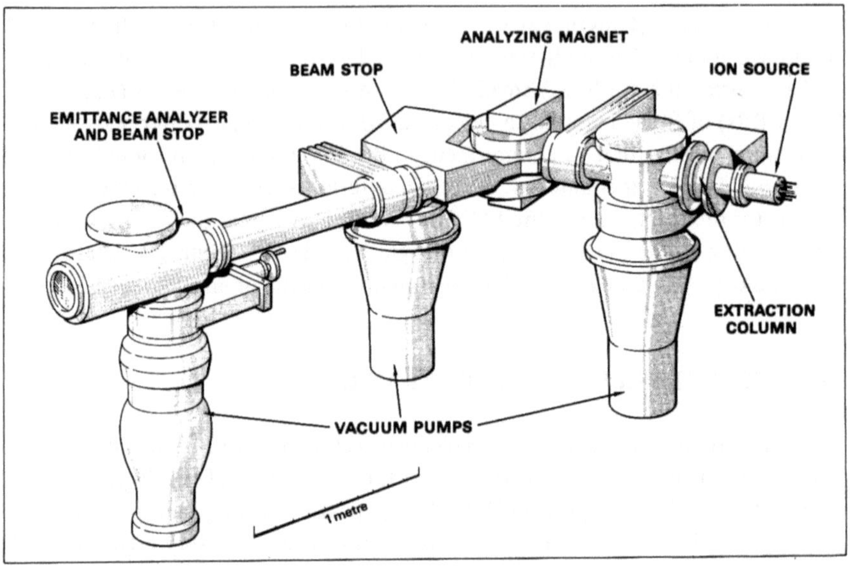

FIGURE 2 The Chalk River Nuclear Laboratories injector test experiment for experiments on ion sources and beam-formation columns for the accelerator breeder.

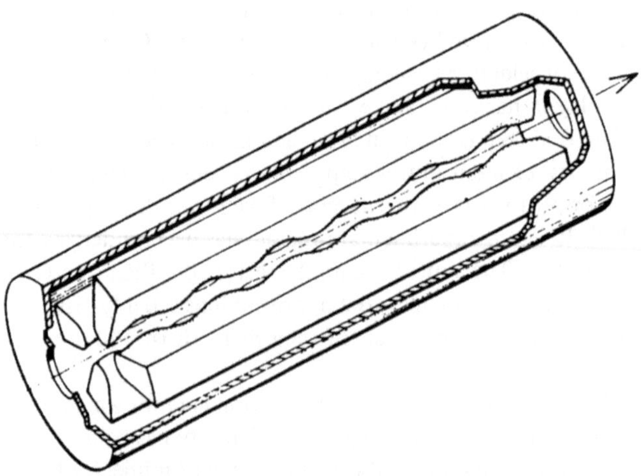

FIGURE 3 A cutaway view of an RFQ accelerating structure (reprinted with permission from Crandall et al., 1979). The RFQ would be the low-energy accelerating structure in an accelerator breeder.

Potter et al. (1979) and a theoretical treatment in Crandall et al. (1979). The very valuable feature of the RFQ that makes it of interest here is its capacity to accept a relatively low-energy (50 keV) beam, focus it, accelerate it, and bunch it with very low losses. At the same time it permits the use of a more efficient version of the drift-tube linac which follows the RFQ.

The drift-tube linac and coupled-cavity linac structures shown in Figure 1 have been the subject of study at Chalk River for several years. A drift-tube linac for operation at 100% duty factor (Figure 4) has been built and tests have begun with modest beam currents. A coupled-cavity test accelerator for electrons (Figure 5) has been operated successfully at Chalk River for a number of years in the continuous mode. Although these structures are well developed for the conventional pulsed mode of operation which is typical of research accelerators, the continuous mode of operation, which will be an essential feature of accelerators used in the accelerator breeder, requires the development

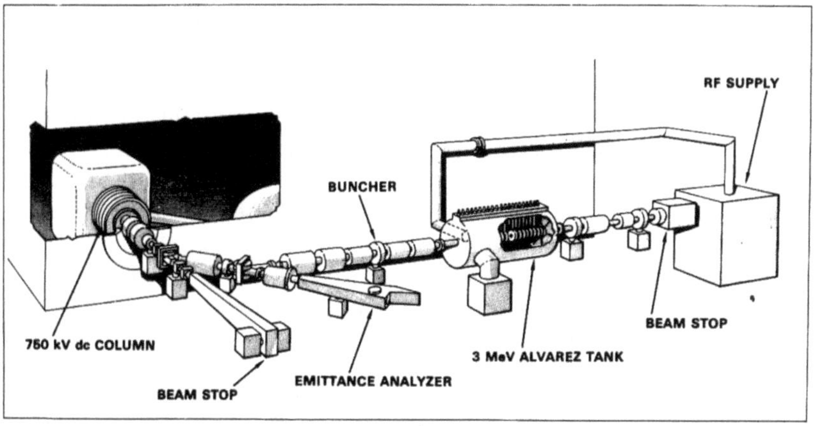

FIGURE 4 The Chalk River Nuclear Laboratories high-current test facility for experiments on the acceleration of intense proton beams at 100% duty factor in a drift-tube linac typical of the intermediate-energy section of an accelerator breeder.

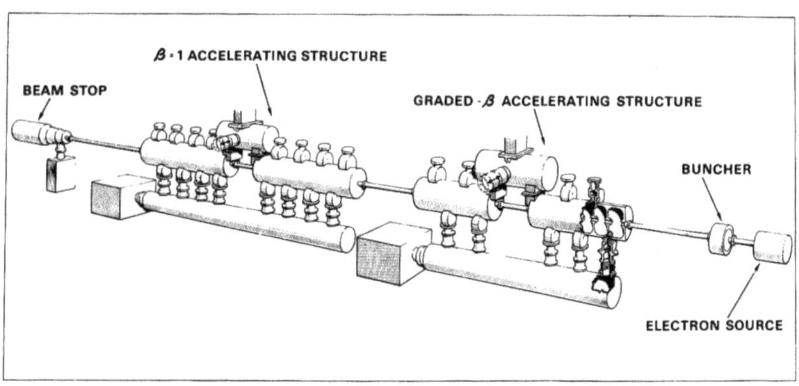

FIGURE 5 The Chalk River Nuclear Laboratories electron test accelerator for experiments on the acceleration of intense beams with a large accelerating gradient at 100% duty factor in a coupled-cavity linac typical of the high-energy section of an accelerator breeder.

of techniques for cooling, for control of the flow of large amounts of RF power, for control of multitank systems, and for beam handling.

A coherent program to tackle some of these problems has been proposed at Chalk River. Because it is anticipated that the important energy range in which operation with high beam current has to be demonstrated is below 10 MeV, a test accelerator has been proposed to deliver a full-current beam of 300 mA at 10 MeV. Since this energy is a very small fraction of the nominal output of an accelerator breeder (1000 MeV), the convenient acronym ZEBRA for Zero-Energy BReeder Accelerator has been adopted. ZEBRA would in fact consist of the first three components of Figure 1 but with the drift-tube linac truncated at 10 MeV instead of 200 MeV. This proposal has the very important advantage that it is possible to test, at full scale, all of the components in the low-energy section of the accelerator breeder for a relatively modest expenditure.

Beyond the 10-MeV energy, one can envisage further development stages which could be mounted. For example, a 200-MeV facility with 70 mA and a liquid lead target would provide an intense fast neutron source for materials damage studies of interest for the accelerator breeder project, and, with a moderator, a neutron source for a broad program of condensed-matter research. Slow neutron fluxes of the order of 10^{15} cm^{-2} s^{-1} would be achievable. Such a facility would also provide a test bed for the engineering development of high-power liquid-metal target systems.

At a still later stage, with the addition of coupled-cavity structures and their associated RF power supplies, the energy would be raised to the nominal 1000 MeV of the accelerator breeder, with the current initially restricted to 70 mA. This would permit initial experimentation with fissile-fuel breeding in a realistic environment. Such a large facility might be located on the site of a large nuclear power station. The prototype accelerator breeder would initially present a load to the power grid of about 200 MW. With a power gain of between two and five, which is typical of what is expected with a uranium blanket with enrichments up to 4%, the substantial amount of energy produced in the target assembly could profitably be recovered as high-quality steam and could thereby largely cover the operating cost of the device.

To date, considerations of the target and blanket problems have been fragmentary. The approaches have been on two very different levels. On the one hand, several groups of workers have studied simple target systems both experimentally and theoretically (Fraser et al., 1980; Russell et al., 1980; Alsmiller et al., 1981). With the aim of establishing the validity of Monte Carlo codes such as NMTC (Coleman and Armstrong, 1970), HETC (Chandler and Armstrong, 1972), and MORSE (Straker et al., 1970) as predictors of the neutron production and fertile-to-fissile conversion rates, experiments have been carried out on depleted uranium and thorium metal targets. The targets were thick enough to stop the proton beams of 480 and 800 MeV. The overall agreement of the measurements with the predictions is fairly good but in some experiments the observed neutron production exceeded the predictions by over 30% (Fraser et al., 1980).

On the other hand, some theoretical studies have been carried out on full-scale target/blanket conceptual designs which are derivatives of various fast-reactor types. In these designs the high-power proton beam either impinges on a liquid-metal central plug in an LMFBR core (Mynatt et al., 1977), directly on a molten salt (Mynatt et al., 1977; Furukawa et al., 1981), or directly onto fuel pins in a gas-cooled fast-reactor variant (Mynatt et al., 1977).

The heat generated by the beam and, more importantly, by the fast-neutron-induced fission is much more concentrated than is the case in a reactor. A number of target/blanket designs, which are not so closely linked to existing fast-reactor cores, have been produced to alleviate this serious problem. Multiple liquid Pb–Bi jets have been suggested (Grand, 1979) to distribute the neutron source and hence the power density in the surrounding blanket.

The enhancement of the fissile-fuel production rate close to the neutron source, a feature common to all accelerator breeder designs, means that a fuel-management scheme would be required to keep the power density under control (Burns et al., 1979). With a more diffuse primary spallation neutron source and a small resulting loss in the fissile-fuel production rate, power densities can be brought down to the levels found in the LMFBR.

Clearly the development of target/blanket designs needs a considerable amount of effort to balance the progress that has already been made in accelerator technology. It seems likely that liquid-metal loop technology will have to be developed (with materials such as Pb–Bi and lithium) that is beyond the present experience in fast reactors. Furthermore, the materials problems, especially in the beam impact zone, are somewhat more severe than those in fast reactors (Horak, 1977).

6 SUMMARY AND CONCLUSIONS

Since the earliest proposals for the accelerator-based breeding of fissile material over 30 years ago, developments in accelerator physics and technology have greatly enhanced the prospects for the technical feasibility of the accelerator breeder. Continuing investigations of the economic outlook for the accelerator breeder reaffirm that it will have an economic role to play in the field of thermal fission power when the cost of uranium in constant dollars rises above the current level by a factor of about three. In spite of the present low cost of uranium the probability of a cost rise of that magnitude is high if potential alternative sources of electrical energy such as the FBR, fusion, and photovoltaics prove to be costly.

It would appear to be timely to begin a modest development program directed towards the practical demonstration of key components of the accelerator breeder. The starting point clearly has to be the injector and initial acceleration of the total beam current that is forseen. To this end a design study for a 300-mA, 10-MeV proton accelerator has been initiated at the Chalk River Nuclear Laboratories.

REFERENCES

Alsmiller, F.S., Alsmiller, R.G., Jr., Gabriel, T.A., Lillie, R.A., and Barish, J. (1981). A phenomenological model for particle production from the collisions of nucleons and pions with fissile elements at medium energies. Report ORNL/TM-7528. Oak Ridge National Laboratory, Oak Ridge, Tennessee.

Bartholomew, G.A. and Tunnicliffe, P.R. (Editors) (1966). The AECL Study for an Intense Neutron Generator. Report AECL-2600. Atomic Energy of Canada Limited, Chalk River, Ontario.

Bell, R.E. and Skarsgard, H.M. (1956). Cross sections of (p, xn) reactions in the isotopes of lead and bismuth. Canadian Journal of Physics, 34:745.

Bercovitch, M., Carmichael, H., Hanna, G.C., and Hincks, E.P. (1960). Yield of neutrons per interaction in U, Pb, W, and Sn by protons of six energies between 250 and 900 MeV selected from cosmic radiation. Physical Review, 119:412.

Burns, T.J., Bartine, D.E., and Renier, J.P. (1979). Concept evaluation of a nuclear design for electronuclear fuel production: evaluation of ORNL's proposed TMF–ENFP. Report ORNL/TM-6828. Oak Ridge National Laboratory, Oak Ridge, Tennessee.

Chandler, K.C. and Armstrong, T.W. (1972). Operating instructions for the high energy nucleon meson transport code HETC. Report ORNL-4744. Oak Ridge National Laboratory, Oak Ridge, Tennessee.

Church, T.G. (Editor) (1967) ING Status Report July 1967. Report AECL-2750. Atomic Energy of Canada Limited, Chalk River, Ontario.

Coleman, W.A. and Armstrong, T.W. (1970). The nucleon–meson transport code NMTC. Report ORNL-4606. Oak Ridge National Laboratory, Oak Ridge, Tennessee.

Crandall, K.R., Stokes, R.H., and Wangler, T.P. (1979). RF quadrupole beam dynamics design studies. In Proceedings of the 1979 Linear Accelerator Conference. BNL-51134. Brookhaven National Laboratory, Upton, New York, p. 205.

Critoph, E., Bannerjee, S., Barclay, F.W., Hamel, D., Milgram, M.S., and Veeder, J.I. (1976). Prospects of self-sufficient equilibrium thorium cycles in CANDU reactors. Report AECL-5501. Atomic Energy of Canada Limited, Chalk River, Ontario.

Dreyfuss, D.J., Augenstein, B.W., Mooz, W.E., and Sher, P.A. (1978). An examination of alternative nuclear breeding methods. Report R-2267-DOE. US Department of Energy, Washington, D.C.

Fraser, J.S., Garvey, P.M., Milton, J.C.D., Kiely, F.M., Thorson, I.M., and Pate, B.D. (1980). Neutron production in thick targets of lead, thorium and uranium bombarded by 480 MeV protons. In Proceedings of the Symposium on Neutron Cross-Sections from 10 to 50 MeV. BNL-NCS-51245. Brookhaven National Laboratory, Upton, New York, p. 155.

Furukawa, K., Nakahara, Y., and Tsukada, K. (1981). Single-fluid-type accelerator molten-salt breeder concept. Journal of Nuclear Science and Technology, 18:79.

Giebeler, R.H. (1969). High power microwave generation. Journal of Microwave Power, 4:79.

Goeckerman, R.H. and Perlman, I. (1948). Characteristics of bismuth fission and high energy particles. Physical Review, 73:1127.

Grand, P. (1979). Accelerator breeding, or the use of high energy accelerators in the nuclear fuel cycle. Nature, 278:693.

Hagerman, D.C. (1979). LAMPF operations at 500 μA. In Proceedings of the 1979 Linear Accelerator Conference. BNL-51134. Brookhaven National Laboratory, Upton, New York, p. 78.

Horak, J. (1977). Materials technology for accelerator production of fissile isotopes. In the Proceedings of the Information Meeting on Accelerator Breeding. CONF-770107. Brookhaven National Laboratory, Upton, New York, p. 232.

Kapchinskii, I.M. and Teplyakov, V.A. (1970). Linear ion accelerator with spatially homogeneous strong focusing. Pribou i Tekhnika Eksperimenta, 119(2):19.

Lewis, W.B. (1952). The significance of the yield of neutrons from heavy nuclei excited to high energies. Report AECL-968. Atomic Energy of Canada Limited, Chalk River, Ontario.

Lien, E.L. (1970). High efficiency klystron amplifiers. In Proceedings of the International Conference on Microwave and Optical Generation and Amplification, 8th, Amsterdam, September 1970. Kluwer, Deventer, The Netherlands, p. 11.

Livermore Research Laboratory (1954). Status of the MTA (Materials Testing Accelerator) process. Report LRL-102. Livermore Research Laboratory, Livermore, California.

Mynatt, F.R., Alsmiller, R.G., Jr., Barish, J., Gabriel, T.A., Bartine, D.E., Burns, T.J., Martin, J.A., Saltmarsh, M.J., and Bettis, E.S. (1977). Preliminary report on the promise of Accelerator Breeding and Converter Reactor Symbiosis (ABACS) as an alternative energy system. Report ORNL/TM-5750. Oak Ridge National Laboratory, Oak Ridge, Tennessee.

Nucleonics Week (1981). Nucleonics Week, 22 (8, February 26).

O'Connor, P.R. and Seaborg, G.T. (1948). High energy spallation and fission products of uranium. Physical Review, 74:1189.

Potter, J.M., Williams, S.W., Humphry, F.J., and Rodenz, G.W. (1979). Radio frequency quadrupole accelerating structure research at Los Alamos. IEEE Transactions on Nuclear Science, 26(3):3745.

Russell, G.J., Gilmore, J.S., Proel, R.E., Robinson, H., and Howe, S.D. (1980). Spallation target–moderator–reflector studies at the Weapons Research Facility. In Proceedings of the Symposium on Neutron Cross-Sections from 10 to 50 MeV. BNL-NCS-51245. Brookhaven National Laboratory, Upton, New York, p. 169.

Schriber, S.O., Fraser, J.S., and Tunnicliffe, P.R. (1977). Future of high intensity accelerators in nuclear energy. In Proceedings of the International Conference on High Energy Accelerators, Xth. CERN, Geneva, Vol. 2, p. 408.

Steinberg, M., Powell, J.R., Takahashi, H., Grand, P., and Kouts, H.J.C. (1979). The linear accelerator fuel enricher–regenerator (LAFER) and fission product transmuter (APEX). IEEE Transactions on Nuclear Science 26 (3):3002.

Straker, E.A., Stevens, P.N., Irving, D.C., and Cain, V.R. (1970). The MORSE code – A multigroup neutron and gamma ray Monte Carlo transport code. Report ORNL-4585. Oak Ridge National Laboratory, Oak Ridge, Tennessee.

Tunnicliffe, P.R., Chidley, B.G., and Fraser, J.S. (1976). High current proton linear accelerator and nuclear power. In Proceedings of the 1976 Proton Linear Accelerator Conference. Report AECL-5677. Atomic Energy of Canada Limited, Chalk River, Ontario.

Warner, D.J. (1972). Accelerator research and development with the CERN 3-MeV linac. In Proceedings of the 1972 Proton Linear Accelerator Conference. Report LA-5115. Los Alamos Scientific Laboratory, Los Alamos, California.

Weinberg, A.M. (1981). The future of nuclear energy. Physics Today, 34(3):48.

THE FRG PROJECT FOR A HIGH-POWER SPALLATION NEUTRON SOURCE FOR FUNDAMENTAL RESEARCH

G.S. Bauer
Institut für Festkörperforschung der Kernforschungsanlage Jülich GmbH, D-5170 Jülich, Postfach 1913 (FRG)

J.E. Vetter
Institut für Kernphysik des Kernforschungszentrums Karlsruhe GmbH, D-7500 Karlsruhe, Postfach 3640 (FRG)

ABSTRACT

A concept for a new neutron source for fundamental research has been developed in the FRG, taking advantage of the low heat dissipation per neutron of the spallation process and the fact that almost any required kind of time structure can be imposed on the proton beam driving the reaction. In its first stage the source is designed to reach a time-average thermal neutron flux of 7×10^{14} cm^{-2} s^{-1} and a peak flux of 1.3×10^{16} cm^{-2} s^{-1} at 100-Hz repetition rate, which would make it superior to all existing neutron sources for beam hole research. The proposal includes a 1.1-GeV, 5-mA time-average linear accelerator consisting of an Alvarez-type low-energy part and a disk-and-washer (daw) high-energy accelerating structure which operates at a peak current of 100 mA and at RF frequencies of 108 (Alvarez) and 324 MHz (daw). The target material is proposed to be lead initially. It will be arranged on a 2.5-m diameter wheel rotating at 0.5 rps to reduce the heat load and radiation damage to the target. The proton beam will impinge on the circumference of the wheel at right angles to its axis. Provisions are proposed for neutron scattering (thermal and cold neutron source) irradiation facilities, neutrino physics, nuclear physics, and meson research. With its time-average beam power of 5.5 MW and peak proton current of 100 mA the facility would also be an important step in the development of the very high power machines needed for future applications in nuclear technology. Options for future extensions include: an increase in proton beam power up to 11 MW time average, the use of uranium as target material, the addition of a proton pulse compressor to provide pulses of 0.7 μs duration with a peak current of 66 A, and the construction of a second target designed for pulsed neutron work mainly in the epithermal energy range.

1 INTRODUCTION

"A world of neutron abundance" – what a prospect this is for those struggling for stepwise improvements in the performance of the neutron sources presently available! One of the most intriguing features of the neutron's role in modern science and technology is the large variety of its applications. In a meeting which emphasizes the extraordinary versatility of this particle in future energy concepts it seems appropriate to recall also its role in fundamental research which has the prospect of becoming even more important in the future. The availability of high-performance neutron sources is an essential condition for further progress in this field. In this context, a high-power spallation neutron source has been proposed in the Federal Republic of Germany (FRG). Such a facility, although more than an order of magnitude below the power levels discussed for technical applications, is to be considered as an important step in the development of these plants.

2 THE ROLE AND AVAILABILITY OF NEUTRONS IN CONDENSED-MATTER RESEARCH

Through scattering experiments, using thermal and slow neutrons as a probe, a large body of scientific knowledge has been accumulated which has affected our technological progress and will continue to affect it as requirements on newly developed materials become more and more stringent.

The neutron shares its role as a probe for the investigation of matter with other types of radiation such as X rays, visible, ultraviolet, and infrared radiation, γ rays, electrons, and even mesons and ions. However, none of these has as wide a variety of applications as the neutron. Without going into details, the following list of some milestones on the neutron's progress to its present role in the science of condensed matter may serve as an illustration: crystal structure, magnetic structure, lattice dynamics, magnetic excitation, the structure and dynamics of liquids, the dynamics of quantum fluids, the dynamics of molecules and molecular crystals, phase transitions, metal–hydrogen systems, diffusion in solids, type-II superconductors, polymer physics, disordered structures, protein crystallography, biological structures. Undoubtedly this list of topics will continue to grow in the future.

It is not only the variety of physical problems that can be studied with neutron scattering which makes the neutron an indispensable tool for modern science, but also the quality and uniqueness of the information that can be obtained. This is due to the particular combination of important properties of the neutron, a combination which is not found with any other particle. These properties include the following:

- The neutron has no charge. Therefore it is able to penetrate into thick samples without being affected by electrons.
- The neutron interacts with atomic nuclei. The strength of the interaction (scattering) depends not only on the kind of target atom but also on its isotopic mass. By substituting isotopes in a defined way, a useful contrast variation can be achieved. It is very important, and may become even more important in the future, that hydrogen, which is hardly detectable with X rays, has a rather large scattering length for neutrons.
- The wavelength of thermal neutrons is of the order of the atomic distances in condensed matter. This makes thermal neutrons particularly well suited for structure determinations.
- The kinetic energy of thermal neutrons is comparable to the oscillation energy

of the atoms. This makes the dynamics of condensed matter accessible to investigation by neutron scattering as well.

- The neutron has a magnetic moment. Therefore it interacts with the magnetic moments of electrons and nuclei. The structure and dynamics of magnetic materials are important fields of neutron-scattering research.
- The neutron has a nuclear spin. Thus the scattering amplitude will in general depend on the orientation of the neutron's spin relative to that of the atom from which it is scattered. This is an important fact not only for the production of polarized neutrons but also for the study of oriented nuclei especially at low temperatures. Spin-incoherent nuclear scattering has developed into a powerful tool for studying single-particle motion (e.g. self-diffusion and molecular rotation).
- Because of their zero charge and low energy, thermal neutron beams do very little damage to organic substances. This is very important in the rapidly growing application of neutrons in biological research.

An increasingly large number of scientists of more and more different disciplines are starting to take advantage of neutron-scattering techniques and their unique possibilities.

Some of these techniques do, however, require very strong restrictions with respect to the energy, momentum, and spin state of the neutrons before as well as after the scattering process. As a consequence and because of the limited strength of the thermal neutron sources currently available, the present precision and information content of neutron-scattering experiments are limited by experimental rather than fundamental considerations, and improvements of orders of magnitude are still conceivable.

Neutrons whose energies are appropriate for scattering applications cannot be generated directly but have to be obtained from much more energetic neutrons by moderation. As indicated in Figure 1, a primary source of fast neutrons is surrounded by a moderator where the neutrons lose energy in a number of collisions with the moderator atoms until thermal equilibrium is reached. Beam holes, which reach into the region of maximum thermal neutron flux, serve to extract the moderated neutrons through the biological shield for use in various types of spectrometers. The design goal for such a neutron source should therefore be to obtain as high a flux of thermal neutrons as possible at the end of the beam tubes.

The most powerful sources so far devised for neutron-scattering research are the High-Flux Reactors (HFRs) at Oak Ridge, at Brookhaven, and at the Institute Laue–Langevin (ILL) in Grenoble. This last source has been especially optimized for neutron-scattering applications and has a thermal neutron flux of 10^{15} cm^{-2} s^{-1}. It operates at a power of 57 MW and the number of neutrons produced by fissions in the highly enriched uranium core is about 4×10^{18} s^{-1}. It is generally believed that, owing to the power density in the core, no substantial further increase in the thermal neutron flux will be possible.

As will be shown in the next section, advantage can be taken of a suitable time structure to increase considerably the performance of a neutron source. The new Soviet IBR II reactor, which has been built at Dubna, is designed to operate in a pulsed mode. As can be seen from Table 1, where important data are listed for the HFR at Grenoble, the IBR II, and the first stage of the FRG project for a spallation neutron

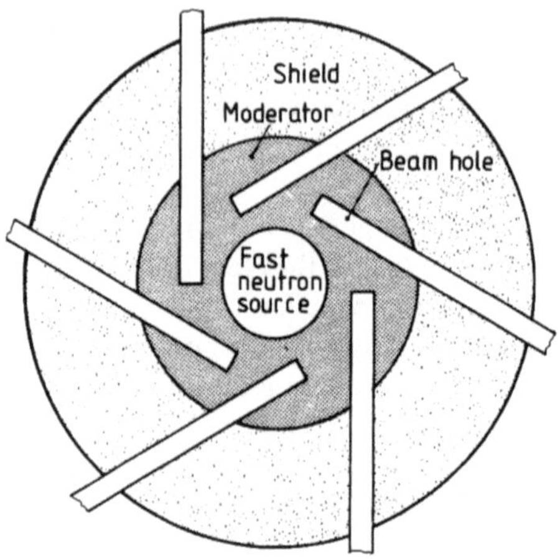

FIGURE 1 A schematic representation of a neutron source for beam-hole research.

TABLE 1 A comparison of characteristic data for the HFR at Grenoble (operational), the IBR II in Dubna (starting operation), and the intensity-modulated station DIANE of the SNQ project proposed in the FRG.

Characteristic	HFR (ILL)	IBR II[a]	DIANE
Peak thermal flux (cm^{-2} s^{-1})	10^{15}	10^{16}	1.3×10^{16}
Average thermal flux (cm^{-2} s^{-1})	10^{15}	7.5×10^{12}	7×10^{14}
Pulse repetition rate (s^{-1})	–	5	100
Pulse duration (μs)	–	150	500
Number of beam holes	14 + 11 N.G.[b]	14	14 + 12 N.G.[b]
Fuel or target material	U-235 (HEU)[c]	Pu	Pb (W)
Mode of operation	Critical	Periodically supercritical	Noncritical
Coolant	D_2O	Na	H_2O
Average thermal power (MW)	57	4	2.9
Peak thermal power (MW)	57	6000	58
Moving part	–	Reflector	Target
Extensions	Second cold neutron source	Electron-induction linac to give source pulse of 7 μs	Compressor ring to give source pulse of 0.7 μs U target to increase flux

[a] Data for the IBR II are taken from Rudolph and Wrobel (1981).
[b] N.G., neutron guides.
[c] HEU, highly enriched uranium.

source, the IBR II surpasses the HFR by a factor of 10 in peak flux but is very low in average flux despite considerable technical efforts and enormous peak thermal power.

In this situation the possibility of using the spallation reaction rather than nuclear fission as a primary neutron source has been investigated owing to the almost tenfold lower heat dissipation per useful neutron for spallation. This idea, which was originally followed in the Canadian Intense Neutron Generator (ING) Project (Bartholomew and Tunnicliffe, 1966) and which has recently been put forward again in conjuction with the upgrading of the cyclotron at the Swiss Institute for Nuclear Research (SIN) (Fischer et al., 1979), was the subject of an extensive study carried out jointly by the two FRG nuclear research centers at Jülich and at Karlsruhe between May 1979 and May 1981. The project is referred to as SNQ from the German word Spallations-Neutronenquelle (spallation neutron source).

3 GOALS AND DESIGN BASE OF THE SNQ PROJECT

The SNQ was conceived as an FRG national facility to substitute for presently used low- and medium-flux reactors, most of which are now more than 20 years old and outdated in design. It was felt that it should provide neutron-scattering capabilities comparable to those of an HFR but with an improved performance due to a well-developed time structure of its neutron flux. A suitable time structure, with a peak thermal neutron flux that is more than an order of magnitude higher than the time-average flux, constitutes an advantage in thermal neutron-scattering work for the following reasons (Figure 2).

- Spectrometers requiring a chopped beam to use the neutron's flight time over a well-defined distance to determine its energy can be synchronized to the time structure of the source and hence can profit in full from the peak flux of the source.
- A large fraction of the experimental background is correlated with the fast neutron pulse and can be eliminated by gating the detectors in a suitable way to count only the signal neutrons which arrive at the detector at a later time.
- Crystal monochromators, which are frequently used for energy selection, will also transmit integer fractions of the desired neutron wavelength. These can also be eliminated by a suitable time gate, thus avoiding a parasitic effect which is difficult to correct for in many experiments.

If skillfully exploited, a suitable time structure can result in substantial gains in data-collection rates as compared to a continuously operating neutron source. (For this reason several new neutron sources based on the spallation reaction have been or are being built (Table 2), and these, despite a moderate time-average flux, are potentially competitive with much stronger reactor sources for certain classes of experiments. These sources, which are referred to as pulsed neutron sources, are trimmed for short intense pulses of neutrons. They are especially advantageous if neutrons above the thermal energy range, i.e. of several hundred millielectronvolts in energy, are to be used.)

However, there exist – and will continue to exist in the future – a number of applications of thermal neutron sources which require a high time-average neutron flux. This is true for isotope production, irradiation experiments and certain classes of neutron-scattering spectrometers. Therefore, in addition to the time structure, the goal in the SNQ design was to reach a thermal neutron flux of at least 6×10^{14} cm^{-2} s^{-1} to be superior to all existing neutron sources in the FRG and to come within a factor of two of the highest fluxes presently available.

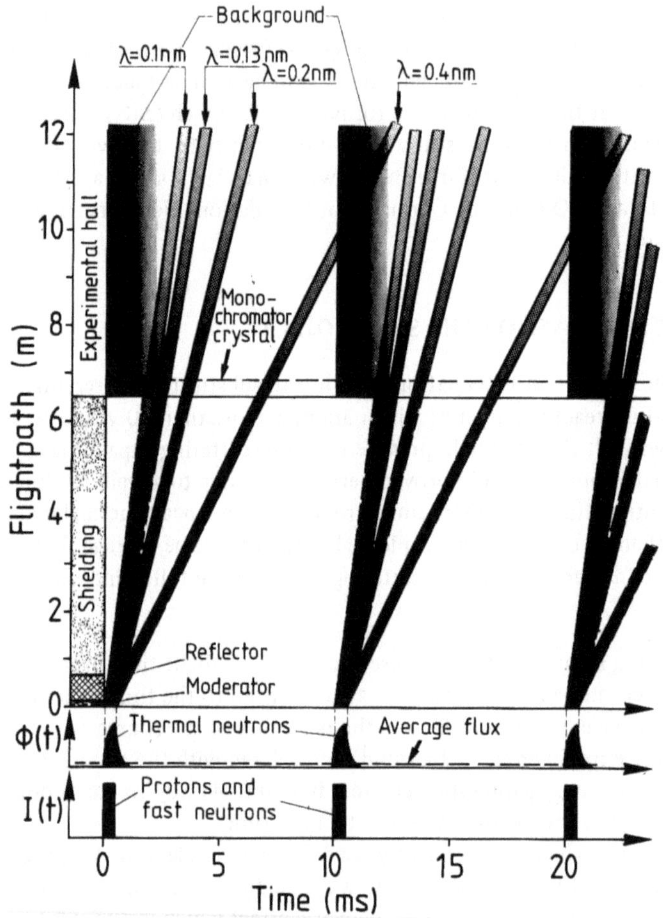

FIGURE 2 A space–time diagram of thermal neutrons in an intensity-modulated source. The sequence of proton and resulting fast neutron pulses (0.5 ms long and separated by 9.5 ms) is indicated on the bottom line. Above it, the shape of the resulting thermal neutron pulses in an H_2O moderator is shown. Whilst all neutron energies in the thermal range are present in the moderator at the same time, they separate according to their different velocities as they travel along the beam holes penetrating the shield (as indicated at the right-hand side of the diagram). Four neutron velocities corresponding to the various orders of reflection of a graphite crystal adjusted to 0.4 nm are shown. Also indicated is the time dependence of the background caused by fast neutrons penetrating the shield; this is of course correlated with the primary proton pulses.

In order to reach these goals a basic concept (Figure 3) was developed which is characterized by the following parameters: type of accelerator, linac; proton energy, 1.1 GeV; maximum proton current, 100 mA; pulse repetition rate, 100 Hz; proton pulse width, 500 μs; target material, lead; thermal neutron source, H_2O moderator, lead reflector; cold neutron source, liquid D_2 source in large D_2O tank. With these parameters the target station is characterized as "intensity modulated" rather than as a "pulsed source" because the width of the neutron pulses will be 500 μs and this is much too long to be

TABLE 2 Characteristic data[a] for pulsed spallation neutron sources.

Facility[b]	Laboratory[c]	$\bar{I}_p$ (μA)	t_p (μs)	$\bar{\Phi}_{th}$ ($cm^{-2}\ s^{-1}$)	$\hat{\Phi}_{th}$ ($cm^{-2}\ s^{-1}$)	Status (May 1981)
ZING-P'	ANL, USA	4.8	0.15	3.6×10^{10}	10^{14}	Shut down
WNR	LANL, USA	6	10	1×10^{12}	3×10^{14}	Operating
KENS	KEK, Japan	1.5	0.05	2×10^{10}	7×10^{13}	Operating
IPNS-I	ANL, USA	24	0.15	3×10^{11}	9×10^{14}	Operating
SNS	RAL, UK	212	0.22	7.2×10^{12}	4.5×10^{15}	Under construction

[a] $\bar{I}_p$, average proton current; t_p, pulse duration; $\bar{\Phi}_{th}$, average thermal neutron flux; $\hat{\Phi}_{th}$, peak thermal neutron flux.
[b] The various facilities are abbreviated as follows: ZING-P', Zero-Gradient Synchrotron Intense Neutron Generator – Prototype II; WNR, Weapons Neutron Research Facility; KENS, KEK Neutron Source; IPNS-I, Intense Pulsed Neutron Source I; SNS, Spallation Neutron Source.
[c] The various laboratories are abbreviated as follows: ANL, Argonne National Laboratory; LANL, Los Alamos National Laboratory; KEK, Laboratory for High Energy Physics; RAL, Rutherford-Appleton Laboratory.

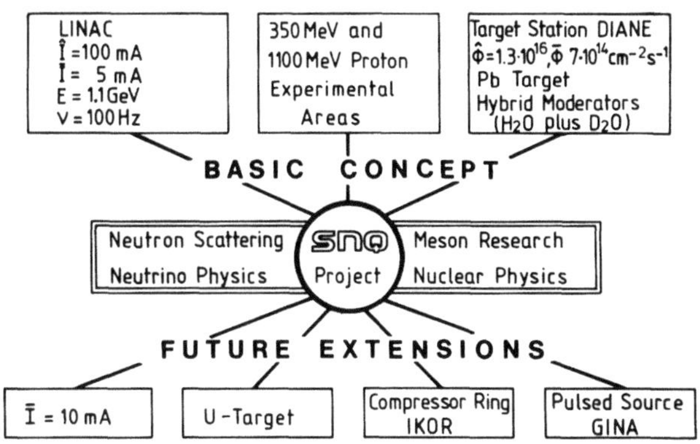

FIGURE 3 The basic concept, fields of research, and optional future extensions of the SNQ project.

used as the primary resolution element in time-of-flight measurements over reasonable flight paths. The target station has therefore been called DIANE which stands for "Deutsche Intensitätsmodulierte Anlage für Neutronen-Experimente" (German intensity-modulated facility for neutron experiments). A model view of the arrangement of buildings for the basic concept is shown in Figure 4.

It is obvious that such a facility will be attractive not only for neutron-scattering experiments but also for other scientific disciplines. Irradiation facilities for low- and ambient-temperature irradiation will be located in the D_2O tank. Uses of the proton beam itself are foreseen as well as exploitation of the large number of neutrinos originating from the decay of pions and muons produced in the target. This facility will in fact be the world's most exciting neutrino source if a suitable time structure can be provided (pulses less than 200 ns long separated by at least 10 μs) to distinguish between ν_μ and

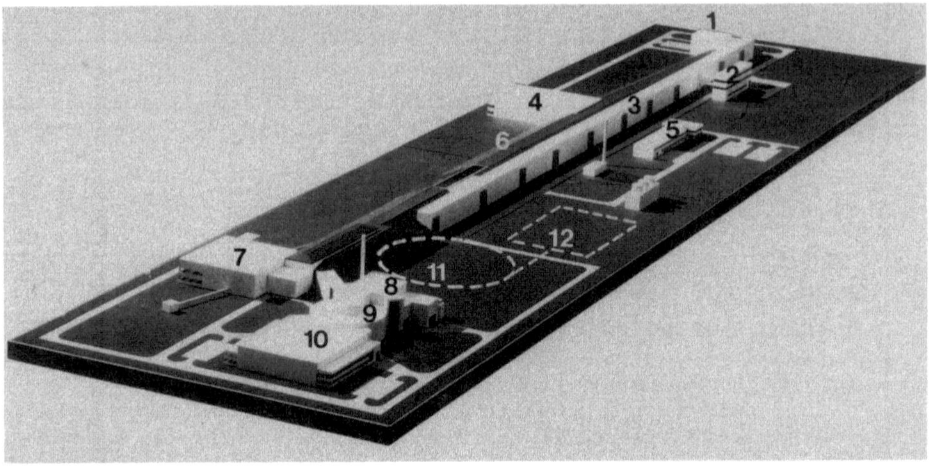

FIGURE 4 A model view of the SNQ facility. The buildings of the basic concept are shown together with the locations foreseen for the compressor ring and a future pulsed neutron source: 1, injector building; 2, test buildings; 3, RF gallery; 4, 350-MeV experimental area; 5, operations building; 6, accelerator tunnel; 7, 1100-MeV experimental area; 8, maintenance building; 9, target building; 10, neutron-guide hall; 11, future compressor ring; 12, future pulsed source.

ν_e through the vastly different lifetimes of their parent particles, the pions ($\tau_{\pi^+} = 2.6 \times 10^{-8}$ s) and the muons ($\tau_{\mu^+} = 2.2 \times 10^{-6}$ s). Provisions to make these additional uses of the facility possible have been foreseen in the concept. However, in order to exploit the potential of the facility in full, a number of future options have been conceived as indicated in Figure 3. They will be discussed below.

4 THE RATIONALE FOR THE MAIN DESIGN PARAMETERS

Protons were chosen as the particles to be accelerated despite an expected increase in neutron yield from deuterons (Figure 5) because the lower speed of deuterons at a given energy makes it more difficult and costly to overcome space charge effects in the low-energy part of the accelerator. However, the main reason was that beam losses in the low-energy part may lead to undesirable neutron production via d–d reactions (self-target effect).

Lead was selected as target material, at least for the initial phase of SNQ operation, because of its good neutron yield (Figure 6), low heat release per neutron (Figure 7), low thermal neutron absorption cross section, and the relaxed safety requirements due to the absence of fissionable material. It was, however, recognized that a considerable gain in performance could be expected through the use of uranium. The possibility of a later change to uranium as target material was therefore an important factor in all design decisions.

The average proton beam power required to obtain the desired thermal neutron flux of more than $6 \times 10^{14}\,cm^{-2}\,s^{-1}$ was determined to be 5.5–6 MW from mock-up

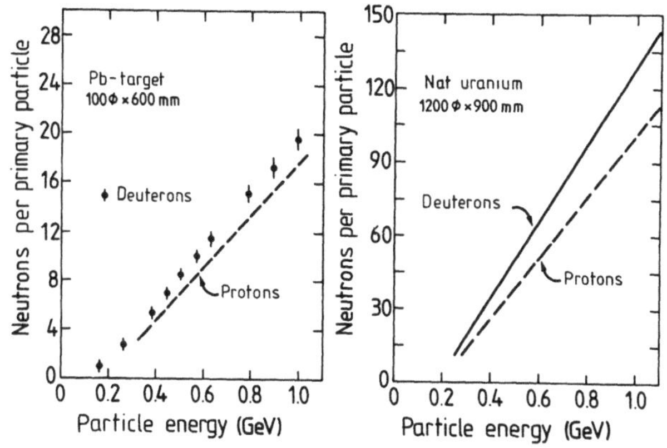

FIGURE 5 Neutron yield per primary particle for protons and deuterons on lead (100ϕ X 600 mm) and natural uranium (1200ϕ X 900 mm) targets (after Barashenkow et al., 1974); ϕ denotes diameter (mm).

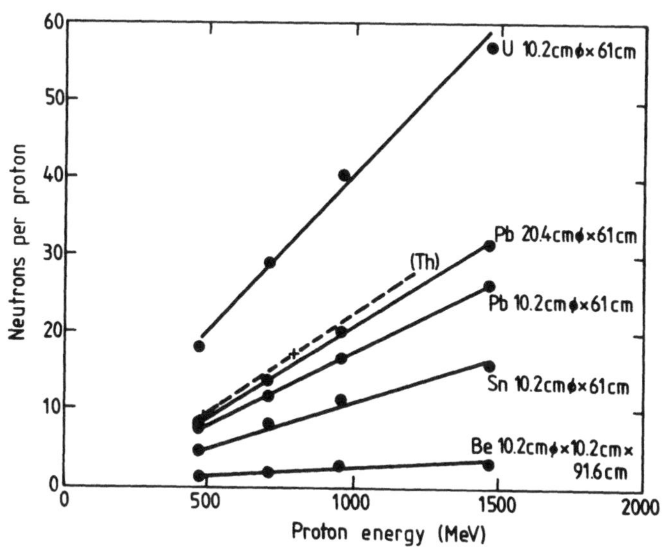

FIGURE 6 Neutron yield per incident proton as a function of proton energy for various target materials (from Bartholomew and Tunnicliffe, 1966; the data for thorium are taken from Garvey, 1980, and Russell et al., 1981).

experiments performed at the SIN proton cyclotron at Villigen, Switzerland, and at the proton synchrotron SATURNE at Saclay, France.

The pulse width and repetition rate of the thermal neutron pulses were selected after consideration of both experimental requirements and technical limitations. From Figure 2 it can be seen that, if neutrons of wavelengths up to 0.4 nm or more are to be used, a repetition rate ν of 80–100 Hz is desirable. Also, in order to achieve a reasonable

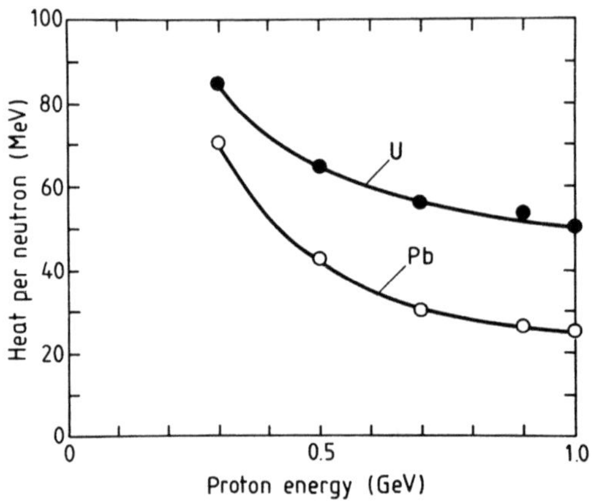

FIGURE 7 Heat release per neutron for lead and uranium as a function of proton energy (from Bartholomew and Tunnicliffe, 1966).

separation between the various orders of reflections from a crystal monochromator, a tolerable pulse width t_p (Full Width at Half-Maximum, FWHM) will be 500 μs, taking into account the fact that a 10-m distance between source and detector will result from the shield thickness of the source and the dimensions of the spectrometers. These two values, $t_p = 500\,\mu s$ and $\nu = 100$ Hz, have been found to be also compatible with other requirements, as will be seen in the following sections. The effect on the thermal neutron peak flux and the pulse width of deviations from a 500-μs proton pulse width at fixed frequency and fast neutron source strength is visualized in Figure 8.

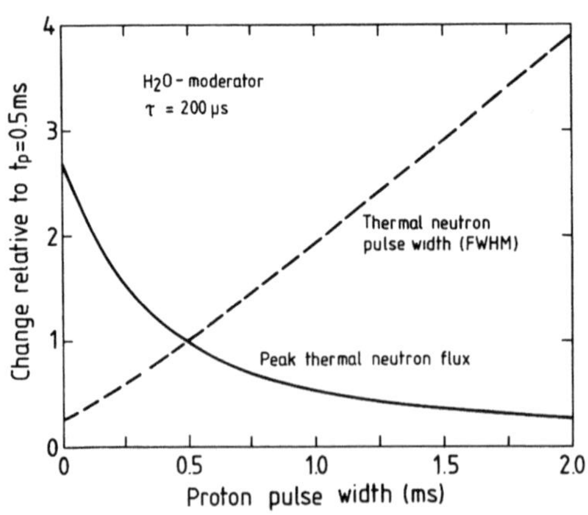

FIGURE 8 The change in peak thermal neutron flux and thermal neutron pulse width with proton pulse duration at constant source strength and repetition rate (H_2O moderator, with a decay time τ of the thermal neutron field of 200 μs). The FWHM at $t_p = 0.5$ ms is 0.516 ms.

The linac-type accelerator was found to be the only one able to satisfy the need for 6-MW beam power and the need for a 100-Hz 500-μs time structure at the same time. Synchrotrons can deliver much shorter pulses but are at present limited to beam powers of the order of 0.5 MW if the energy is in the range of 1 GeV as desired for a spallation neutron source. Cyclotrons can reach suitable energies and proton currents but would allow a suitable time structure only at the expense of a drastically reduced beam power. Also, in view of the option of adding a proton pulse compressor to the system (see later) it should be possible to accelerate also H^- ions to an energy of at least 1 GeV. This is not the case with circular orbit accelerators because of Lorentz stripping.

The choice of the proton energy was affected by a large number of parameters. From Figure 7 it is obvious that in order to reduce the power density in the target an energy above 1 GeV is desirable. Also the increase in neutron yield per proton with energy (Figure 6) is in favor of a higher energy because it allows shorter pulses at a given peak proton current. Another aspect of considerable importance was the possibility of providing extremely short proton pulses of less than 1-μs duration while retaining almost the full time-average beam power at a later date. This can be done by incorporating a compressor ring in the system (see below). At energies below 1.1 GeV severe limitations arise concerning the feasibility of such a compressor ring. Large-aperture magnets become necessary to allow for the space charge effects which increase with decreasing relativistic mass γ and velocity β as $\beta^{-2}\gamma^{-3}$. Also, the isochronous mode of operation of such a ring, as it has been envisaged for reasons of simplicity, becomes more and more difficult to establish below 1 GeV. Since, in contrast, the length and cost of a linac increase in proportion to the energy, it was decided to fix the proton energy at the lower limit compatible with the foregoing requirements, namely 1.1 GeV.

The peak proton current was chosen to be 100 mA for the following reasons. It is the minimum value compatible with a 0.5-ms pulse duration at a 10-ms pulse separation if the average power and proton energy are fixed at 5.5 MW and 1.1 GeV, respectively. Although a high proton current is desirable for good linac efficiency, beam losses which are likely to impede hands-on maintenance of the accelerator (or of parts of it) will increase with peak current. Higher currents would allow pulses shorter than 500 μs but, in view of the fact that the accelerator structures have to be excited before the beam can be injected (to ensure stable operation), the gain in efficiency on acceleration would be more and more counterbalanced by the increasing fraction of filling time.

5 THE ACCELERATOR

5.1 Design Aspects

High peak and high average beam power are the specific characteristics of the SNQ linac. A comparison with existing machines or projects shows that an extrapolation of about one order of magnitude has to be made in either quantity (Table 3). Two design requirements follow: (1) economically operating systems are needed to produce the Radio-Frequency (RF) power required and to transfer it to the beam with high efficiency; (2) beam transport has to be optimized for minimum beam spill, particularly at higher energies, to reduce activation of machine components.

TABLE 3 Linear accelerators with high intensities or high duty factors as compared to the SNQ linac.

Machine or project	Laboratory[a]	Particle	Final energy (MeV)	Peak beam current (mA)	Duty factor of the beam	Status (1981)
New linac	CERN	p	50	150	10^{-4}	Operating
p linac	FNAL	p	200	300	10^{-4}	Operating
		H^-		46	10^{-3}	
LAMPF	LANL	p		12	6–12%	Operating
		H^-	800	0.1		
Unilac	GSI	Heavy ions	10[b]	μA[c]	25%	Operating
FMIT	HEDL, LANL	d	40	100	100%	Design, prototyping
Electrobreeder	CRNL	p	1000	300	100%	Study, prototyping
Proton linac	New England Nuclear	p	45	60	10%	Under construction
SNQ linac		p (H^-)	1100	100	5% (10%)	Study

[a] The various laboratories and projects are abbreviated as follows: CERN, Centre Européen pour la Recherche Nucléaire; LAMPF, Los Alamos Meson Physics Facility; FNAL, Fermi National Accelerator Laboratory; LANL, Los Alamos National Laboratory; GSI, Gesellschaft für Schwerionenforschung; FMIT, Fusion Materials Irradiation Test Facility; HEDL, Hanford Engineering Development Laboratory; CRNL, Chalk River Nuclear Laboratory.
[b] 10 MeV per nucleon, upgrading to 20 MeV per nucleon in 1982.
[c] Peak beam current of the order of microamperes.

Apart from these arguments, which hold for any high-power accelerator, a high degree of flexibility in the machine concept was desired for the SNQ. For the basic machine special time structures are required which depend on the individual experiments. In addition to feeding the neutron target, as discussed later, fractions of beam pulses will be split off to serve experiments in nuclear physics in simultaneous operation. Furthermore, the options to increase the beam power to 11 MW and to provide a 100-mA H^- beam for injection into a compressor ring have been included in the linac design.

The main parameters were discussed earlier. Table 4 gives a more detailed summary. The schematic arrangement of the linac is shown in Figure 9. Protons are extracted from a magnetic multipole ion source and are preaccelerated to an energy of 450 keV by a Direct-Current (DC) preinjector. In the transport path of the low-energy beam the beam quality is measured and the beam is defined by collimators. Macropulses up to 500 μs long are cut out of the beam by a fast deflector, which can, at the same time, impose on the pulses any substructure desired. RF bunches are formed by a three-gap buncher system. The beam is matched in the transverse phase space to the accelerator entrance by quadrupole triplets. In the Alvarez accelerator, operated at 108 MHz, the beam is accelerated to 105 MeV over a distance of 85 m. After a matching section, a Disk-And-Washer (DAW) accelerator provides acceleration to the final energy. When the DAW accelerator is operated at three times the Alvarez frequency, its acceleration efficiency is higher from about 100 MeV on. Not including a beam-switching section at 350 MeV, the DAW accelerator has a total length of 400 m, which results in an energy gain of 2.5

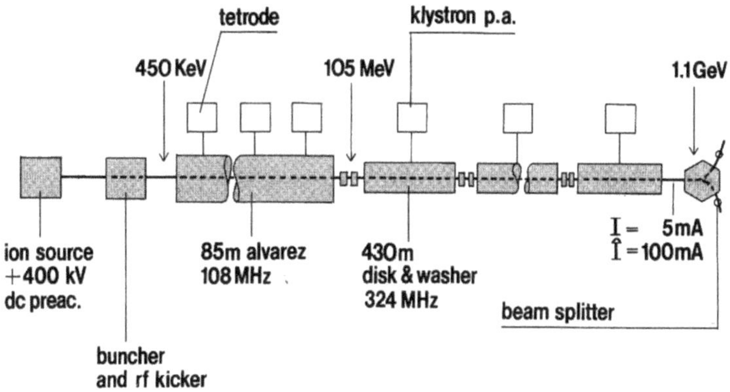

FIGURE 9 The schematic arrangement of the SNQ linac.

MeV per meter of technical length. This gradient was found to be close to the optimum of capital and operation costs. When it reaches its final energy the main beam is transferred to the neutron target, but fractions of the pulse can also be split off to serve the nuclear physics experiments. A beam dump is provided in the forward direction for testing purposes.

5.2 Beam Dynamics

The choice of essential accelerator parameters such as operation frequencies and injection energies was affected by beam dynamics and, in particular, by space charge effects. From analytical calculations (Mittag, 1979) which complement data from multiparticle tracing codes, the following guidelines can be formulated.

TABLE 4 Parameters of the SNQ accelerator.

Final energy (MeV)	1100	
Peak beam current (mA)	100	
Pulse length (μs)	$\leqslant$ 500	
Repetition frequency (Hz)	$\leqslant$ 100	
Power consumption (total facility) (MW)	50	
	Alvarez accelerator	DAW accelerator
Frequency (MHz)	108	324
Input energy (MeV)	0.45	105
Technical length (m)	85	400
Minimum transmission of beam current (%)	85	99.65
Peak power input into accelerator (MW)	21	267
Peak beam power (MW)	10.5	100
RF duty factor (%)	6.5	5.5

– The injection energy into the RF accelerator should be made as high as possible. The upper limit is, however, given by the reliability of high-voltage preinjectors. As an alternative, the RF-quadrupole structure may be used for preacceleration (Stokes, 1981). It was found to be difficult to combine this injection scheme with the fast deflector needed for precise pulse formation in the SNQ accelerator.
– If transport of the maximum beam current (optimum transmission) rather than high brightness (sharp focus at the target) of the beam is required, a low operation frequency must be chosen. In this way higher transverse acceptances and better compression of the beam buckets in time are obtained. (The bunches must be compressed to fit into the stable range of phase of the DAW accelerator which is reduced as compared to that of the Alvarez accelerator because of the higher frequency.)

Multiparticle calculations have shown that, for the parameters chosen, a 115-mA beam with an emittance of 3π mm mrad can be transported through the Alvarez accelerator with 87% transmission. The 13% of particles which are not captured are lost in the first Alvarez section at energies where cross sections for activation are still low. From the second Alvarez tank on, 100% transmission is obtained up to the final energy.

In a real accelerator, beam loss cannot be completely suppressed. Unfortunately even the most elaborate beam-dynamics codes are not able to predict losses to better than 1%. Beam loss is caused by various effects: misalignments, fluctuations of parameters such as beam current or setting of the electric and magnetic fields, and, particularly for pulsed machines, transient effects. Beam loss can therefore be reduced by proper choice of parameters (e.g. generous apertures), by stabilization of the guiding and accelerating fields, and by suitable alignment procedures. To determine the optimum

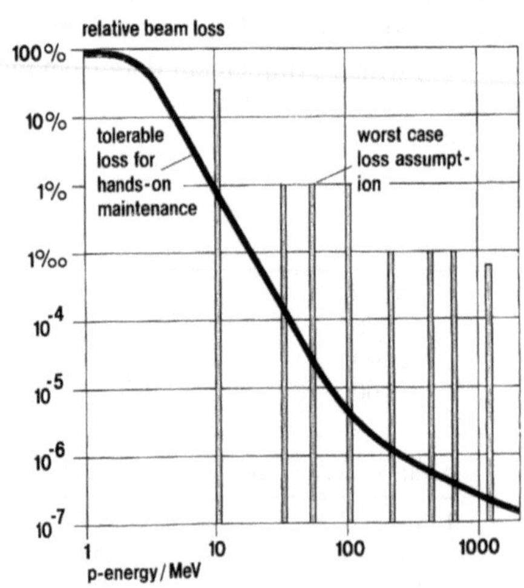

FIGURE 10 Beam loss versus proton energy. The upper limit of beam loss (fractions of total beam per meter of length) permissible for hands-on maintenance of the accelerator decreases steeply with proton energy (curve). The shaded bars show sources of point losses assumed as worst cases for the design of buildings, accelerator components, and handling systems.

values for beam operation and to maintain them over periods of at least weeks, extended beam diagnosis has to be applied. Data have to be stored and corrected by computer. It is to be expected that, with some experience from beam tests at a reduced repetition rate, it will be possible to tune the accelerator in such a way that beam losses remain below the limit for hands-on maintenance (Figure 10).

For the design of the buildings and the choice of materials and handling schemes, higher losses were assumed; these are shown as point sources of loss in Figure 10 and represent the worst-case situation. Under these assumptions, the accelerator building was divided into different zones of accessibility (Figure 11). Accelerator components in the beam tunnel are handled remotely from the maintenance tunnel (access after shutdown). A wall of concrete blocks which is set up as needed separates these two areas. Radioactive components are handled by an overhead crane system which, if required, can be supplied with machine tools or manipulators. A detailed study of handling systems and suitable component design did not show basic difficulties or an excessive increase in installation cost. Components of activated water circuits and supplies which must not be exposed to permanent radiation are housed in the equipment tunnel. As the activation of cooling water decays rapidly this area is accessible after shutdown. The RF gallery and its basement are of unlimited access.

5.3 Accelerator Components

To obtain a beam which is reproducible and free of noise, a magnetic multipole (cusp-field) ion source is proposed. From a source developed for the SNQ project 220 mA of total ion current could be extracted in DC operation at 50 keV. The beam quality and stability were found to be adequate. Studies of beam transport with space charge neutralization are being performed at present. In the ion source the beam will be only prepulsed; the exact shaping is done by the fast deflector located in the low-energy beam transport line at the 450-keV level. The deflector consists of pairs of deflecting plates connected to a

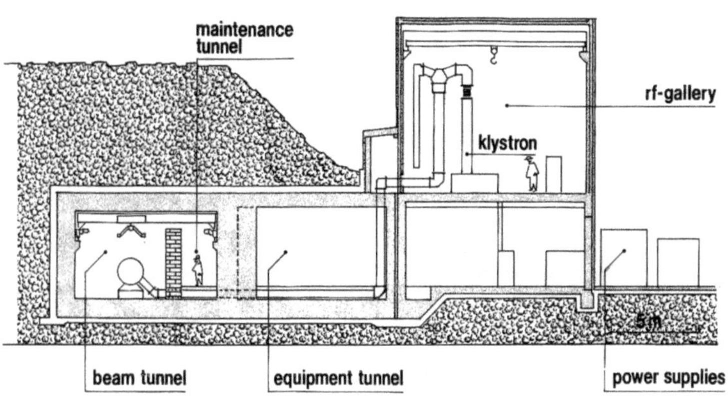

FIGURE 11 A cross section of the DAW accelerator. Zones of different accessibility are delimitated from each other; the building of the RF gallery is freely accessible during machine operation.

meander line which adapt the pulse velocity to the velocity of the beam. With broadband high-power amplifiers that have recently become available the beam can be deflected at a speed of a few nanoseconds per sweep, i.e. between two of the micropulses formed by the buncher for acceleration. Beam pulses with a wide variety of time structures can be formed this way.

The Alvarez accelerator consists of seven tanks, each 12 m long (Figure 12). Focusing of the beam is obtained from quadrupole doublets included in the drift tubes. For better field stability and better transfer of energy the structure is postcoupled (Figure 13). Drift tubes are mounted on girders (three per tank) so that they can be preadjusted and inspected with the tanks remaining in place. Alvarez accelerators with similar specifications are already in operation, so no essential problems are anticipated.

The performance of DAW accelerators has only been demonstrated in laboratory tests of full-scale prototypes (Andreev et al., 1976) but the main principles of construction and operation resemble those of other high-energy structures. The essential advantages of the DAW structure are high shunt impedance (i.e. high acceleration efficiency) and good propagation of energy (stable fields with beam loading, reduced requirements on tolerances). Focusing in the DAW accelerator is obtained by quadrupole doublets in between tanks where beam diagnosis and steering elements are also placed. By this scheme of separate functions the accelerator can be built in a modular way which facilitates handling. The lengths of the 57 individual tanks vary from 3.1 m to 7.2 m along the accelerator. From 350 MeV on, pairs of resonators which are separately fed with RF but which form one unit mechanically are used (Figure 14).

5.4 RF Power Supplies

21 MW of peak RF power have to be produced for the Alvarez accelerator at a frequency of 108 MHz. Power amplifiers equipped by a tetrode delivering up to 2 MW

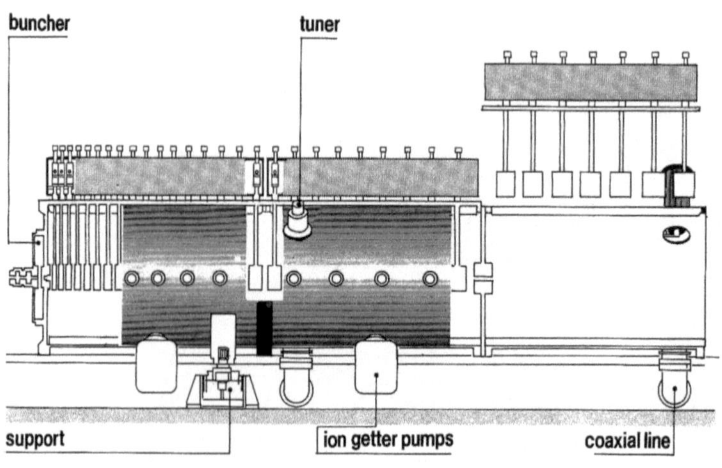

FIGURE 12 The first tank of the Alvarez accelerator. The buncher is directly flanged to the front end plate. Sets of drift tubes are suspended from girders.

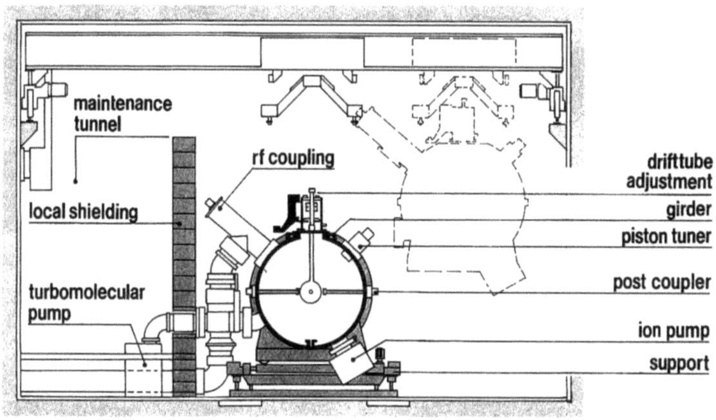

FIGURE 13 A cross section of the beam tunnel in the Alvarez accelerator.

of peak power at a 10% duty cycle have recently been developed. The output of two such amplifiers is combined by a hybrid to feed one Alvarez tank. More than 90% of the RF power required (267 MW total) is produced by the 89 klystron power amplifiers for the DAW accelerator. The peak power required (3.5 MW of saturated power) per amplifier exceeds the state of the art which is at present limited to 600 kW of continuous-wave power in this frequency range. Development of a suitable high-efficiency high-power klystron was therefore started within the framework of the SNQ study. As a first step, the V107SK klystron (Figure 15) was built and tested. Tests at high power showed that an efficiency of 70% (RF power to electron beam power) can be obtained and that extrapolations to the power level proposed can be made with confidence. Properties relevant for pulsed operation were examined and suitable control and protecting circuits were designed for accurate setting of the amplitude and phase of the accelerating fields during the beam pulse (Vetter et al., 1981).

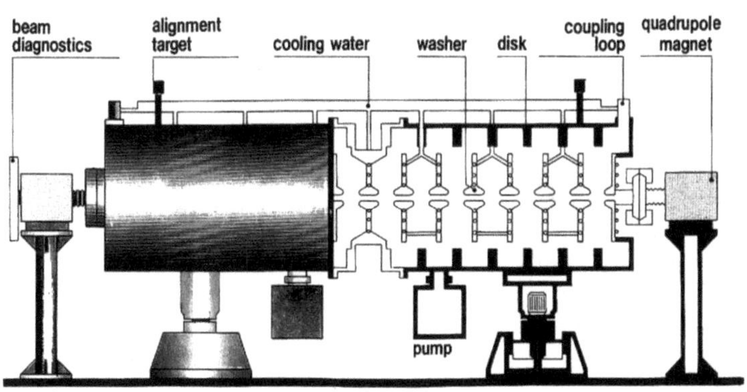

FIGURE 14 A twin tank of the DAW accelerator. Accelerating cells are formed by pairs of washers, while the ring-shaped slit between disks and washers provides a high flow of energy.

FIGURE 15 The test klystron V107SK. With a power capability of 1.5 MW and an efficiency of 70% the klystron is a development step towards the SNQ series production klystron.

5.5 Conclusions for Accelerator Construction

For the accelerator proposed, feasibility and cost were analyzed in collaboration with industry. The system and its components were found to be feasible. Development is needed in particular for high-power klystrons and beam diagnostic equipment. The performance of essential components (the Alvarez tank, the DAW tank, fast magnets, power amplifiers, and injection equipment) has to be demonstrated early in the construction phase. Eight years are required for construction and putting into operation. RF systems including auxiliary equipment were found to be the dominant part of the construction cost. The operating cost is essentially determined by the power consumption. At full operation 50 MW of mains power are needed to produce 5.5 MW of beam power. Higher efficiencies, as would be needed for technical applications, can be obtained in principle by a modified design but were not compatible with the design goals of the SNQ.

6 THE TARGET

With a time-average beam power of 5.5 MW and assuming a gaussian beam profile of 4 cm FWHM as chosen to obtain a compact source of primary neutrons, the maximum heat density in the lead target is 6 MW l^{-1}. In an uncooled target this would result in a temperature rise of 4000 K s^{-1}. Cooling of the target under these conditions is only possible, without excessive dilution by coolant channels, if the target material hit by the proton pulse is moved out of the beam and is allowed to cool down elsewhere. Two concepts to achieve this have been discussed: a flowing liquid-metal target (Pb–Bi eutectic, Hoffmann, 1980) similar to the one proposed for the ING concept (Bartholomew and Tunnicliffe, 1966) and a rotating solid target (Bauer, 1980). The rotating-target concept was selected for the following reasons:

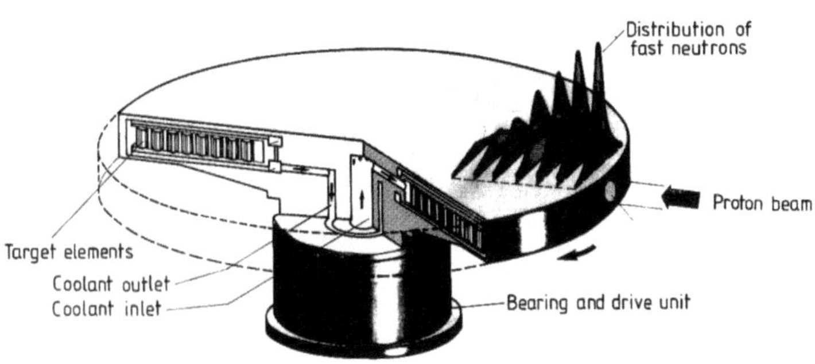

FIGURE 16 A schematic representation of the rotating target arrangement with individually canned target pins cooled by water which flows in and out through the hub of the wheel. The diameter of the wheel is 2.5 m and the height of the target material is 10 cm. The distribution of fast neutron leakage as calculated with the HET code is indicated for the upper surface.

TABLE 5 Radiation effects and thermomechanical data for components of the fast breeder reactor, the fusion reactor, and the SNQ.

Reactor	Average damage rate[a] (dpa d^{-1})	He production rate[b] (appm d^{-1})	H production rate[b] (appm d^{-1})	Operating temperature (°C)	Thermal cycles (s^{-1})	Thermomechanical stress level (MPa)
Fast breeder (material: steel)	0.1–0.2	0.03		300–600	3×10^{-7}	60–120
Fusion reactor (inertial confinement)	0.2–0.1	0.07–1.4	0.3–2.1	300–1000	0.3–30	100–200
SNQ						
Stationary window (material: Mo)	1.9	170	1530	800	100	100
Rotating window (material: Al)	0.01	0.6	1.6	50	0.5	80
Structure of rotating target (Al)	< 0.01	< 0.3	< 0.3	50–70	4×10^{-7}	50–80

[a] dpa, displacements per atom.
[b] appm, atomic parts per million.

– It enables the use of a solid target material and therefore leaves open the possibility of changing to uranium at a later date.
– It does not require a stationary beam window.
– Radiation damage is reduced not only in the target itself but also in the structural material, especially the proton beam window, which rotates with the target.

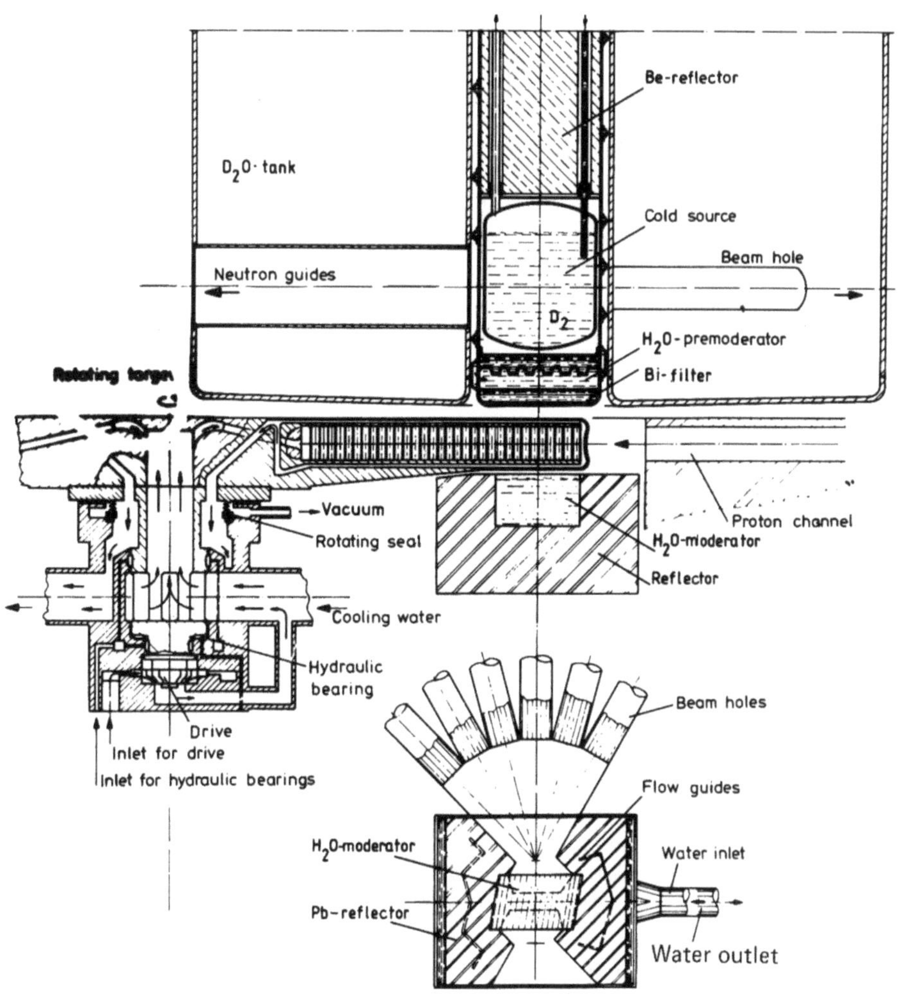

FIGURE 17 The hybrid moderator concept proposed for the target station DIANE. A large volume of high neutron flux, housing the cold neutron source and irradiation facilities (not shown), is arranged above the target to facilitate access from above. The H_2O moderator below the target is surrounded by a lead reflector on three sides and is viewed by 12 beam tubes.

– It works with a horizontal proton beam line which can be built more cheaply and easily than a vertical one.
– Further advantages of this geometry are that it is well suited for a hybrid moderator concept which could provide moderators with both a large volume of high flux and a good time structure at the same time.

The target material (a total of 4000 kg) is mounted on the outer 70 cm of a wheel 2.5 m in diameter (Figure 16) and is subdivided into some 9000 individually canned pins about 24 mm in diameter and 10 cm high which are directly cooled by water flowing between them. The wheel structure is of aluminum alloy and the cooling water is brought in and out through the hub of the wheel. The proton beam enters at the outer periphery through an aluminum window 0.5 cm thick which is heated up by 12° C each time it is hit by a proton pulse. The wheel's speed of rotation is adjusted in such a way that adjacent areas of the circumference are hit by successive beam pulses. This leads to a frequency of approximately 0.5 Hz (4 m s^{-1} circumferential speed). The load on the structural material due to the proton beam is greatly reduced in this way and very favorable operating conditions are achieved, as can be seen from Table 5 where data are compared for a fast breeder reactor, a fusion reactor, and a stationary as well as a rotating beam window in the SNQ. The hot spot of the target itself is heated by 42° C once every 2 s and operates at a maximum temperature of 63° C above the coolant temperature, i.e. at about 120° C.

The mechanical and thermal stresses occurring in the target and in the canning, as well as those in the structural parts, were very thoroughly analyzed during the study and were found to be tolerable at all points. Based on an analysis of the radiation effects, a minimum lifetime of 12,000 h at full power (approximately 2 years) is expected. Research is going on to employ other materials such as SAP (Sintered Aluminum Product), metallic glasses, glassy carbon, or refractory metals for special applications.

7 THE MODERATORS

Depending on the kind of use that is to be made of the thermal neutrons, different moderator characteristics are required. Irradiation facilities and a cold neutron source are served best with a large volume of high flux. A large tank of heavy water, with its extremely low absorption cross section, is best suited for this purpose. However, in order to obtain a good time structure a small volume of material with high slowing-down density is preferable. Light water excels in terms of its superior slowing-down properties and is to be preferred for a good time structure.

Taking advantage of the natural separation of the two regions of high neutron leakage above and below the rotating target, it was decided to arrange for a hybrid moderator concept with a D_2O tank above the target and a suitable volume of H_2O below, as indicated in Figure 17. It was found in the mock-up experiments mentioned earlier that a grooved surface on the H_2O moderator results in an increased thermal neutron leakage. It was also shown that a reflector of lead is equivalent to a beryllium reflector with respect to peak flux and superior with respect to time structure.

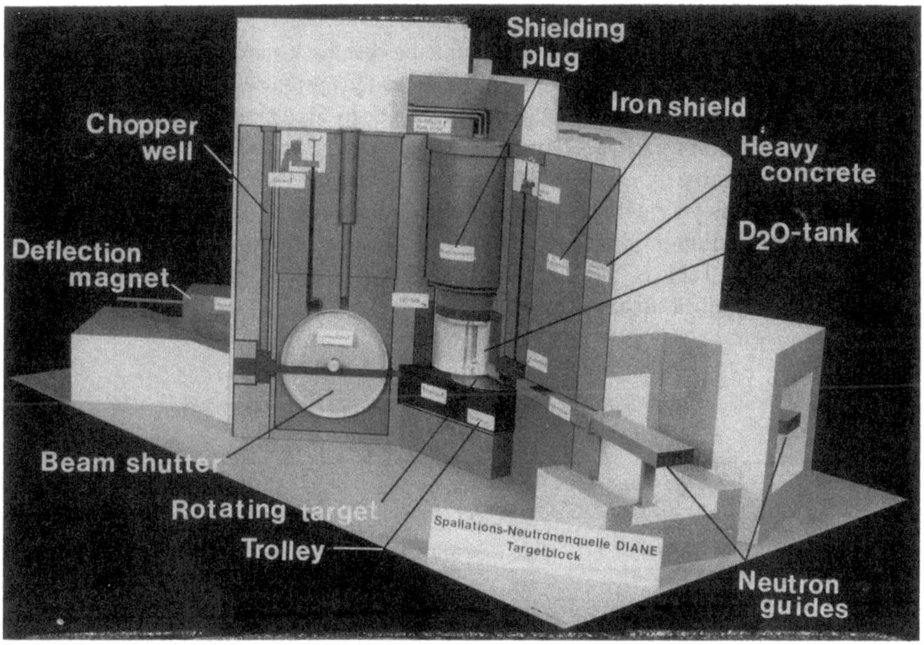

FIGURE 18 A cutaway model view of the target block of the target station DIANE.

8 THE TARGET BLOCK

A model of the target block is displayed in Figure 18. Calculations have shown that a total of 6–6.5 m of iron and heavy concrete is needed to bring the radiation dose down to tolerable levels on the outside of the shield. The beam holes which penetrate the shield can be shut off safely while the source is operating by means of rotating shutters, 3 m in diameter, which place the beam channel vertical when closed. In this position, collimator inserts can be exchanged from above. The heavy-water tank is mounted in the shield and can be removed from above if necessary. The target itself and the H_2O moderator below it are mounted on a heavy trolley which can be moved on rails back into the service area adjacent to the target block. For this purpose the last bending magnet, which is also located in the service area, has to be retracted.

A total of 12 beam holes viewing the H_2O moderator and two beam holes viewing the cold neutron source will penetrate the shield and end in the experimental hall (40 m in diameter) surrounding the target block. In addition, two bundles of six neutron guides each will lead to an external cold neutron laboratory (neutron-guide hall) which provides space for about 30 spectrometers.

It has been proposed to provide two suitably shielded caverns below the target block for neutrino experiments (Zeitnitz, 1981).

9 PERFORMANCE ESTIMATES

The expected performance of the source has been estimated both on the basis of extensive mock-up experiments and by a sophisticated system of Monte Carlo computer programs where the HET and MORSE codes have been supplemented by suitable analysis routines and geometry packages (Armstrong et al., 1981). In this way it was possible to simulate the full three-dimensional geometry of the target–moderator–shielding arrangement and to obtain very detailed data on neutron flux, energy deposition rates, particle spectra, and spallation products. A summary of expected thermal neutron flux levels for DIANE is given in Table 6. The calculated data are in excellent agreement with those derived from the mock-up experiment.

TABLE 6 Thermal neutron flux data for the moderators of DIANE as derived from measurements and as calculated[a].

Moderator	Average thermal flux $\bar{\Phi}$ ($\times 10^{14}$ cm^{-2} s^{-1})		Peak thermal flux $\hat{\Phi}$ ($\times 10^{14}$ cm^{-2} s^{-1})	Pulse width FWHM (μs)
	Measurement	Calculation	Measurement	Measurement
H_2O	7	7.8	130	515
D_2O	6.8	7	56	850

[a] For the case of the H_2O moderator "equivalent fluxes" (i.e. isotropic fluxes which would result in the same neutron current at the beam-hole exit) are given. The data for the D_2O tank are unperturbed flux data; i.e. the effect of experimental installations and beam holes in the tank is not taken into account.

The expected time structure of the neutron flux in the two moderators, also derived from measured data, is shown in Figure 19. Figure 20 gives the energy-dependent peak flux per unit energy interval in comparison to the mean flux of the HFR at ILL, Grenoble. The pulse width at DIANE will be 500 μs for practically all energies and therefore the time-average flux is a factor of 18–20 below the peak flux everywhere.

With its time-average flux of 7×10^{14} cm^{-2} s^{-1} and its peak flux of 1.3×10^{16} cm^{-2} s^{-1} at 100 Hz DIANE will be the most powerful neutron source in the world. Some data on the HFR at Grenoble and on the IBR II at Dubna are given in Table 1 for comparison. It is also interesting to compare the technical information given in the lower part of the table.

10 OPTIONS FOR FUTURE EXTENSIONS

The basic SNQ concept, as described in the foregoing, was selected also with a view to its potential for future improvements. Four possible extensions have been considered.

– Doubling of the proton pulse length from 0.5 to 1 ms. This would make it possible to supply all experimental facilities with protons in a satisfactory way simultaneously

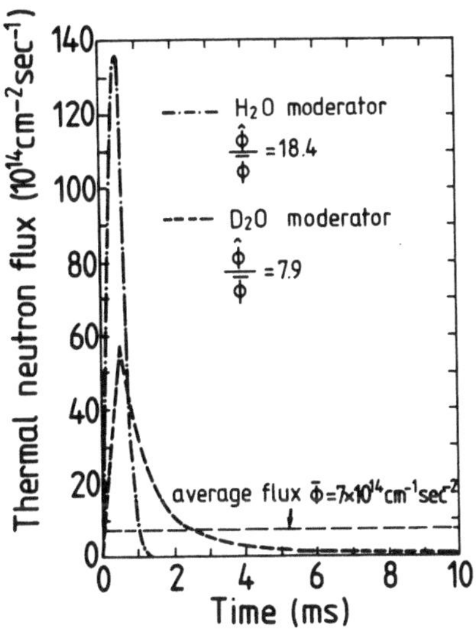

FIGURE 19 The time structure of the thermal neutron flux in the moderators of DIANE as derived from measured data.

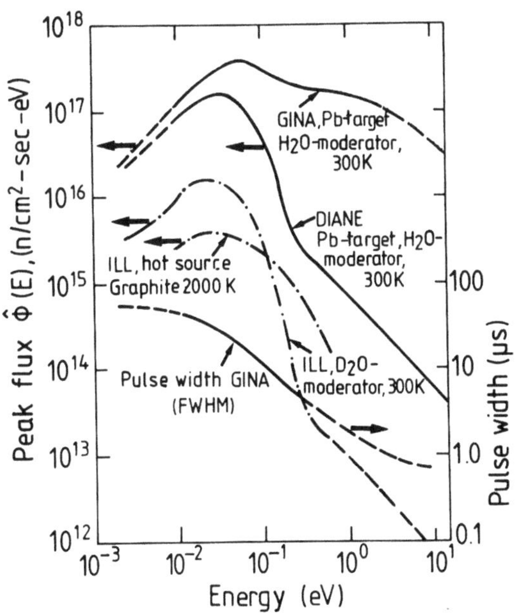

FIGURE 20 Peak neutron flux at DIANE as a function of energy in comparison to the steady-state flux at the HFR, Grenoble. The pulse width at DIANE is about 500 μs for all energies and the time-average flux is a factor of 18–20 below the peak flux. The curve labeled GINA is for the option of a pulsed source as outlined in the text.

(350-MeV proton beam research area, 1100-MeV proton beam research area, low-duty-cycle pulse precursor for neutrino experiments (see Zeitnitz, 1981), neutron-production pulse on target). It would be indispensable if two target stations were to be operated simultaneously (see later).

- Change to uranium as target material. In the mock-up experiments it was shown that this would increase the time-average neutron flux by a factor of 1.8 in the D_2O tank and by a factor of 2.1 in the H_2O moderator. Since the operational cost of the facility would remain practically unchanged this is a very attractive option.
- Addition of a proton pulse compressor to supply proton pulses of only 700 ns duration and 66 A beam current. If used on the target station DIANE this would raise the thermal peak flux by a factor of approximately 2.3 in the H_2O moderator. It would, in particular, open up the possibility of providing a time structure suitable for type-selective neutrino experiments without cutting down too much on the time-average beam intensity (three bunches of 150-ns length in the ring). Another advantage would be that the neutron pulses above thermal energies (slowing-down regime) could be made short enough for these energies to be used for time-of-flight experiments. This is a necessary technique because deflection angles from crystals are often extremely small. However, this technique requires specially designed moderators and reflectors and therefore a fourth option has been considered:
- Construction of a second target station, GINA (from the German: Gepulste Intensive Neutronen-Anlage), (pulsed intense neutron facility), which would be designed as a pulsed source to be exploited by time-of-flight techniques in the thermal and epithermal energy range. For this purpose the moderators would be poisoned with an absorber for thermal neutrons and decoupled from the reflector against low-energy neutrons to provide suitably short pulses. The expected energy-dependent peak flux and pulse width of this facility are also shown in Figure 20.

These prospects for future extensions add considerably to the attraction of the SNQ concept. Because of its novelty, the proton pulse compressor has been studied most thoroughly. The proposal is to operate an 11-segment ring, 64 m in diameter, in an isochronous mode, i.e. at the transition energy. Although this mode is inherently unstable it was found that the rise time of the instabilities could be made sufficiently long to have no serious effect on the beam during the filling time (0.5 ms) of the ring. The beam is ejected immediately after the filling is complete. In order to make the necessary stripping–injection into the ring possible (to overcome the conservation of phase-space density) the accelerator has to supply an H^- beam of 100-mA peak current. This is possible with a linac although it poses severe restrictions on the vacuum quality and the magnetic guide fields in the accelerator.

11 CONCLUSIONS

The low heat output and high neutron yield of the spallation process make it possible to build thermal neutron sources with flux levels above those obtainable in steady-state reactors. When used as a facility for fundamental research, the economic advantage of operating a linear accelerator at high current levels can be exploited beneficially to build an intensity-modulated neutron source which is clearly superior in performance to a steady-state source of the same average flux. The proposed facility, which would operate at a time-average proton beam power of 5.5 MW and a peak power of 110 MW, can be built using existing technological know-how. If funded, this spallation neutron source would not only have to be considered as an outstanding center for numerous disciplines in fundamental research but would also make possible indispensable research work (and would constitute a valuable and necessary step) on the way to more powerful plants to be exploited in the various concepts for energy-supply systems discussed in this meeting.

REFERENCES

Andreev, V.G., Guslitskov, I.K., Mirochnik, E.A., Pirozhenko, V.M., and Polyakov, B.I. (1976). Proceedings of the Proton Linear Accelerator Conference, Chalk River. AECL 6584. Atomic Energy of Canada Limited, Chalk River, Ontario, pp. 269–271.

Armstrong, T.W., Cloth, P., Filges, D., and Neef, R.D. (1981). Results of nuclear calculations on the spallation neutron sources SNQ. Jül-Spez 120. Jülich Report Series, Kernforschungsanlage Jülich GmbH, Jülich, FRG.

Barashenkow, V.S., Toneev, V.D., and Chigrinow, S.E. (1974). Interaction of high energy deuteron beams in matter. Atomnaya Energiya, 37:480–483. (English translation in Soviet Journal of Atomic Energy, 37:1256.)

Bartholomew, G.A. and Tunnicliffe, P.R. (Editors) (1966). The AECL Study for an Intense Neutron Generator. AECL 2600. Atomic Energy of Canada Limited, Chalk River, Ontario.

Bauer, G.S. (1980). Design studies for a rotating target arrangement for a spallation neutron source. In G.S. Bauer (Editor), Proceedings of the Meeting on Targets for Neutron Beam Spallation Sources. Jül-Conf 34. Kernforschungsanlage Jülich GmbH, Jülich, FRG, pp. 155–183.

Fischer, W.E., Joho, W., Tschalär, Ch., Sigg, B., and Rauch, B. (1979). Studie über eine kontinuierliche Spallations-Neutronenquelle am SIN. 200 Dok. -Inf. Schweizerisches Institut für Nuklearforschung, Villigen, Switzerland.

Garvey, P.M. (1980). Neutron production by spallation in heavy metal targets. In G.S. Bauer (Editor), Proceedings of the Meeting on Targets for Neutron Beam Spallation Sources. Jül-Conf 34, Kernforschungsanlage Jülich GmbH, Jülich, FRG, pp. 1–16.

Hoffmann, H. (1980). Prospects of a liquid metal target system for a spallation neutron source. In G.S. Bauer (Editor), Proceedings of the Meeting on Targets for Neutron Beam Spallation Sources. Jül-Conf. 34. Kernforschungsanlage Jülich GmbH, Jülich, FRG, pp. 127–153.

Mittag, K. (1979). In Proceedings of the Proton Linear Accelerator Conference, Montauk BNL 51134. Brookhaven National Laboratory, Upton, New York, p. 253.

Rudolph, H. and Wrobel, D. (1981). Alltag des Einmaligen. Wissenschaft und Fortschritt, 31:101–105.

Russell, G.J., Meier, N.M., Gilmore, J.S., Prael, R.E., Robinson, H., and Taylor, A.D. (1981). Measurement of spallation target–moderator–reflector neutronics at the weapons neutron research facility. In Y. Ishikawa et al. (Editors), ICANS-IV: Proceedings of the Meeting of the International Collaboration on Advanced Neutron Sources, 4th. KENS Report II, KEK Laboratory for High Energy Physics, Tsukuba, Japan, pp. 210–223.

Stokes, R.H., Wangler, T.P., and Crandall, K.R. (1981). The radio frequency quadrupole – a new linear accelerator. IEEE Transactions on Nuclear Science, NS-28(3):1999–2003.

Vetter, J.E., and the Linear Accelerator Working Group, Kernforschungszentrum Karlsruhe (1981). A high intensity proton linear accelerator for the German spallation neutron source (SNQ). IEEE Transactions on Nuclear Science, NS-28(3):3455–3457.

Zeitnitz, B. (1981). Neutrino physics at the spallation neutron source. KfK 3155. Kernforschungszentrum Karlsruhe, FRG.

ON THE NEED FOR ACCELERATOR BREEDERS AND FISSION CONVERTER REACTORS

P. Jansen
Institut für Energiewirtschaft, Technische Universität Wien, Vienna (Austria)

In the following discussion, the term "accelerator breeder" simply denotes a machine producing one metric ton of fissile plutonium per year. The question is whether developing and implementing such machines would have significant advantages with respect to the worldwide need for natural uranium. In considering this question we start by elaborating some yardsticks to judge what "advantageous" may mean and this leads us to an estimate of the long-term nuclear energy supply. We will restrict the discussion to the uranium–plutonium fuel cycle since it is the only one that has already been demonstrated.

Essentially, no one knows what amount of nuclear energy might be used in the future. However, the incentive for discussing the needs for accelerator breeders stems from the possibility that mankind might be well advised to prepare for a large contribution from nuclear energy to the world's energy needs. For the WOCE region (the World Outside the Centrally-planned Economies), a figure of 600 GWe in the year 2000 and double that in 2025 is not unreasonable. Furthermore, the use of nuclear energy in developing countries might be only just beginning then, so in order to assess the role of accelerator breeders quantitatively we go on adding 600 GWe every 25 years. This leads to 2400 GWe in the year 2075, representing about 5 GWe for every 14 million people in the WOCE region. For industrialized economies this would not be very much.

Starting with the case of the Light-Water Reactor (LWR) once-through (OT) fuel cycle, which requires 135 tons of natural uranium per gigawatt-year of electricity generated, we arrive at 17 million tons of natural uranium used by the year 2075. There is a chance that future improvements in LWR technology (starting in 2000) might lower the figure to 12 million tons. Though it is not impossible that 12 million or even 17 million tons of natural uranium may be available, scientists and politicians generally agree that it would be highly risky to rely on this amount. In particular, there might be severe ecological obstacles to supplying so much uranium. International institutions agree on 5 million tons of natural uranium as the upper limit for policy-oriented analysis. This means that we need some basic improvement in the efficiency of uranium *usage*. Here the concept that is technically most developed is the Fast Breeder Reactor (FBR).

Assuming present-day design goals for FBRs (with a breeding ratio of 1.2 and a cycle inventory of 6.4 tons of fissile plutonium) can be met, and starting them in 2000

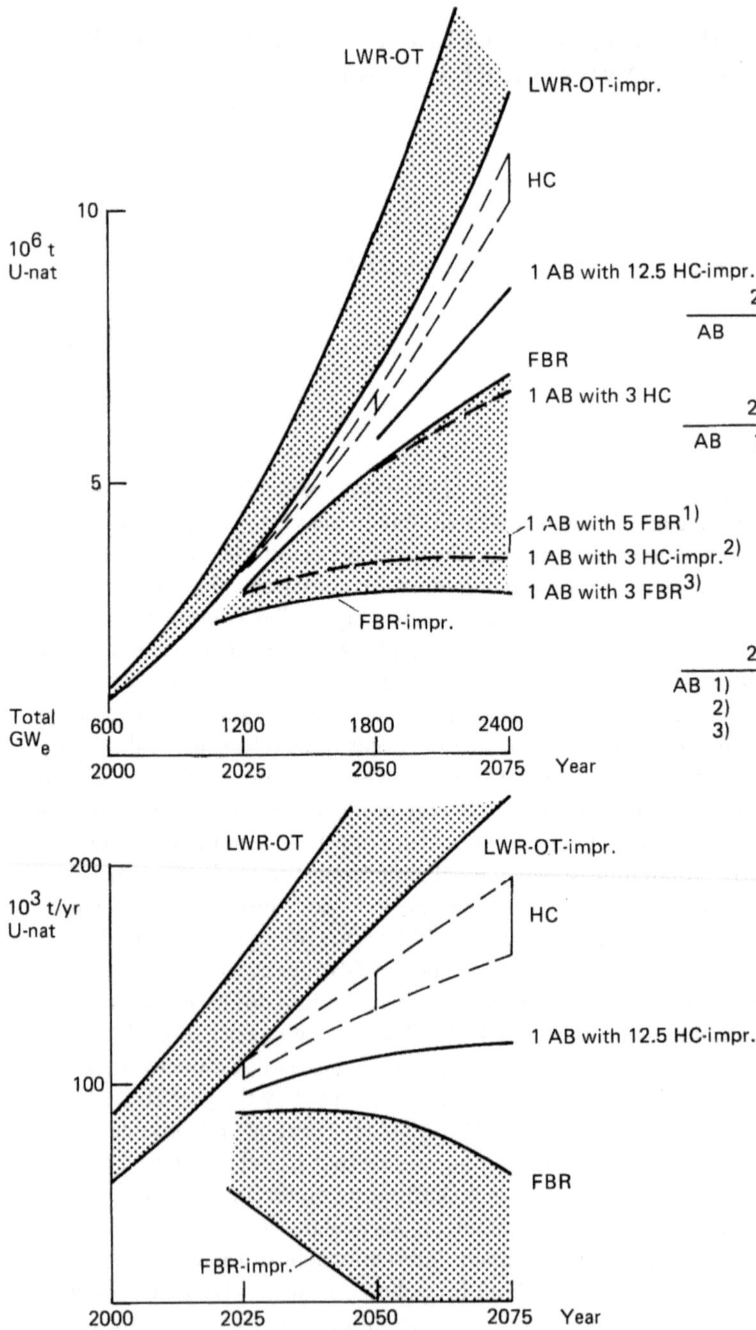

	2025	2050	2075
AB	40	80	120

	2025	2050	2075
AB	145	370	265

	2025	2050	2075
AB 1)	280	0	0
2)	245	145	190
3)	120	0	0

FIGURE 1 Natural-uranium demand for several breeder strategies. The following abbreviations are used: HC, high converter; AB, accelerator breeder; impr., improved; U-nat, natural uranium.

and using up the plutonium stock accumulated by then, the cumulative natural-uranium need until the year 2075 could be brought down to 7 million tons. By then the annual natural uranium requirement will be decreasing, having already fallen to 50,000 tons per year, whereas in the LWR-OT case it would be increasing and have reached a level of 230–330 thousand tons per year by 2075, even assuming that the LWR is of the improved type (with a fissile-plutonium surplus of 110 kg per GWe-year). Further improvements of FBR technology producing a breeding rate of 1:3 and a 1-yr out-of-pile time (i.e., a system inventory of 4.5 tons of fissile plutonium) might even make possible a cumulative consumption of only 3 million tons of natural uranium up to the year 2050 with no further uranium needs thereafter. With a cumulative natural-uranium requirement of 3–7 million tons, according to these calculations, the use of FBRs may meet the availability restrictions on natural uranium.

The FBR is an expensive and somewhat complicated machine. Much thought has been given to alternatives which achieve similar goals using cheaper and easier technologies. One step in this direction is an LWR variant with a tight lattice, plutonium fuel, and a high conversion rate of about 0.9 (with a system inventory of 6 t of fissile plutonium) or 0.95 (with a system inventory of about 8 t). Using these LWRs alone leads to a cumulative uranium demand of 10–11 million tons by the year 2075. The second step, however, is to support the high converter by accelerator breeders. Various support strategies are possible. Some illustrative examples are presented below and are depicted in Figure 1.

If we were to install one accelerator breeder for every nine to twelve high converters there would be no need for plutonium refueling from outside except for the inventory buildup. In this case the growth of our power-demand projection requires 110–160 thousand tons of uranium in 2025, reaching a cumulative total of 8.5–10 million tons by 2075.

If on the other hand we were to install one accelerator breeder for every three high converters, the results would be similar to those for the FBR strategies; i.e. we would arrive at a cumulative natural-uranium need up to 2075 of 3.5–6.5 million tons, and a zero or declining natural-uranium need thereafter. Thus we have a substitute for FBRs if one accelerator breeder is provided for every three high converters. The reason that we need so many accelerator breeders is that we have to provide for the buildup of a significant plutonium inventory. It is questionable whether a system of three high converters plus one accelerator breeder would be cheaper than three FBRs and whether an accelerator breeder is technologically any simpler than an FBR. In the above schemes for accelerator breeding, the fuel cycle is the same anyway. Thus, when deciding which strategy is appropriate for the future, the potential advantages of an accelerator breeder system (or perhaps a fusion breeder system which is not considered here) will have to be weighed against the fact that at present the FBR is technically the most developed system of the three and does not cause any reduction in uranium efficiency. In these considerations we have assumed that the accelerator breeders will be available starting from the year 2000. If they are introduced later, their importance would further decline.

Section Five

Handling the Waste

FUTURE PROCESSING OF SPENT REACTOR FUELS

F. Culler and R.C. Vogel
Electric Power Research Institute, Palo Alto, California (USA)

ABSTRACT

The probable future for the reprocessing of nuclear fuels over the next 50 years is discussed in the light of four main factors: the processing of light water reactor (LWR) fuel on schedules to be determined by individual national policies; the eventual processing of fast reactor fuels; present and future degrees of concern on arms proliferation issues and the vulnerability of processing plants to terrorist activities; and general public acceptance of reprocessing. An overriding factor may well be the institutional and financial arrangements necessary for reprocessing (and therefore the whole fuel cycle) to progress. The general principles of reprocessing technology have already been established although improvements may still be necessary in meeting new safeguard requirements and developing methods of waste disposal that are acceptable to the public. Use of the "sphere pac" fuel fabrication process colocated with appropriately designed reprocessing plants may allay the serious concern about possible diversion of plutonium.

1 INTRODUCTION

In this paper we will consider the speculative subject of the future of reprocessing during the next 50 years. Some of the factors involved in the future may be (a) processing of Light-Water Reactor (LWR) fuel, perhaps on a delayed schedule as determined by individual national policies, (b) eventual processing of fast reactor fuels, (c) continued pressure on proliferation issues, (d) concern about the possible vulnerability of reprocessing plants to terrorist activities, and (e) public acceptance of reprocessing. The last factor depends on a well-formulated radioactive-waste disposal program. An overriding factor may be the institutional and financial arrangements which permit reprocessing and the rest of the fuel cycle to move ahead. In spite of the foregoing challenges to moving ahead with reprocessing it is highly important to do so to alleviate the accumulation of spent fuel. In considering reprocessing the large expense of plants is considered to be a disadvantage. However, the cost of a single reactor is very much greater than the reactor's share of the reprocessing cost so that the problem is really an institutional one of organizing the reprocessing activity for a group of reactors. The same reasoning is of course valid for the waste-disposal component of reprocessing. Thus only large "collectives" of perhaps 50 reactors can achieve the maximum cost savings of reprocessing using a large plant rather than a number of smaller reprocessing plants. However, the fractional cost increase in the fuel cycle if a small reprocessing plant is used is not so great that most groups could not still afford to reprocess in a small plant if they so desired.

Reprocessing technology is established in a general way except that possibly it may not adequately meet new requirements for safeguards and waste disposal. There will very possibly be a need for near-complete containment of wastes through a recycle within the plant with a single high-level waste stream to be handled. The reprocessing of fast reactor fuels will also require an approach that is somewhat different from that for LWR reprocessing. For example, the higher plutonium concentrations can cause dissolution difficulties and different criticality problems. The mechanical head end, if it is used, will have to be modified. In our view, however, it does not seem likely that a completely new reprocessing technology will be commercialized within the next 50 years, although the evolution of significant improvements in waste-treatment techniques and perhaps in the head end can be expected.

With regard to the public attitude on reprocessing and its associated waste-disposal problems, we feel that a less negative posture may develop. The certain disappointment in "soft energy" approaches and enhanced concerns about (a) the environmental effects of CO_2, (b) toxic chemical wastes, and (c) the dependability of oil supplies could all help to develop a more balanced attitude about radioactive wastes.

In this paper it is appropriate to take a world prospective. However, each country's situation will differ depending on the political environment, the availability of other energy sources, the degree of industrialization, and the international institutional arrangements available to undertake reprocessing and the balance of the fuel cycle.

2 FACTORS INFLUENCING FUTURE REPROCESSING

2.1 Loads

The future reprocessing loads to be expected will depend on the world growth of nuclear energy. One of the current sources of information is the International Nuclear Fuel Cycle Evaluation (INFCE, 1980a). Table 1 presents estimates from this document which can be used as a basis for discussion.

In addition to the worldwide nuclear energy demand, the reprocessing loads will depend on the variables involved in reactor operation such as burnup, the attitudes of various countries concerning the long-term storage of fuel instead of reprocessing, the time of introduction of fast reactors, and the rate of growth of fast-reactor technology.

The data from the INFCE(1980b) report on reprocessing shown in Table 2 give the assumed worldwide spent-fuel discharges to the year 2000. These totals exclude 6,500 tons of spent fuel, mostly in storage in North America at the end of 1977. The data of Table 2 indicate an increasing amount of spent fuel in storage; it is estimated that there will be 40,000 tons in 1985, 86,000 tons in 1990, and 225,000 tons in 2000. If the worldwide generating capacity in the year 2000 is 850 GWe (see Table 1) and the discharge rate is 30 tons of fuel per 1000 MWe, the spent-fuel discharge rate will be approximately 25,000 tons per year. Thus, without additional reprocessing capability, not only will there be no opportunity to reduce the amount of spent fuel in storage

TABLE 1 Projected nuclear generating capacity in the world outside the centrally planned economies (in gigawatts of net electric power).

Year	Low case		High case	
	Total	FBR	Total	FBR
1985	245	2.1	274	2.4
1990	373	5.1	462	6.6
1995	550	12.8	770	15.9
2000	850	21.8	1200	38.3
2005	1100		1650	
2010	1300		2150	
2015	1450		2700	
2020	1650		3350	
2025	1800		3900	

but it will be accumulating at an ever-increasing rate. Over 70% of the cumulative world spent-fuel discharges will *not* have been reprocessed by the year 2000. Table 2 clearly shows that the United States has the greatest accumulation of spent fuel in storage. The data in INFCE and presented in Table 2 were developed on the assumption that the United States will not undertake reprocessing before 2000. We hope that this will be proved to be an erroneous assumption.

A future need for reprocessing arises since there will not be sufficient plutonium to undertake the planned fast-reactor programs in the world outside the centrally planned economic areas on the schedule anticipated unless reprocessing is accelerated. There will be a shortfall of about 130 tons of fissile plutonium. This 130 tons could be readily recovered from the approximately 1,540 tons of plutonium in the spent fuel that will be in storage by the year 2000 if additional reprocessing capacity were made available.

Using the nuclear generating capacity shown in Table 1 it is possible to estimate the loads beyond 2000. This is of course quite speculative. Some of the uncertainties involved in converting the data of Table 1 into reprocessing loads for 2025 concern the fraction of the total which will be generated by fast reactors in 2025, the characteristics of the fast reactors, and the characteristics of the thermal reactors in use at that time. INFCE (1980c), perhaps with some optimism concerning the rate at which breeders will be introduced, predicts a breeder capacity of 50 GWe in 2000 and 200 GWe in 2005. The 50 GWe for the year 2000 does not agree with the data presented in Table 1 which come from Working Group 1 rather than Working Group 5 of INFCE. The predicted capacities indicate an addition of 30 GWe year^{-1} from 2000 to 2005. If one assumes the same continuing rate of introduction of fast reactors, by 2025 there would be perhaps 800 GWe generated by fast reactors. There would then be 1000 GWe from LWRs adding up to 1800 GWe for the low-range prediction. The data just presented may seem to be optimistic concerning the rate of introduction of the breeder. However, we believe that it is now a cheaper power source than coal and oil. Furthermore, the very low uranium requirements and the lack of dependence on enrichment are powerful arguments for pressing forward with an aggressive breeder program.

If we assume that the core burnup of fast reactor fuel is about 50 MWd kg^{-1} then about 25–30 tons year^{-1} of blanket plus core must be reprocessed in the fast-reactor

TABLE 2 Spent fuel arising from thermal reactors in the world outside the centrally planned economies (in tons of heavy metal at end of year)[a].

Region	1985			1990			2000		
	Cumulative spent-fuel discharge	Spent fuel in storage at year end	Cumulative spent fuel reprocessed	Cumulative spent fuel discharge	Spent fuel in storage at year end	Cumulative spent fuel reprocessed	Cumulative spent-fuel discharge	Spent fuel in storage at year end	Cumulative spent fuel reprocessed
European Economic Community	8,200	3,200	5,000	22,100	10,500	11,600	> 57,300	> 8,050	49,250
Total Europe	12,000	6,100	5,900	30,200	16,900	13,300	> 70,200	> 11,100	59,100
North America	27,100	27,100	0	56,450	56,450	0	166,700	166,700	0
Pacific	3,900	1,500	2,400	8,300	2,700	5,600	30,000	0	30,000
Total OECD	43,000	34,700	8,300	94,950	76,050	18,900	> 266,900	> 177,800	89,100
Non-OECD	4,100	3,850	250	11,600	9,950	1,650	> 54,500	47,100	7,400
Total	47,100	38,550	8,550	106,550	86,000	20,500	> 321,400	> 244,900	96,500

[a] See INFCE (1980b) for assumptions.

case. If higher burnups are achieved, the reprocessing rate of course decreases. Thus within wide limits the annual amount of material to be reprocessed does not depend strongly on the mix of fast versus thermal reactors but rather depends more on the total growth of nuclear power. Whether fast or thermal reactor fuel is being reprocessed is of course a matter of concern to the reprocessor because of the differing procedures required.

The rate of discharge of spent fuel beyond 2000 was estimated by using the low-range growth curve. The very rough figures are presented in Table 3. These figures can also be checked against those in the report of the Organization for Economic Cooperation and Development on Nuclear Fuel Cycle Requirements (OECD, 1978) which are slightly higher.

TABLE 3 Estimated spent-fuel discharge from LWRs + fast reactors in the world outside the centrally planned economies (in tons per year).

Year	Discharge
2000	26,000
2005	33,000
2010	37,000
2015	42,000
2020	44,000
2025	46,000

The estimated worldwide reprocessing capacity as used by INFCE (1980b) is also needed to give a reasonably quantitative picture of the status of reprocessing. These data are presented in Table 4. A comparison of the spent-fuel discharges of Table 3 with the reprocessing capacity as presented in Table 4 shows that a serious accumulation of spent fuel will occur and will become worse each year. In addition, the data of Table 2 show that as we enter the twenty-first century there will be 250,000 tons of spent fuel already in storage.

TABLE 4 Estimated reprocessing capability in the world outside the centrally planned economies (in tons per year).

Year	Schedule	
	High	Low
1980	750	750
1985	1200	1200
1990	5900	4100
1995	9500	8100
2000	9500	8100

In summary, there will be an ever-increasing shortfall of reprocessing capacity unless many plants are built during the next 50 years. We do not believe that storage of increasingly large amounts of spent fuel is prudent from considerations of (a) the economical use of fissionable and fertile material and (b) the increasing ease of diversion of plutonium from the spent fuel as its associated fission products decay.

2.2 Potential New Safeguards Criteria

In considering the future of reprocessing it is necessary to extrapolate the safeguards criteria over the next few decades. This has two features. One feature centers on the diversion of weapons material by subnational groups. The other feature is to ensure that countries acquiring nuclear power facilities do not use these facilities in national nuclear weapons programs.

As one considers future trends it is difficult to anticipate whether criteria related to safeguards will be more or less strict. If during the next few decades the number of states with nuclear weapons significantly increases, concerns regarding the proliferation of nuclear weapons may decrease since the proliferation will be a fait accompli. It does not seem likely that the trend will be in the other direction.

Considering the present trend of events with regard to the possible diversion of fissionable material by terrorist (subnational) groups, it seems likely that these activities will increase. Therefore we guess that the protection of fuel-cycle facilities from terrorists may require measures for safeguards and physical protection that are even more stringent than those that are now envisaged.

In addition to effective safeguards measures being in place, the public must have confidence that such measures are indeed effective if nuclear energy is to be accepted.

A number of papers have been written on the subject of procedures which can be used to evaluate fuel cycles from the viewpoint of safeguards. It is not within the scope of this paper to go into this subject in any detail. However, the following brief comments may be of interest. The easier plutonium is to convert to a weapon, the more attractive it will be to a potential divertor. Four attributes of plutonium-bearing materials might be considered for a safeguards evaluation: (1) the concentration of plutonium in the material; (2) the radiation intensity of the material per unit weight of plutonium; (3) the time required to extract the plutonium in metallic form; (4) the plutonium detection sensitivity at the specific point in the process under consideration.

An additional group of important factors relate to both the facility design and the process itself. These are (1) access denial provided by the facility design, (2) inventory capability, (3) safeguard-instrument reliability, (4) personnel access to plutonium-bearing material, (5) penetrations between personnel work areas and process vessels, (6) unidirectional macroflow of plutonium (lack of recycle), (7) maintenance procedures, (8) surveillance procedures, and (9) operations procedures.

A further important factor is the cost and benefit of each step taken to achieve an enhanced degree of safeguards. The installation of overly conservative and expensive safeguards procedures represents a cost to the user of electricity almost in perpetuity; indeed, once installed, such procedures are awkward to discontinue even though generally recognized as not cost-effective.

The Electric Power Research Institute (EPRI) is currently supporting work by Exxon Nuclear on the development of evaluation methodology to assess the diversion resistance of fuel cycles. The factors involved in this methodology are those mentioned in the preceding paragraphs. However, an attempt is being made to evaluate fuel cycles in a more quantitative fashion than can be described here.

A rough statement of the tentative conclusions is that within the reprocessing segment of the fuel cycle the more vulnerable areas are (1) the PuO_2-conversion area, (2) the PuO_2-rework area, and (3) the laboratory. Common concerns at these locations were the need for improved control of personnel movement and reduced hands-on access to PuO_2. A recently issued study by Cameron and Bleck (1980) based on the AGNS (Allied-General Nuclear Services) facility at Barnwell has also identified the analytical laboratory and PuO_2-conversion areas as having the greatest diversion vulnerability.

Another important conclusion which appears to be emerging from the EPRI–Exxon Nuclear study is that the traditional method of fabricating plutonium-bearing fuels, i.e. pelletizing, is a particularly vulnerable operation because of hands-on access to PuO_2 and the difficulty of controlling personnel movement. This is leading to the recommendation that fuel fabrication should be carried out by a process which is less vulnerable to these criticisms. The fuel-fabrication procedure which will be recommended is "sphere pac". This process appears to offer appreciably fewer opportunities for diversion.

The opinion just stated is sufficiently important to deserve further elaboration. The Exxon Nuclear group has tentatively concluded (final reports are not yet published) that a "sphere pac" fuel-fabrication process for Liquid-Metal Fast Breeder Reactor (LMFBR) fuel would require six enclosures as compared with 23 enclosures for the standard pellet process. This means that the safeguards problems are simpler for a "sphere pac" process. Furthermore, powder preparation, pelletizing, and grinding operations are eliminated. The "sphere pac" process can eliminate the scrap-recovery operation by sharing it with reprocessing. This does not seem to be such a reasonable suggestion for the pellet process since the scrap for the pellet process is more likely to be contaminated with extraneous materials than that from a "sphere pac" process. Finally an analysis of the "accountability points" and/or potential scrap sources (an inverse figure of merit used in evaluating the ability to safeguard) suggests about 50 for the pellet process compared with 16 for the "sphere pac" process. Another overall aspect of the "sphere pac" process is that it appears to be possible to automate it more readily than the conventional pellet process. The automation will lead to lower radiation exposure for workers. The final report will expand on these areas.

The foregoing recommendation for "sphere pac" fuel fabrication also has the interesting effect of mitigating the more serious diversion concerns for fuel reprocessing since it is possible (a) to coprocess, i.e. to keep plutonium diluted with uranium, and (b) to avoid conversion to PuO_2. Thus it becomes clear that in matters of safeguards it is necessary to consider the whole fuel cycle. Therefore in this paper it is appropriate that we discuss not only fuel reprocessing but also fuel fabrication.

Another study, supported jointly by the Swiss Federal Institute for Reactor Research and EPRI and being carried out by Professor B.I. Spinrad of Oregon State University has reached a similar conclusion concerning fuel fabrication of plutonium-bearing fuels and recommends that "sphere pac" be considered a more diversion-resistant process than the conventional pelletizing process.

There are other possible approaches to the safeguards problem. For example, it is possible to use radioactivity to protect plutonium from diversion. This can be done by "spiking" the plutonium with fission products. There are several ways of accomplishing this objective (IAEA, 1977a, b; Meckani et al., 1977). These are more extreme measures which may not be necessary.

2.3 Institutional Considerations

As the need for reprocessing of spent fuel becomes more pressing for each nation it will become necessary to resolve the institutional problems which are currently impeding the reprocessing of spent fuel. The problems will differ from one country to another and indeed will change with time. Each country will, however, have a licensing group which has to be satisfied. There is a general need for licensing requirements based on reasonable technical requirements. These institutional problems appear to be more intractable than the technical problems.

2.3.1 Industry Support

In the United States the negative government attitude concerning reprocessing during the Carter administration is so well known that further elaboration in this paper is not required. It now appears that the current administration is developing a more positive attitude toward reprocessing. However, the development of another problem is now apparent in the United States. The disappointments which US industry has suffered involving investments in the fuel cycle are well known. Recent losses now total more than US$400 million. US industry is therefore thoroughly discouraged on further investment in the nuclear fuel cycle to compensate for dwindling governmental support, despite the more positive attitude toward doing the fuel-cycle job that seems to be developing. At this point we can do no more than identify the US problem and hope that a solution develops.

2.3.2 The Time Cycle

The difficulties cited in the previous section are somewhat peculiar to the United States. However, there are always inherent difficulties in a technical activity whose gestation period is perhaps of the order of ten years and which interacts with a political system operating on a significantly shorter time cycle (i.e. often four years in the United States between administrative changes). For long-term technical projects it is necessary to have transadministration stability if investment is to take place. This represents an institutional problem which is not peculiar to the United States and which needs to be addressed.

2.3.3 Regional Fuel-Cycle Centers

There are at least two recent studies considering regional fuel-cycle centers as a solution to some of the institutional problems. Both INFCE (1980b) and the International Atomic Energy Agency (IAEA, 1977a, b; Meckani, et al., 1977) have offered some significant ideas. The advantages stated for a regional fuel-cycle center are as follows. Reductions in the number of national facilities constructed and reductions in shipment of special nuclear materials will give a nonproliferation advantage. In addition, the intergovernmental agreements envisaged for the regional fuel-cycle center will enhance control of the transfer and use of nuclear materials. Radioactive-waste management will probably be facilitated by the provision of a centralized large-scale facility incorporating temporary liquid storage and solidification-plant operation followed by further storage and eventual disposal. There will also be economic advantages due to scale. Several countries with small nuclear programs could cooperate in the establishment of a regional fuel-cycle center in order to achieve the economies of the large plants.

Two complicating factors are involved in the consideration of the regional fuel-cycle center. It appears to us that the legal problems will be very complicated. It will be difficult to negotiate contracts between the many entities involved in the construction and operation. Another complicating factor is the assurance of some competition between centers so that a utility (or nation) is not the captive customer of a single fuel-cycle center.

2.3.4 Colocation (Reprocessing and Fabrication)

Another type of institutional problem is that of colocation of processing and fabrication facilities. We feel that colocation is advantageous from technical considerations. It is of course possible to envisage colocation of fuel fabrication and reprocessing plants on a national or international scale. Whatever the eventual arrangement, advantages would accrue owing to minimization of the risk of diversion through elimination of external shipments of fissile material. There would in addition be cost savings due to the use of common facilities. The colocation concept is inherent in the fuel-cycle flow sheet which is presented in Section 3 since scrap from fabrication is recycled back through reprocessing.

3 FUTURE REPROCESSING TECHNOLOGY

3.1 Guidelines

In this section we will outline our ideas of what the future reprocessing technology may be beyond 2000. In Section 2.1 we described the potential reprocessing loads. It appears from this information that a significant number of new reprocessing plants will be required. This will allow and stimulate innovation in the field. We have also pointed out that safeguards requirements, particularly with reference to defense against terrorists, may become stricter and that colocation of fuel reprocessing and fabrication facilities is attractive because it offers both enhanced safeguards capabilities and potential economic incentives.

Additional waste-management requirements will arise for reprocessing. It will be necessary to control tritium and krypton, to recycle low-level wastes internally, and to blend the low-level wastes with the high-level wastes. We believe that there will be a need for near-complete containment of the lower-level wastes within the plant with a single high-level waste stream to be handled.

The spent-fuel cooling time is another variable for consideration. For the next few decades it will certainly not be necessary to reprocess fuel that has only cooled for a short time. However, by 2025 if there is an aggressive fast-reactor program a plutonium shortage may develop. If this situation comes about there may be interest in decreasing the plutonium out-of-reactor inventory in the fuel cycle. The more pressing problems introduced by going from 180 days or more of cooling to 30–60 days of cooling would be self-heating of the fuel during the mechanical-head-end steps and iodine containment. At this point it does not seem necessary to address these problems.

From preliminary reports it appears that the EPRI–Exxon Nuclear study will suggest that a colocated fuel processing–fabrication plant based on solvent extraction with coprocessing followed by "sphere pac" fuel fabrication is more diversion resistant than the conventional solvent extraction plus pellet fuel fabrication. Furthermore it is our intuitive judgment that such a fuel cycle is possibly less costly than the conventional one. We will therefore take the liberty of assuming that this process is the reference process and will describe it in general terms. Since in this case fuel fabrication and fuel processing are closely related we will describe the complete fuel cycle.

3.2 The Process

There are three fuels which should be considered as future candidates for reprocessing: UO_2, LWR-MOX (i.e. mixed PuO_2–UO_2 fuel) and LMFBR fuel (presumably UO_2–PuO_2 but perhaps UC–PuC at some future time). We feel that it is likely that the core and blanket materials for fast reactors will be mixed and processed together. It seems likely that UO_2 and LWR-MOX fuels can be smoothly reprocessed in similar if not identical equipment. However, in the case of LMFBR fuels it may be desirable to have dedicated plants. This is because LMFBR fuels are different from LWR fuels since (1) the fuel may be wet with sodium or sodium logged, (2) the cladding is stainless steel instead of Zircaloy, (3) the fuel pins are of smaller diameter, (4) the dissolution characteristics may be different owing to the higher plutonium and fission-product concentrations, and (5) the plutonium concentrations in the various streams will be higher, leading to more stringent criticality restrictions.

The two fuels which we will consider as products of the "sphere pac" fabrication plant are the LWR-MOX and LMFBR fuels. It is very likely that these two fuels would be fabricated in separate lines or perhaps separate plants because of the more stringent criticality problems for the fast reactor fuel.

Figure 1 presents a generalized fuel-cycle flow sheet. Since the flow sheet is presented for three different spent fuels and two different product fuels various steps will differ depending on the situation. However, a generalized description of the process follows. For each stage, the numbers in parentheses refer to the boxes in Figure 1.

Fuel Reception and Storage (1)

Spent fuel is received in heavily shielded casks shipped either by rail or truck. The spent fuel is removed from the casks and stored in the fuel-storage pool. The water in the pool is kept clean by circulation through ion-exchange columns and filters. After cooling to the appropriate radiation level, the fuel is sent to the head-end preparation area for processing.

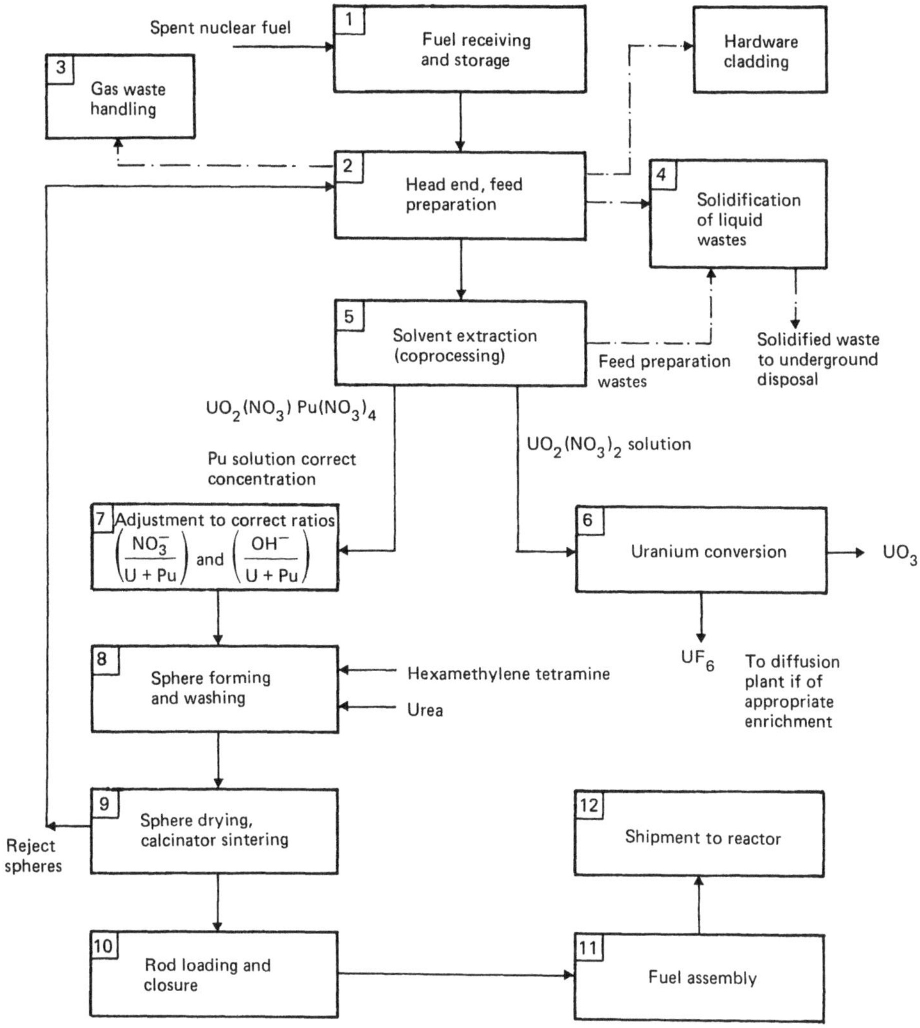

FIGURE 1 Generalized fuel-cycle flow sheet.

Head End, Feed Preparation (2)

Processing starts by cutting end fittings from the fuel assemblies. The remaining portions of the fuel assemblies are mechanically sheared into small pieces to expose the uranium and plutonium oxide. The fuel is then dissolved in nitric acid. Oxygen is added to the dissolver. This permits more efficient recovery of the evolved nitrogen oxides and allows control of fission-product off-gases (fumeless dissolution). The undissolved cladding is washed with water, removed from the dissolver, and sent to solid-waste treatment. The solution from the dissolver will contain cladding fines and some undissolved fission products (principally, molybdenum, ruthenium, rhodium, palladium, and technetium) and plutonium. This solution is clarified by centrifuge to separate the solids for additional treatment.

Gaseous-Waste Handling (3)

Off-gases are treated to remove radioactive iodine (if present), ruthenium, krypton, and particulates to give an effluent that is nearly free of radioactive materials. Iodine can be removed by adsorption on a silica-based material impregnated with silver. Kr-85 can be recovered from the off-gas by cryogenic distillation and packaged in pressurized gas cylinders. Ruthenium can be removed by adsorption on a bed of ferric oxide. All off-gases that have been freed of radioactive materials are passed through Fiberglas filters before discharge. The flow sheet shows a radioactive off-gas stream only from the head end whereas in fact there are also radioactive off-gas streams from other operations in the plant. Furthermore, the details of off-gas handling are considerably more complicated than this discussion may imply.

Solidification of Liquid Wastes (4)

This subject has been so controversial that a few general comments are appropriate. We believe that there are several quite adequately safe procedures for storing high-level radioactive wastes. It is not a profitable exercise to try to determine which is the safest of the safe procedures. A step which would help to resolve the waste question would be to establish reasonable criteria that were well supported by technical data. EPRI is establishing a systems-based methodology for determining these criteria through the Retention Quotient (RQ) procedure (Rodger and Tripathi, 1981). Physically, the reciprocal of the RQ of a repository system is the fraction of the source inventory which must be ingested by a receptor to give a selected radiation dose. A breakdown of the RQ for the components of the system is helpful in determining the required retention capabilities of various barriers such as waste form, repository structure, the geosphere, and the biosphere.

We have shown in our reference flow sheet that we would prepare a glass from the waste and store the glass in a deep dry (salt or granite) underground repository. The vitrification process which could very possibly be used would be similar to the French continuous AVM (Atelier de Vitrification de Marcoule) process or the Battelle spray calciner and joule melter.

Solvent Extraction (Coprocessing) (5)

The solvent-extraction system under consideration at this step requires further investigation. The "coprocessing" system involves two product streams: a decontaminated uranium–plutonium nitrate solution of the proper plutonium-to-uranium ratio to correspond to the specifications of the fuel being manufactured, and a decontaminated uranyl nitrate solution. The system presented has not yet been demonstrated and indeed further investigation would be required to establish it. However, there seems little doubt that such a procedure could be established.

Uranium Conversion (6)

The uranyl nitrate solution from the solvent-extraction system is concentrated to approximately 100% uranyl nitrate heptahydrate and is fed to a fluidized-bed calciner to form UO_3. The UF_6 is formed by fluorination with F_2 gas and is given a final purification by using sorbent beds containing sodium fluoride, magnesium fluoride, and cobaltous fluoride. Distillation can also be used for final UF_6 purification. The UF_6 can be fed to isotope enrichment if it is appropriate and if UF_6 is the desired compound.

It should be kept in mind that with uranium recycling undesirable isotopes such as U-232 may build up in the gaseous diffusion cascades, making the cascades radioactive, and that this is a problem which requires evaluation.

Adjustment to Proper $NO_3^-/(U + Pu)$ and $OH^-/(U + Pu)$ Ratios (7)

There is inadequate information in the open literature to prepare a definitive flow sheet for steps 7–9 when a mixture of uranium and plutonium is involved. A pilot plant is now in operation at Exxon Nuclear in Richland, Washington, in a jointly funded program between the Department of Energy and Exxon. This pilot plant uses the "internal gelation" procedure (employing hexamethylene tetramine (HMTA) and urea) with pure uranium and is directed to the fabrication of UO_2 spheres only. Such a procedure would undoubtedly have to be modified to prepare uranium–plutonium fuels. However, we suspect that the operations represented by box 7, i.e. adjustment of the ratios of the constituents of the solutions to specific values, will indeed be necessary to give appropriate gelation conditions.

Sphere Forming and Washing (8)

Boxes 8 and 9 in fact represent three separate lines for fabricating three separate sphere sizes. For LWR-MOX fuels the diameters of these sintered spheres would presumably be 1200, 300, and 33 μm. These three sphere sizes, with gentle vibration, pack to a smear density of 86–87%, the same smear density achieved by the use of pellets. In this case the plutonium concentration in the product spheres would be about 4% by weight.

If LMFBR fuel were being fabricated it might be desirable to use different sphere sizes because the fuel rod has a smaller diameter than the LWR rod. The plutonium concentration would be perhaps 15% by weight.

The sphere-forming step will possibly consist of mixing the adjusted and cooled uranium–plutonium nitrate solution with HMTA and urea. The HMTA decomposes to form NH_3 which precipitates ammonium diuranate gel in spherical drops. These drops are formed by passing the solution through a vibrating nozzle into a hot immiscible organic liquid. The heating of the solution brings about the HMTA decomposition. The spheres thus formed are washed, dried, calcined, and sintered to form spheres of near-theoretical density.

It should be noted that reject spheres can be recycled to the dissolver at the beginning of the reprocessing operations. The spheres are expected to be pure material as compared to the scrap from a pellet line which contains, for example, grinding sludge from wet-pellet grinding. The potential elimination of a dedicated scrap-recovery operation for fuel fabrication is a special advantage of colocation and of the "sphere pac" process.

The procedure that has been briefly sketched is general because the final process for plutonium-bearing fuels after development is complete may be rather different. However, there is a great deal of related work in the open literature on which investigators can draw.

Sphere Drying, Calcination, and Sintering (9)

The drying and calcination steps (if pure uranium is any indication) are completed when the temperature of the spheres has been increased from ambient temperatures to

400–600° C. The sintering is carried out at 1400–1600° C in a hydrogen atmosphere to produce UO_2. The conditions of the drying, calcination, and sintering steps have to be carefully established through systematic development work and will vary with the conditions in the sphere-forming step and of course with the plutonium concentration.

Rod Loading and Closure (10)

It has been found to be possible to vibrate a typical LWR rod gently while simultaneously feeding 1200-μm, 300-μm, and 33-μm diameter UO_2 spheres in the proportions of 56%, 27%, and 17%, respectively, to give a fuel rod of 86–87% smear density. These rods are performing satisfactorily in various in-reactor tests. The current designs of the LMFBR fuel call for smaller-diameter rods. Work specially directed at the loading of these rods may have to be undertaken.

Fuel Assembly (11)

The completed fuel rods are assembled in the traditional manner.

4 CONCLUDING COMMENTS

To institute the fuel cycle just described, traditional development work will be necessary. The following are examples of work to be done. (1) Waste management for the reprocessing including confinement of tritium and krypton will have to be improved. Improved containment of iodine may be necessary if it is felt desirable, in 50 years time, to process short-cooled fast reactor fuel, for example. (2) The coprocessing steps need to be established and demonstrated. (3) The sphere-forming flow sheets need to be developed and demonstrated. (4) Rod loading for fast reactor fuel needs developing.

We feel that increased attention must be given to new and improved reprocessing schemes over the next decades since it seems certain that fuel reprocessing and closure of the fuel cycle are necessities.

REFERENCES

Cameron, C.P. and Bleck, M.E. (1980). International safeguards for a light water reactor fuels reprocessing plant: containment and surveillance concepts. DOE Document SAND 80-0160, December. US Department of Energy, Washington, D.C.

IAEA (1977a). Regional Nuclear Fuel Cycle Center, Vol. 1. International Atomic Energy Agency, Vienna.

IAEA (1977b). Summary 1977 Report of the IAEA Study Project, Vol. 2. International Atomic Energy Agency, Vienna.

INFCE (1980a). Fuel and heavy water availability. Report of INFCE Working Group 1. International Nuclear Fuel Cycle Evaluation, Vienna.

INFCE (1980b). Reprocessing, plutonium handling, recycle. Report of INFCE Working Group 4. INFCE/PC/2/4. International Nuclear Fuel Cycle Evaluation, Vienna.

INFCE (1980c). Fast breeders. Report of INFCE Working Group 5. International Nuclear Fuel Cycle Evaluation, Vienna.

Meckani, V., Catlin, K.J., and Bennet, L.L. (1977). Regional Nuclear Fuel Cycle Centers, IAEA Study Project. IAEA-CN-36/487. International Atomic Energy Agency, Vienna. (Presented at the International Conference on Nuclear Power and its Fuel Cycles, Salzburg, Austria, May 2–3, 1977.)

OECD (1978). Nuclear Fuel Cycle Requirements. Organization for Economic Cooperation and Development, Paris.

Rodger, W.A. and Tripathi, R.R. (1981). On the use of Retention Quotients (RQ) for evaluation of repository performance. EPRI-NP Project RP1579-2, Interim Report, May. Electric Power Research Institute, Palo Alto, California.

A PROPOSED CONCEPT FOR ACTINIDE-WASTE TRANSMUTATION

Hiroshi Murata and Hideo Kuroi
Fast Reactor Physics Laboratory, Japan Atomic Energy Research Institute, 1-1-13 Shinbashi, Minato-ku, Toyko (Japan)

ABSTRACT

Underground nuclear waste disposal in stable geological formations requires predictions over periods of the order of 100,000 years; although these periods are not long by geological standards, it is difficult to get public acceptance for strategies based on such seemingly long-term forecasts. A concept for the transmutation of higher-actinide wastes from reactors is considered which might contribute to a significant reduction of the biological hazard of the wastes and thus help public acceptance of an expanding nuclear-power economy. Our present knowledge of the nuclear cross-sections of actinides is not sufficiently accurate for reliable forecasts of the behavior of the proposed Actinide-Burning Fast Reactor (ABFR) which will be characterized by a very hard neutron spectrum. Accordingly, a series of measurements is underway at the Japan Atomic Energy Research Institute to check seemingly favorable calculations on the possibility of operating an ABFR entirely on higher-actinide mixed-oxide fuel.

1 INTRODUCTION

Debates on the energy problem usually address two different aspects, namely resources issues and environmental issues. The environmental issues are related to the subjects of (1) fuel materials (i.e. safeguards and security), (2) energy-producing plants (plant safety), and (3) waste management (ultimate disposal).

During the early stages of nuclear development emphasis was placed primarily on providing safety, safeguards, and security for fuel materials and energy-production plants; more effort is still needed in areas (1) and (2) to define criteria of what is safe and what is not that are technically as well as publicly acceptable. During this period, however, comparatively little attention was given to ultimate means of disposing of high-level radioactive waste; only concepts for the temporary storage of waste were considered. It is now widely accepted that the most serious problem to be solved before the large-scale use of nuclear energy can proceed is the safe disposal of radioactive waste. Strong efforts have been made in many countries with nuclear power programs to establish acceptable strategies for radioactive-waste disposal. Present strategies involve the solidification of high-level liquid waste in glassy or ceramic structures and storage of the containers in underground geological formations that are known to be stable, in order to isolate the waste from the biosphere for about 100,000 years, until it decays to an innocuous hazard level.

Technical assessment studies on the safety of this method of disposal are being performed in many places using risk-analysis techniques. However, in implementing the waste-disposal policy we cannot overlook current trends in public opinion since it is extremely difficult to implement any large-scale project without public acceptance. Current public opinion shows the following two features: (1) there is a perception gap between the "specialist" and the public with regard to the risks of novel large-scale technology; (2) there are skeptics who suspect that the "specialist" is willing to emphasize the advantages of the technology while sweeping the disadvantages under the carpet for future consideration.

We now need a general means of convincing the public that they can rely on the "specialist". In order to cope with the difficulties of public acceptance we are convinced that it is essential to show the whole picture of the energy problem, starting with the relationships between the various alternatives, to deal with technical and social issues on different time spans, and to explain any specific technology under discussion as a part of the whole picture. It is also essential for the public to see that the whole picture consists of different alternative technologies, each supported by visible evidence.

With regard to disposal in stable geological formations, the risk analyses that have been made so far suggest that this will be quite safe. However, since there is no way to predict reliably the extraordinary events that might happen over the next 100,000 years, there will still be many controversial discussions from both the technical and the social standpoints in the assessment of this concept. It is therefore important to at least demonstrate that an alternative *well-defined* disposal option *does exist,* regardless of whether or not we implement it in the near future. In this context transmutation, using neutron-induced reactions, of the higher actinides contained in radioactive waste is considered.

The Japan Atomic Energy Research Institute (JAERI) has just started to touch on the basic research for demonstrating this transmutation concept. This paper is a summary of current activities in this field carried out at JAERI.

2 IMPLICATIONS OF ACTINIDE-WASTE TRANSMUTATION IN THE NUCLEAR ENERGY OPTION

To carry out any technical project related to the nuclear energy option usually requires enormous funds and long-range planning that is necessarily somewhat speculative. Project activities tend to acquire a large "momentum". Continual assessment of the project is therefore essential in order to clarify what has been achieved and what remains to be done. Any such assessment should take into account other alternatives, social considerations, and the resources problem as described next.

2.1 Energy Consumption and Fission Energy Resources

In illustrating the energy resources and consumption that are available to mankind, various numbers have been introduced as energy units. In order to help the public visualize information more easily, we introduce a new informative unit, the Sunshine Energy Unit (SEU), where 1 SEU is defined as 1% of the total energy that the earth receives from the sun per year. Estimated world energy resources reported by IIASA (Häfele et al., 1977) have been rewritten in the new sunshine energy units and are shown in Table 1.

TABLE 1 Estimated world energy resources in "sunshine energy units".

Resource	Amount
Fossil	1.47
Lithium up to \$60 kg^{-1}	140.0
Uranium up to \$250 kg^{-1}	166.0
Thorium	120.0
Uranium in ocean	4.4×10^3
Lithium in ocean	0.2×10^6
World energy consumption in 1975	5×10^{-3}

A model to estimate the global climatological change due to energy consumption will be the subject of a future study, but we should note that a disruption of the global climate is quite within the bounds of possibility if the global energy release (on top of what the earth receives from the sun) is somewhere between 0.1 and 1 SUE. Therefore it is reasonable that world consumption of nonsolar energy should not exceed 0.1 SEU. As can be seen in Table 1, fission energy resources will last over 3000 years even at the maximum consumption rate of 0.1 SEU y^{-1} and the resources are considered to be essentially inexhaustible. The value of 0.1 SEU is about 20 times the world energy consumption in 1975 (0.005 SEU).

Assuming a simple extrapolation of a 3% annual increase in world energy consumption, we will reach 0.1 SEU in 2080, about 100 years from now. We will then enter into a new era where world energy consumption is constrained by climatological considerations and any increment of energy consumption must, in principle, be supplied by the solar energy. In the period of transition to the new era, energy consumption will be constrained by the technical and social problems of establishing a new large-scale technological system to utilize nuclear energy.

2.2 Radioactive Waste Generated by Fission and Fusion

Radioactive wastes produced from fission and fusion reactors have different characteristics. Per unit energy released, there are about five times more neutrons which are seven times more energetic in a fusion reactor than in a fission reactor, but there are no radioactive "ashes" from D–T fusion. Radioactive wastes from fusion reactors are therefore characterized by radioactivity of structural material induced by high-energy neutrons. Considering differences in time spans for decay to a harmless level, we can broadly divide radioactive wastes from fission reactors into two groups: radioactive "ashes" of fission products and the so-called actinide waste produced in the reactor's own fuel cycle.

Häfele et al. (1977) evaluated radioactive wastes generated by fission and fusion reactors. Their results are used in Figure 1 to compare the decay and hazard characteristics of radioactive wastes from a fusion reactor and from a fission reactor (both fission products and actinide waste); a parameter of Biological Hazard Potential (BHP) is used. It should be noted that the BHP resulting from fusion is nearly the same as that of fission products after about 1000 years. As far as the concept of radioactive-waste disposal in a stable geological formation is concerned, confinement for a long enough period for decay to a harmless level to take place (in this case the residual radioactivity after about 1000 years) is essential. Therefore, as can be seen in Figure 1, if the

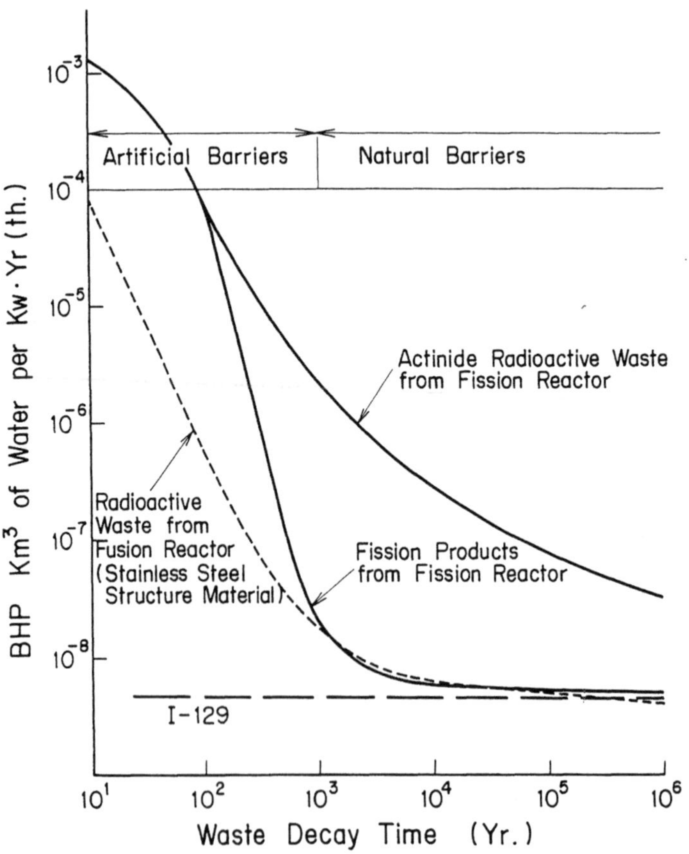

FIGURE 1 Decay and hazard characteristics of radioactive waste from fusion and fission reactors.

actinides can be separated from the waste and can be transmuted to fission products by neutron-induced fission, the severe requirements imposed on the waste-disposal concept will be alleviated significantly.

2.3 Radioactive-Waste Disposal in a Stable Geological Formation

The lifetimes of some nuclides in radioactive waste are very long by human standards. However, they are short in comparison with geological times. Consequently a concept of ultimate confinement of radioactive waste in geological formations is being evaluated by many countries and it is a common view among a number of geologists that suitable geological formations do exist in the world.

In November 1977 the International Atomic Energy Agency (IAEA) published the report "Site Selection Factors for Repositories of Solid High-Level and Alpha-Bearing Waste in Land Geological Formations" (IAEA, 1977) to provide useful background information to those authorities responsible for national radioactive-waste-management programs.

Many factors must be considered in selecting the site for a waste depository: the time span for confinement, the heat generation from the waste, the corrosive nature of the waste, the course of any underground water, the sorption coefficient, the type of borehole, etc. The time span of confinement of waste with a long half-life is the most important problem to be solved. The general conclusion of the IAEA report is that disposal of waste with a long half-life must be appropriately assured on the basis of a confinement time of the order of 100,000 years at least. In guaranteeing this confinement time, it should be noted that many geological formations which have been stable for hundreds of thousands of years in the past are likely to remain so for another 100,000 years in the future.

Given the length of the confinement time required, especially for α-bearing wastes, a storage-facility concept cannot be considered as a satisfactory alternative to waste disposal, and we should be very cautious in placing much reliance on artificial barriers for long-term confinement.

The IAEA report also mentioned that countries engaged in nuclear power generation should develop disposal sites for long-lived radioactive waste *within their own countries.* The trend observed in many discussions in the IAEA report and in some others on waste disposal is towards continental geological structures or salt formations as prospective sites for ultimate waste disposal.

In the problem of waste disposal in a stable geological formation the major question from the point of view of public acceptance may be the explanation and assurance of 100,000 years' confinement in a geological formation; a time span of 100,000 years is not easily comprehensible to the public because only a few thousand years of human history are well known and it is debatable whether any model used for risk analysis can reliably predict extraordinary events that might occur over the next 100,000 years.

3 ACTINIDE-WASTE TRANSMUTATION: PRELIMINARY CONSIDERATIONS

Since Claiborne's transmutation study was published (Claiborne, 1972), much work has been carried out and published by numerous organizations. These efforts indicate that the study of transmutation still has some speculative features as shown by the variety of technical options and in the plethora of possible conclusions. Taking account of results which have been reported so far, we consider the following points in the present study.

(1) Many reports have concluded that the harder the neutron spectrum, the better is the transmutation system for the actinide waste. The general reason for this conclusion is that the fission-to-capture cross-section ratio of actinides increases with increasing neutron energy. However, a more important reason which should be emphasized is that various dominant actinides in the transmutation chain (Np-237, Am-241, and Am-243) possess an appreciable fission cross-section of a threshold type in a high-energy region, as shown in Figure 2. This contributes to the achievement of a high transmutation rate and a high multiplication of neutrons if such a spectrum is available.

(2) Fast reactors, fusion reactors, and accelerator reactors are considered as transmutation candidates because of their ability to produce fast neutrons in large quantities. Typical examples of neutron spectra in fast reactors and in an accelerator reactor are shown in Figure 2. The hardest neutron spectrum achieved in a fast reactor, as represented by the JEZEBEL critical assembly at Los Angeles Scientific Laboratory, is shown together with the neutron spectrum of a Liquid-Metal Fast Breeder Reactor (LMFBR). A fast reactor used to transmute the actinide waste would most likely have a neutron spectrum somewhere between those of the LMFBR and JEZEBEL. As can be

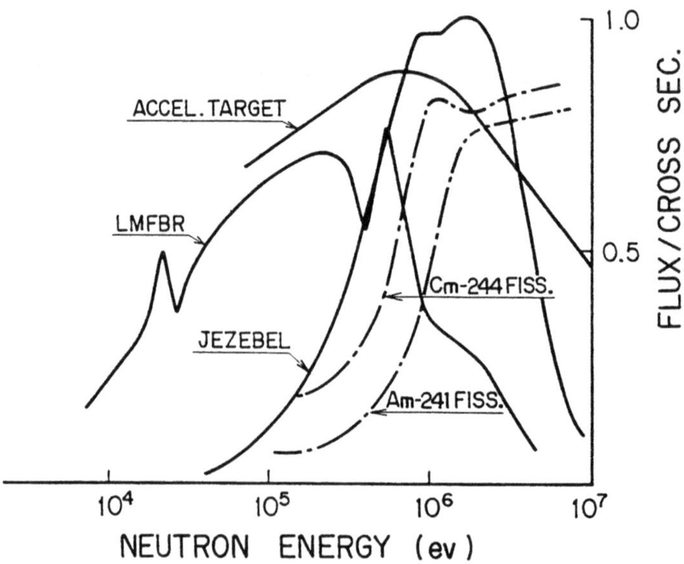

FIGURE 2 Neutron spectra and fission cross-sections.

seen in Figure 2, fast reactors may provide a sufficiently hard neutron spectrum for the efficient transmutation of actinides.

A fast fission reactor with a hard neutron spectrum is the primary option considered here, since the engineering feasibility of controlled thermonuclear fusion reactors and accelerator reactors are somewhat uncertain at present. As a long-term prospect for actinide transmutation, however, controlled-fusion or accelerator reactors seem to be alternatives well worth exploring, especially for easing difficult "out-of-reactor" problems associated with fuel reprocessing, partitioning, and fabrication processes.

(3) Problems to be solved in achieving transmutation of actinide waste arise in two sectors, the "in-core" sector and the "out-of-core" sector. With regard to the in-core sector, many studies have proposed the spiking of actinide target subassemblies in a host reactor for transmutation. The conventional LMFBR has been considered as a host reactor. The inclusion of target subassemblies in a host reactor might have an adverse impact on the optimum operating conditions of the host reactor whose main purpose is power generation. The main effects would be problems of power peaking and decay-heat removal. To eliminate any impact on conventional power-producing reactors it would be preferable to design a special Actinide-Burning Fast Reactor (ABFR) system with the primary purpose of transmutation, and with power production of only secondary concern.

Since the ABFR would deal with only 4% of the total fuel cycle, the impact of its investments and running cost on the total power-generation cost will be minor, and since power generation is not its major concern, its operating conditions can be easily chosen to satisfy various safety considerations. The introduction of a special burning reactor that is completely separate from the power reactors is essential for realizing an on-site fuel reprocessing and fabrication concept, which has been demonstrated once at EBR-II.

(4) Within the framework of the complete fuel cycle for utilizing nuclear energy, actinide partitioning and transmutation will introduce a new fuel-cycle approach, namely the "strata fuel-cycle concept". The strata of fuel cycles should be constructed so as to minimize the impacts of a new cycle on the conventional fuel cycle for power generation. It could be dealt with best by considering two separate strata whose interface is radioactive waste from the fuel reprocessing and fabrication plants. As schematically shown in Figure 3, the conventional fuel-cycle facilities of Light-Water Reactors (LWRs) and LMFBRs are not perturbed by higher actinides at all and will carry about 96% of the total fuel cycle. The higher-actinide fuel cycle consists of actinide-fuel reprocessing and fabrication plants and of the ABFR. This latter fuel cycle will carry about 4% of the total if the ABFR can be operated entirely on the higher-actinide "waste".

Assuming 99.5% recovery of uranium and plutonium in fuel-reprocessing plants, ten 3000-MWth LMFBRs produce radioactive wastes containing about 9000 kg of fission products and about 1600 kg of actinide waste per year, as shown in Figure 3. After separation of about 1200 kg of uranium from the actinide waste, the total amount of hazardous actinides to be transmuted per year is about 400 kg. Consequently one 1000-MWth ABFR operated entirely on the actinide waste would be needed to transmute the actinide waste produced from ten power stations of conventional size.

High radiation levels due to higher-actinide fuels impose the need for remote-handling devices in a shielded facility for the fabrication of the fuel elements of the ABFR. These devices will be similar to those that will be used for an advanced fabrication

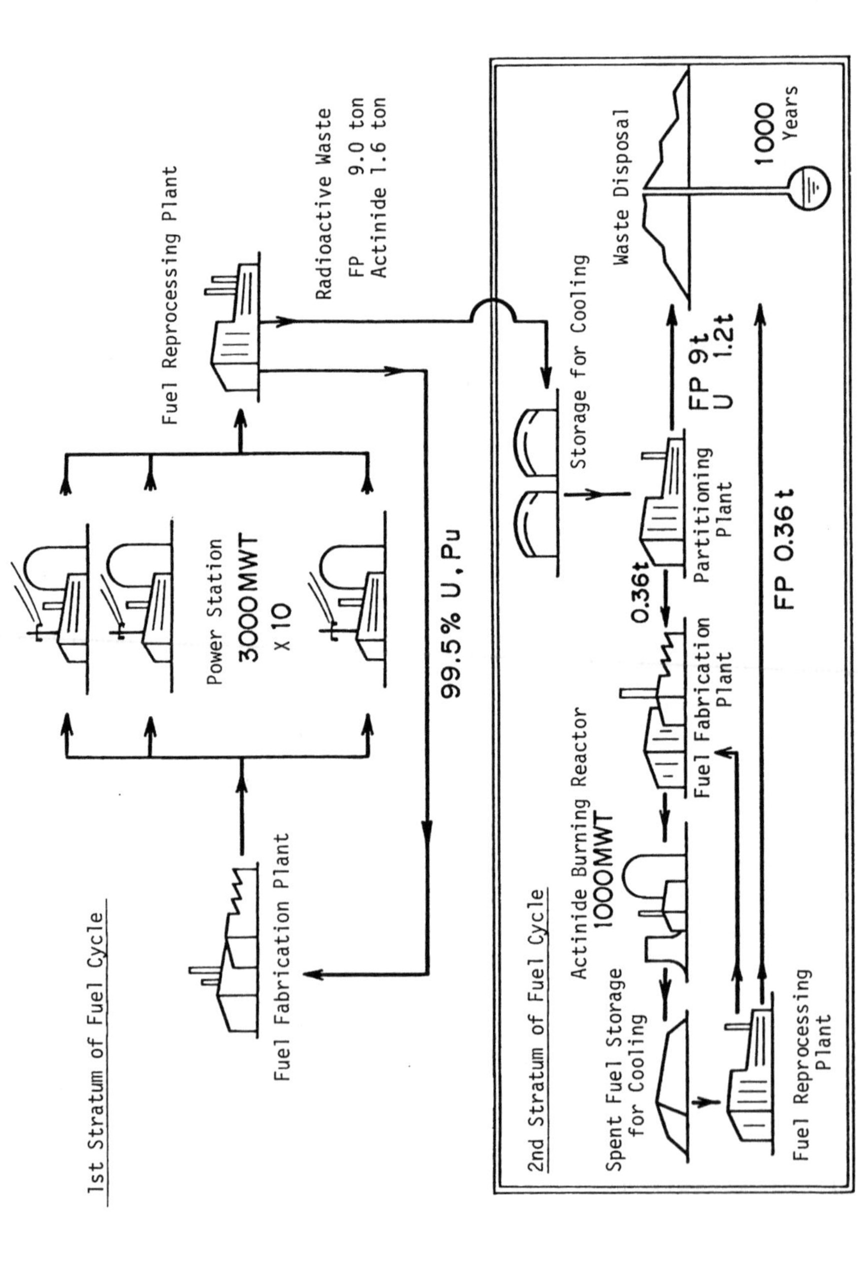

FIGURE 3 The flow of radioactive waste per year.

plant for LMFBR mixed-oxide fuels. The release of decay heat will necessitate a new procedure for handling the actinide in vented facilities in order to dissipate the heat generated.

4 PREDICTED PERFORMANCE OF THE ABFR

In this section we discuss the performance of the ABFR operated entirely on actinides contained in the waste from fuel-reprocessing plants. The uranium and plutonium recovery process in the reprocessing plant is assumed to be 99.5% efficient. Typical weights of actinides in the waste from reprocessing of fast reactor fuels are given in Table 2. For a reactor that is operated entirely on a mixture of actinides in the waste it is necessary to separate uranium from the other actinides because the presence of a large proportion of U-238 is unfavorable in view of the reactor's criticality and the hard neutron spectrum necessary for efficient actinide burning.

The actinide mixture given in Table 2, excluding the uranium isotopes, is used as a "fresh fuel" for the ABFR; the ABFR is assumed to be operated entirely on this fuel in oxide form with a sodium coolant, and the volume ratio fuel:coolant:structure in the core is assumed to be 40:30:30.

Reactor core performances have been predicted on the basis of present knowledge of actinide cross-sections. The predictions were made using the 25-group ABBN-type cross-sections obtained from ENDL and ENDF/B data files.

TABLE 2 Actinide waste from LMFBRs per 1000 MWd.

Nuclide	Waste (g)	Nuclide	Waste (g)
U-235	0.215	Pu-241	0.797
U-236	0.0057	Pu-242	0.494
U-238	133.0	Am-241	14.7
Np-237	3.82	Am-243	7.72
Pu-238	0.101	Cm-242	0.600
Pu-239	8.83	Cm-244	0.564
Pu-240	2.92		

(1) The criticality of the ABFR was investigated and k_∞ for the core was found to be as high as 1.85 and the critical radius in bare geometry to be about 32 cm. The ratio of the fission rate to the capture rate in the ABFR core is given in Table 3 together with the ratios for the LMFBR and JEZEBEL. The ratio of actinide reactivity worth to that of Pu-239 in the ABFR is given in Table 4. Owing to the hard neutron spectrum achieved in the ABFR, a fairly large reactivity worth is obtained, even for the actinides whose fission cross-sections are of a threshold type. This high activity, which is achieved by separating U-238 from the actinide waste, offers various advantages in making core design flexible and in relaxing the requirements imposed in the "out-of-core" sector.

(2) To investigate the possibility of abnormal reactivity behavior due to selective burning in certain nuclides, burnup analyses have been made using the computer code

TABLE 3 The average ratio of fission rate to capture rate in the spectra of various reactors.

Nuclide	LMFBR	ABFR	JEZEBEL
Np-237	0.264	1.00	4.47
Pu-238	2.98	9.78	32.1
Pu-239	3.84	8.57	20.3
Pu-240	0.434	1.68	5.55
Pu-241	6.29	9.60	18.2
Pu-242	0.84	3.73	14.1
Am-241	0.42	1.43	9.71
Am-242	7.45	13.1	22.0
Am-243	0.28	1.34	7.9
Cm-243	5.34	10.6	26.4
Cm-244	1.28	5.60	17.6

TABLE 4 The ratio of the sample reactivity worth of a threshold-fission-type nuclide to that of Pu-239.

Nuclide	$\Delta\rho/\Delta\rho$(Pu-239)
Np-237	0.170
Pu-240	0.226
Pu-242	0.170
Am-241	0.191
Am-243	0.107

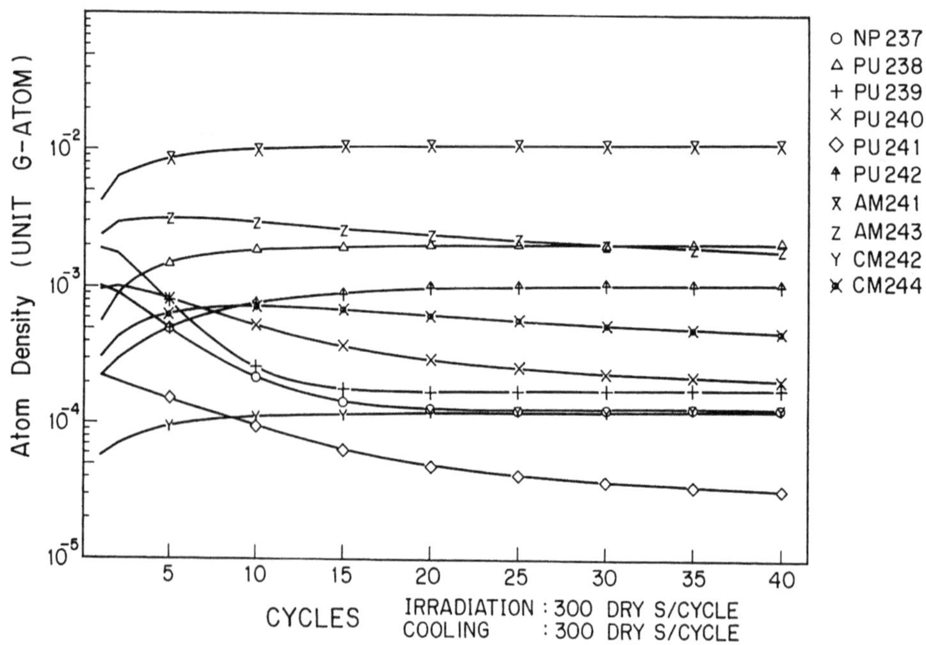

FIGURE 4 The change in atom density at each cycle in the ABFR.

ORIGEN (Bell, 1973). For this purpose the cross-section data library for the LMFBR provided in ORIGEN was replaced with a new one obtained using the neutron spectrum of the ABFR as a weighting function. The recycling calculation was performed up to the 40th cycle, where each cycle consists of irradiation by 1×10^{16} n s^{-1} cm^{-2} for 300 days and cooling for 300 days. At the beginning of each cycle, the criticality is adjusted by feeding fresh actinide mixture from LMFBR fuel reprocessing. Typical actinide concentrations at each cycle are shown in Figure 4 and more detailed values at the first and 20th cycles are given in Table 5. As can be seen in Table 5, even though there is a drastic composition change in the actinide mixture by the 20th cycle, the total atom density of the actinide mixture required to maintain criticality is increased by only 40%. This result indicates that the core criticality depends on the total atom density of the actinide mixture but does not depend strongly on the relative composition of the mixture because of its hard neutron spectrum.

TABLE 5 The actinide atom density in the ABFR core.

Nuclide	Initial	20th Cycle
Np-237	1.38 (− 3)[a]	1.38 (− 4)
Pu-238	3.65 (− 5)	1.99 (− 3)
Pu-239	3.18 (− 3)	1.99 (− 4)
Pu-240	1.06 (− 3)	3.22 (− 4)
Pu-241	2.91 (− 4)	5.19 (− 5)
Pu-242	1.81 (− 4)	6.63 (− 4)
Am-241	5.31 (− 3)	1.34 (− 2)
Am-242M	0.0	3.08 (− 4)
Am-243	2.81 (− 3)	2.78 (− 3)
Cm-242	2.17 (− 4)	1.20 (− 4)
Cm-243	0.0	5.20 (− 5)
Cm-244	2.06 (− 4)	6.49 (− 4)
Cm-245	0.0	4.06 (− 5)

[a] Read as 1.38×10^{-3}, and so on for the other figures.

(3) The effects of fission-product (rare-earth) contamination in the actinide fuel on the criticality of the reactor have been investigated. They can be divided into two effects: one is a reduction of the volume fraction of the actinide mixture and the other is due to the fission product itself. The calculated results are summarized in Table 6. It can be seen that the effect of the fission products themselves is minor even if the actinide fuel pin is contaminated by 50 vol. % of fission products. This means that, whereas conventional reprocessing plants require a high decontamination of fission products from the actinides and a low decontamination of the actinides from aqueous raffinate, the transmutation of actinides requires the reverse, i.e. relatively high decontamination of actinides from the raffinate. This results in a low decontamination of fission products from the actinides.

(4) Zamorani et al. (1980) discussed the fabrication of an actinide-mixture oxide fuel pin for transmutation. In their report they recommend MgO as a dilution material for adjusting the fuel-pin temperature. The effects of the diluent material MgO have been

TABLE 6 The effect of fission-product contamination in fresh actinide fuel.

Case	Fuel/FP/SS/NA/void[a] (vol.%)	Critical radius[b] (cm)
1	40/0/30/30/0	23.6
2	32/8/30/30/0	29.8
3	32/0.01/30/30/8	29.6
4	20/20/30/30/0	48.8
5	20/0/30/30/20	45.6

[a] FP, fission products; SS, stainless steel; NA, sodium coolant.
[b] With a conventional LMFBR blanket 40 cm thick.

TABLE 7 The effect of MgO diluent on fresh actinide fuel.

Case	Fuel/MgO/SS/NA/void[a] (vol.%)	Critical radius[b] (cm)
1	40/8/30/30/0	29.3
2	40/0/30/30/8	29.5
3	20/20/30/30/0	46.0
4	20/0/30/30/20	45.6

[a] SS, stainless steel; NA, sodium coolant.
[b] With a conventional LMFBR blanket 40 cm thick.

TABLE 8 The present uncertainty levels assumed for decay constants and nuclear data.

Nuclide	λ (%)	σ_f (%)	σ_c (%)	$\sigma_{n,2n}$ (%)
U-235	2	5	5	40
U-236	2	15	25	50
U-237	1	25	50	
U-238	3	5	5	40
Np-236	1	30	50	
Np-237	2	20	40	50
Np-238	1	30	50	
Np-239	1	30	50	
Pu-238	1.5	20	50	
Pu-239	1	5	5	40
Pu-240	5	10	20	40
Pu-241	5	10	20	40
Pu-242	5	20	30	50
Am-241	2	30	40	
Am-242M	2	30	50	50
Am-242	1	30	50	50
Am-243	2	30	50	
Cm-242	2	30	50	
Cm-243	3	30	50	50
Cm-244	3	30	50	
Cm-245	3	30	50	

TABLE 9 The effect of cross-section error on the actinide atom density in actinide fuel at the 20th cycle.

Isotope	Uncertainty of atom density (%)	Isotope	Uncertainty of atom density (%)
U-235	18.1	Am-241	28.6
U-238	16.5	Am-242M	30.3
Np-237	25.3	Am-243	26.3
Pu-238	22.5	Cm-242	16.8
Pu-240	11.6	Cm-243	
Pu-241	14.9	Cm-244	31.1
Pu-242	16.6		

investigated in the same manner as the effects of the fission products and with a similar conclusion, as indicated in Table 7.

(5) The uncertainties of the predicted performance of the ABFR have been evaluated through reactor sensitivity analysis together with the current estimated uncertainties of actinide cross-sections given in Table 8. In general it appears that the current reliability of actinide cross-section data is reasonably adequate for predicting the performances of conventional reactors. However, there are considerable uncertainties in making predictions for the entirely new ABFR. Two typical examples are given in Tables 9 and 10. Table 9 gives uncertainties in actinide atom densities at the end of the first and 20th cycles of the ABFR due to cross-section uncertainty. The uncertainties cause fuel volume changes of 3% and 17% at the end of the first and 20th cycles, respectively. In Table 10 uncertainties in the multiplication factor of the ABFR at the first and 20th cycles due to the cross-section uncertainty are given for the following four cases: A, full correlation between errors in different energy groups and between different nuclides; B, full correlation between errors in different energy groups only; C, full correlation between errors in different nuclides only; D, no correlation at all. Thus it is concluded that improvement of cross-section errors by a factor of 2–3 will be essential even for a conceptual study and much greater improvement will be needed for an engineering design.

TABLE 10 The uncertainty in the multiplication factor of the ABFR due to actinide cross-section errors.

Case	1st cycle		20th cycle	
	Capture (%)	Fission (%)	Capture (%)	Fission (%)
A	7.3	11	8.5	17
B	4.6	6.0	7.0	12.0
C	2.4	4.2	3.0	6.5
D	1.6	2.3	2.5	4.4

5 MEASUREMENT OF INTEGRAL EFFECTS OF ACTINIDES IN THE JAERI FAST CRITICAL ASSEMBLY

The integral data which have been measured for conventional reactors are mainly criticalities and reaction-rate ratios. Such measurements are absolutely necessary for improving the prediction of reactor performance. However, most actinides are highly radioactive and expensive at present. Consequently integral measurements that may be realistic for the ABFR are the small-sample worth, the reaction-rate ratio, and the composition change due to irradiation.

In contrast to differential measurements, integral measurements are very accurate, but the neutron energy dependence in integral data is not straightforward so that the choice of an appropriate variety of cores where the integral measurements are made is important in order to obtain successful results. A series of integral measurements of actinides for the ABFR is now underway at the JAERI Fast Critical Assembly (FCA) with the support of the US Department of Energy. The measurements are separated into two phases.

(1) Phase 1 consists of measurements of fission-rate ratio and sample reactivity worth in a series of different cores whose neutron spectra are shifted from hard to soft systematically. The actinide samples used for the reactivity-worth measurements are separated isotopes of 20 g of Np-237, 15 g of Pu-238, 15 g of Pu-240, 20 g of Am-241, and 20 g of Am-243 in oxide form which have been selected as the most relevant isotopes for evaluating the core performance of the ABFR. The fission rates are measured for Np-237, Pu-238, Pu-240, Am-241, Am-243, and Cm-244. The phase-1 measurements were started in September 1980 and will be finished in December 1981. The FCA assembly IX has been built for the phase-1 integral measurements. Assembly IX consists of six different types of cores. The core compositions of these six cores (given in Table 11) have been selected considering the following factors. (a) Each configuration of core should have a simple composition and geometry to simplify analysis. (b) The amount of separated actinide isotope that is available for sample reactivity-worth measurements is limited to 20 g. The core configuration is therefore selected so that its critical mass is less than 150 kg of fissile material. (c) In order to minimize drift of the core during sample reactivity-worth measurements, plutonium is not used as a core fuel; only metallic uranium fuel of either 93% or 20% enrichment (in U-235) is used as a core fuel and either iron or graphite is used as a diluent material to shift the neutron spectrum appropriately. (d) The geometry of the core is cylindrical with a natural uranium blanket and a ratio of core height to core diameter of close to unity. Typical calculated results for integral data to be measured in these cores are given in Table 11 relative to the case of Pu-239. These data show a clear dependence on neutron spectrum.

The contribution of integral data measured in the FCA IX assembly to the improvement of the prediction of ABFR performance can be estimated using the theory of least-squares fitting of cross-section data utilizing integral data (Mukaiyama et al., 1980). Assuming that the integral data are measured in the FCA IX assembly within a 3% error, the uncertainties in the cross-sections of important actinides (see Table 8) in the important energy region for predicting ABFR performance are shown to be improved by the factors given in Table 12. The results, like the uncertainties in the ABFR data given in Tables 9 and 10, are expected to be improved in future by a factor of 2–5.

TABLE 11 Integral measurements of actinides at the FCA.

Assembly:	IX-1	IX-2	IX-3	IX-4	IX-5	IX-6
Material[a] (vol %)						
93% EU	5.29	10.59	15.88	10.59	15.88	15.88
C	79.40	74.10	68.81	0.0	0.0	0.0
SS	10.79	10.79	10.79	84.89	79.60	27.94
Void	4.52	4.52	4.52	4.52	4.52	56.18
Criticality (C/E)	0.9916	–	0.9989	1.0069	–	1.0051
Np-237/Pu-239						
Fission	0.222	0.355	0.396	0.347	0.401	0.466
Worth	– 0.834	– 0.221	0.022	0.059	0.166	0.262
Pu-238/Pu-239						
Fission	0.638	0.812	0.883	0.870	0.912	0.955
Worth	0.704	0.870	0.931	0.890	0.935	0.981
Pu-240/Pu-239						
Fission	0.229	0.341	0.399	0.353	0.400	0.460
Worth	0.221	0.346	0.396	0.309	0.365	0.433
Pu-242/Pu-239						
Fission	0.191	0.292	0.347	0.291	0.339	0.403
Worth	0.161	0.301	0.368	0.283	0.342	0.415
Am-241/Pu-239						
Fission	0.274	0.347	0.378	0.272	0.315	0.387
Worth	– 0.724	0.035	0.274	0.218	0.317	0.431
Am-243/Pu-239						
Fission	0.174	0.258	0.304	0.236	0.280	0.344
Worth	– 0.018	0.244	0.334	0.237	0.303	0.383

[a] EU, enriched uranium; C, carbon; SS, stainless steel.

TABLE 12 The improvement of cross-section uncertainty utilizing integral measurements at the FCA.

Nuclide	Cross section	Important energy region	Improvement factor
Pu-238	Capture	1 MeV–1 keV	0.2
	Fission	3 MeV–10 keV	0.1
Np-237	Capture	1 MeV–1 keV	0.1
	Fission	10 MeV–50 keV	0.1
Am-241	Capture	100 keV–1 keV	0.8
	Fission	10 MeV–800 keV	0.5
Am-243	Capture	1 MeV–5 keV	0.1
	Fission	10 MeV–100 keV	0.1

(2) Phase 2 consists of integral measurements in a simulated ABFR core which is built using 93% enriched metallic uranium and pure plutonium metallic fuels together with diluents to simulate the neutron spectrum and core size of the ABFR. The sample to be used in the phase-2 measurements will be a mixed-actinide sample obtained from a reprocessing plant. The main purpose of the measurements is to confirm whether there is any significant effect on the core performance of the ABFR due to curium and

californium which are contained in the mixed actinide mixture to a certain extent. The timing of phase 2 is not yet fixed and depends on the availability of the mixed-actinide sample.

6 CONCLUDING REMARKS

As indicated by Croft (1976), studies of actinide transmutation to date have mainly emphasized the in-core aspects. Those studies which have investigated the projected out-of-core impacts of transmutation have generally been deficient on in-core matters. The greatest impact of the actinide-transmutation fuel cycle will be felt in the out-of-core sector, i.e. in the partitioning process and in the actinide-fuel reprocessing and fabrication process.

The basic techniques of solvent extraction, oxalate precipitation, and cation-exchange separation for the partitioning of actinides from radioactive waste are available on a laboratory scale. For an engineering assessment of these processes the most significant factor is the allowable amount of fission products in the actinide fuel pin and the allowable amount and type of diluent materials, whether artificial or contamination in the fabrication of the actinide fuel pin. The in-core sector should therefore be studied along with an investigation of how the inherent difficulties in partitioning and fuel-pin fabrication can be alleviated by improving the core design.

The ABFR concept proposed consists of a new stratum of the fuel cycle, completely separated from the conventional cycle, to deal with about 4% of the total fuel cycle. The predicted spectrum of the ABFR is so hard that the effect of fission-product contamination of the actinide fuel pin on reactor performance is calculated to be negligibly small; for example, a fraction of fission products of as much as 50 vol.% mixed with actinide in the actinide fuel pin introduces an effect on reactor performance of only about 3% $\Delta k/k$ reactivity change. Similar trends are found for contamination of the actinide fuel pin due to structural materials or diluent materials like MgO. If these calculated results are confirmed by future measurements, the favorable in-core characteristics may substantially alleviate various severe difficulties in the partitioning and fabrication process for the actinide fuel pin. Consequently it may turn out that the rather pessimistic perception of economic actinide partitioning and reprocessing for transmutation will have to be reconsidered. If so, it will be worthwhile to explore a dry-type process technology such as pyrometallurgy and a liquid tin-nitride separation process for the transmutation fuel cycle.

Continuous dedicated efforts in looking for various technical options to achieve "as low as practicable" risk levels might lead to a reasonable degree of public acceptance in the course of expanding the nuclear power program.

ACKNOWLEDGMENTS

The authors wish to express their thanks to Messrs. K. Koyama, M. Nakano, M. Obu, T. Mukaiyama, and H. Nakamura for providing many useful data for this paper.

REFERENCES

Bell, M.J. (1973). ORNL-4628. Oak Ridge National Laboratory, Oak Ridge, Tennessee.
Claiborne, H.C. (1972). ORNL/TM-3969. Oak Ridge National Laboratory, Oak Ridge, Tennessee.
Croft, A.G. (1976). Actinide transmutation studies: a review. Paper presented at the June Meeting of the American Nuclear Society, Toronto.
Häfele, W., Holdren, J.P., Kessler, G., and Kulcinski, G.L. (1977). Fusion and fast breeder reactors. RR-77-8. International Institute for Applied Systems Analysis, Laxenburg, Austria.
IAEA (1977). Site selection factors for repositories of solid high-level and alpha-bearing waste in land geological formations. IAEA Report. International Atomic Energy Agency, Vienna.
Mukaiyama, T., et al. (1980). In the Proceedings of an International Conference on Nuclear Cross-Sections for Technology, pp. 552–556.
Zamorani, E., et al. (1980). Nuclear Science and Technology, 2(3):681–812.

TRANSMUTATION OF RADIOACTIVE WASTES: AN ASSESSMENT

M.C. Edlund
Department of Mechanical and Nuclear Engineering, Virginia Polytechnic Institute and State University, Blacksburg, Virginia 24061 (USA)

ABSTRACT

The increasing use of nuclear reactors and consequent production of radioactive fission products make necessary an early decision on a reliable, ultimate method for disposal of radioactive wastes. At a cost of about 10% of total energy output it is possible to transmute the actinide component in the wastes using fission reactors but the available neutron fluxes are still too small to transmute the dangerous ^{90}Sr and ^{137}Cs isotopes. The best option at present is burial of the wastes in deep, stable geological formations; this concept has been very thoroughly investigated in many countries. Fusion reactors may in the future attain sufficiently high neutron fluxes to be useful transmutation devices but the waste-disposal problem is too immediate and pressing to await such developments.

1 INTRODUCTION

In this paper I will be concerned with transmutation of radioactive isotopes in the domain of "Nuclear Synergism", as beautifully described by Professors Harms and Häfele in their recent paper (Harms and Häfele, 1981). If nuclear synergism does apply then neutrons produced by fusion reactors or in spallation reactions can be used to reduce the adverse impact of fission products. In brief compass, during the next 30 years or so, nuclear energy will depend on light-water reactors (LWRs), improved water reactors, and possibly fast breeders. To improve the "mass-sustainability" (fuel utilization) we must reprocess and recycle fuels. Thus we have already faced up to the problem of ultimate disposal of radioactive wastes. Also during this period it would not be prudent to depend on the development of fusion reactors or spallation reactors to provide excess neutrons for transmutation of radioactive wastes.

In 30 years from now the development of fusion reactors may, or may not, play a role in our quest for energy. However, assuming that fusion reactors do have a role, what are the incentives to use transmutation technology? A number of transmutation studies have been made in the United States. These studies indicate no particular incentives for using transmutation as compared to ultimate disposal of high-level wastes in deep underground geological formations. This paper will review this assessment of transmutation technology.

2 THE MAGNITUDE OF THE ULTIMATE DISPOSAL PROBLEM FOR FISSION PRODUCTS

The largest potential hazard in the fuel cycle for fission reactors comes from the waste streams produced in chemical reprocessing of spent fuel. After reprocessing, ^{90}Sr is the dominant potential hazard for some 300 years and the actinides present the most danger for 600 to 10 million years. The potential hazard as measured by the ingestion toxicity at 100 and 1000 years after reprocessing is given in Table 1.

TABLE 1 The ingestion toxicity[a] of fission products (in cubic meters of H_2O per metric ton of fuel)[b].

	Years after reprocessing	
Product	100	1000
^{90}Sr	2.18×10^{10}	–
^{137}Cs	5.3×10^{8}	–
^{90}Y	3.3×10^{8}	–
^{129}I	6.2×10^{5}	6.2×10^{5}
^{151}Sm	1.3×10^{6}	–
^{154}Eu	4.4×10^{7}	–
Total actinides[a]	1.7×10^{7}	7.9×10^{6}
Natural uranium ore[c]	4×10^{7}	4×10^{7}

[a] Toxicity is defined as 1/MPC (Maximum Permissible Concentration).
[b] The figures are for 33,000 MWd per million tons of heavy metal (3.3% enriched uranium) and LWRs; 99.5% of uranium and plutonium are removed by extraction.
[c] 3000 Mt of natural uranium ore containing 0.2% uranium.

The current disposal concept is to bury the high-level wastes as a glass or ceramic in stable geological formations. Tectonic stability of the formation for at least 1000 years would allow the potential hazard to decrease to a relatively innocuous level. The area of land required is very modest – about 3000 m^2 per gigawatt-year of electric energy. The depository land area would be equivalent to 3000 metric tons of 0.2% uranium in carnotite ores.

3 BACKGROUND STUDIES

The use of transmutation to reduce the potential hazard of radioactive fission products has been studied over the years by several investigators. The first reported work was done by Steinberg et al. (1958). They reported on the possibility of using high-flux burner fission reactors to reduce the stockpile of ^{85}Kr, ^{90}Sr, and ^{137}Cs. The next transmutation idea was to use a spallation reactor (beam/accelerator technology) (Gregory and Steinberg, 1967). The use of fusion reactors was suggested by Wolkenhauer (1972). All these technologies were summarized and extended in the classic work of Claiborne (1972), who reached the following conclusions.

> "Elimination of the fission products, ^{90}Sr, ^{137}Cs, and ^{85}Kr, by neutron-induced transmutation as a result of recycling in existing or projected designs of power reactors is not possible, since the neutron fluxes are not high enough to lower the effective half-lives of these nuclides by a significant amount. Special burner reactors with neutron fluxes in the order of $10^{17}\,cm^{-2}\,s^{-1}$ are required for that purpose. Spallation reactors and fusion reactors are possibilities, but the latter is certainly not feasible with current technology. The former, at best, would require an extensive development program including, in particular, a method for coping with the potentially severe radiation damage and heat-transfer problems. It seems that ultimate storage in deep geological formations of known characteristics (such as salt mines) remains the best method for fission-product disposal since less than 1000 years is required to reduce the activity to an innocuous level. Assurance of tectonic stability for 1000 years with a very high degree of confidence is quite possible in some geological formations. The actinides and their daughters, of course, with half-lives measured in many thousands of years should be excluded from the biosphere for a length of time for which tectonic stability can be assured with a lesser confidence level. There is, therefore, a stronger motive for disposal or reduction in the accumulation of the actinides by some other method such as by transmutation in nuclear reactors."

Burkholder et al. (1975) attempted to estimate the long-term benefits of removing actinides from high-level wastes before burial in a waste repository. Their model took into account the physical and chemical processes which greatly retard the isotopic mixture ultimately ingested by man. They concluded, ". . . for the situations investigated, the incentives for a special effort to remove any elements, including the transuranics, from high-level wastes are vanishingly small . . ."

More recently Oak Ridge National Laboratory has completed an assessment of the reduction of the proportion of actinides in high-level wastes to be placed in stable geological formations (Croft et al., 1977, 1980). Transmutation studies supported by the Electric Power Research Institute as part of their fusion program have been reported by Lang (1980), Parish and Draper (1980), and Schaffer and Parish (1980).

4 PARTITIONING AND TRANSMUTATION OF ACTINIDES

The reduction of the toxicity of actinides by a factor of 10–100 prior to their burial in deep underground geological formations appears to be technically feasible in LWRs or fast breeders. A scheme for achieving this follows from the suggestion of Claiborne (1972).

The partitioning transmutation scheme involves the partitioning of the actinides from high-level wastes and other material containing a sufficient amount of actinides to be considered as long-lived transuranic (TRU) wastes in a separate reprocessing plant colocated with a Purex reprocessing plant and mixed-oxide (MOX) refabrication plant. The actinides are removed from the high-level wastes by leaching, solvent extraction, or breaking down of organic actinide compounds. The tetravalent and hexavalent actinides are recovered by TBB (tributyl phosphate) extraction. The trivalent actinides and lanthanides are coextracted using a bidentate extractant. They are then separated from the lanthanides by cation-exchange chromatography.

The total loss of actinides from chemical processes and the MOX refabrication plant to the ultimate waste is estimated to be 0.25%. However, Croft et al. (1980) point out that the partitioning flow sheets have considerable internal recycle of plant streams. Thus it is possible that one or more chemical species could form and build up within the plant, resulting in much higher amounts of actinide in the ultimate waste streams. Furthermore, the resins used in the cation-exchange chromatography suffer considerable radiation damage and may have to be replaced after a single use.

The transmutation rate for a liquid-metal fast breeder reactor is about 11% yr^{-1} at full power. Assuming a capacity factor of 0.8 and three years in the reactor and two years in the reprocessing and fabrication plant, we obtain a transmutation rate of 26% per five-year cycle. The maximum reduction of actinides in the ultimate high-level wastes is $Lr/(1-r)$ where L is the loss per cycle and r is one minus the transmutation rate per cycle. The assessment by Croft et al. (1980) gives the reduction of actinides as 0.25% per cycle. Thus the total actinides in the ultimate wastes are reduced to 0.71% of their original amount, i.e. by a factor of 140. Some chemical engineers (Benedict et al., 1981) believe the loss per cycle to be about 1%. This increases the actinide loss to ultimate wastes to 2.85%, i.e. a factor of 35 smaller than the original amount.

The short-term risks have been evaluated to be about three times greater than the risk for reprocessing and burial owing to the large increase in actinides in reprocessing, refabrication, and transportation. The major conclusion given by Croft et al. (1980) is,

> "*There are no incentives for actinide P–T, even if very conservative assumptions are used in the analysis.* The cost of the actinide P–T benefits is $32,400/person-rem if the nonradiological risks are ignored; if the nonradiological risks are included, the short-term risks *exceed* the long-term benefits integrated over 1 million years."

5 FUSION REACTORS

The neutron fluxes in the blankets of various power fusion-reactor concepts are limited by the wall loading to 1–2 MW m^{-2}. The concepts include the tokamak, mirror, θ-pinch, and laser-ignited reactors which rely on a blanket surrounding a plasma chamber to convert fusion energy into heat. The average fast flux in these blankets would be about the same as those in a fast fission breeder. Thus there is no advantage over fission in transmuting fission products (Lang, 1980, and Parish and Draper, 1980).

As pointed out by Claiborne (1972), neutron fluxes of the order of 10^{17} cm^{-2} s^{-1} are required to transmute ^{90}Sr and ^{137}Cs. This requires a wall loading at least two orders of magnitude greater than those to be expected in the concepts just outlined. One such concept is an imploding liner reactor studied by Schaffer and Parish (1980). Although two plasma configurations were studied, only the closed-confinement field-reversed

θ pinch had a sufficiently high neutron flux to possibly transmute ^{90}Sr and ^{137}Cs with an e-fold time of about 6 years. This is so far out that I only mention it in passing. We "only" need breakthroughs in fusion-plasma-confinement techniques, aqueous processing technology, and liquid-metal and fused-salt processing! Finally, to transform ^{90}Sr efficiently would require a process to separate it from the ^{88}Sr isotope.

6 SPALLATION REACTORS

A typical spallation reactor would use electricity to power a high-energy accelerator. In the original work of Gregory and Steinberg (1967) a 500-MW beam of 10-GeV protons impinging on a liquid uranium target was envisaged. The neutron source strength (greater than $10^{20}\,s^{-1}$) would produce a thermal flux of $2 \times 10^{17}\,cm^{-2}\,s^{-1}$ in heavy-water-moderated ^{90}Sr targets. The principal problem would be to find a method to cope with severe radiation damage and heat-transfer problems.

With new devices producing large numbers of neutrons per unit beam current, the reactor might produce a high enough flux to transmute the problem isotopes efficiently.

7 CONCLUSIONS

We must settle on an ultimate, reliable method for the disposal of fission products in the immediate future. Fission reactors can transmute the actinides at a relatively large cost, about 10% of the cost of energy from fission reactors. The neutron fluxes in fission reactors are far too small to transmute ^{90}Sr and ^{137}Cs. The best alternative at this time is disposal in deep geological formations. This has been examined time and again by many countries. Compared to underground disposal there are no incentives to transmute any of the fission products. By the time fusion reactors can play a role as an energy source we will have accumulated a large amount of radioactive fission products. And these will have to be removed from our environment prior to the development of any advanced transmutation concept.

REFERENCES

Benedict, M., Pigford, T.H., and Levi, H.W. (1981). Nuclear Chemical Engineering. McGraw-Hill, New York, p. 601.

Burkholder, H.C., Cloninger, M.O., Baker, D.A., and Jansen, G. (1975). Incentives for partitioning high-level waste. BNWL-1927, Hanford, Washington.

Claiborne, H.C. (1972). Neutron-induced transmutation of high-level radioactive waste. ORNL-TM-3964. Oak Ridge National Laboratory, Oak Ridge, Tennessee,

Croft, A.G., Tedder, D.W., Drago, J.P., Blomeke, J.O., and Perona, J.J. (1977). A preliminary assessment of partitioning and transmutation as a radioactive waste management concept. ORNL/TM-5808. Oak Ridge National Laboratory, Oak Ridge, Tennessee.

Croft, A.G., Blomeke, J.O., and Finney, B.C. (1980). Actinide partitioning–transmutation program final report: I, overall assessment. ORNL-5566. Oak Ridge National Laboratory, Oak Ridge, Tennessee.

Gregory, M.V. and Steinberg, M.V. (1967). A nuclear transmutation system for the disposal of long-lived fission product wastes. BNL-11915. Brookhaven National Laboratory, Upton, New York.

Harms, A.A. and Häfele, W. (1981). Nuclear synergism. American Scientist, 69(3): 310.
Lang, G.P. (1980). Determination of procedures for transmutation of fission product wastes by fusion neutrons. EPRI AP-1642, Vol. 2. Electric Power Research Institute, Palo Alto, California.
Parish, T.A. and Draper, E.L., Jr. (1980). Determination of procedures for transmutation of fission product wastes by fusion neutrons. EPRI AP-1642, Vol. 3. Electric Power Research Institute, Palo Alto, California.
Schaffer, M.J. and Parish, T.A. (1980). Background study of liner fusion systems for transmuting fission reactor wastes. EPRI AP-1644. Electric Power Research Institute, Palo Alto, California.
Steinberg, M., Wotzak, G., and Manowitz, B. (1958). Neutron burning of long-lived fission products for waste disposal. BNL-8558. Brookhaven National Laboratory, Upton, New York.
Wolkenhauer, W.C. (1972). The controlled thermonuclear reactor as a fission product burner. Transactions of the American Nuclear Society, 15:1.

Section Six

Reflections and Summary

QUICK IS BEAUTIFUL*

Freeman J. Dyson
Institute for Advanced Study, Princeton, New Jersey (USA)

When the Princess Rosalba was baptized, her father, King Cavolfiore of Crim Tartary, gave a banquet, and all the royal guests came with expensive presents and flattering speeches (Thackeray, 1855). Then at the end of the line of guests came the Fairy Blackstick. The Fairy Blackstick waved her stick over the baby and said, "As for this little lady, the best thing I can wish her is a little misfortune". The King was furious and ordered his servants to remove the Fairy Blackstick from the hall. But of course the magic worked, and in the end the Fairy Blackstick's present turned out to be more valuable than all the other presents put together. I am sorry I don't have time to tell you the whole story now.

I am very grateful to the organizers of this meeting for giving me the privilege of talking to you at this banquet. I find it a little strange that I should be talking here. I am an ignorant outsider, an amateur, and you are professionals with deep knowledge and long experience in the field of nuclear energy. I think the best I can do for you is to play the role of the Fairy Blackstick. Do not misunderstand me. I am not against nuclear energy. Only I happen to believe, like the Fairy Blackstick, that the best hope of a useful and creative future for nuclear energy lies in a little bit of misfortune. The nuclear power industry, in the United States of America at least, is in deep trouble. Troubles are often good for us, if we can learn from them instead of pretending that they do not exist. If we seriously want to reach the future world of neutron abundance which we have been discussing at this meeting, we must pay careful attention to the causes of our difficulties in the frustrating present. Like the Fairy Blackstick, I will speak frankly. I will try to describe what is wrong with the nuclear power industry, by telling you stories about things that I have seen happen during my life as an observer of energy projects. You can then judge for yourselves whether these stories have any relevance to the problems with which you are faced in your professional lives.

My first story concerns a company called General Atomic which runs a laboratory in La Jolla, California, and manufactures reactors. The company began in the year 1956.

---

* After-dinner remarks at the Workshop, "A Perspective on Adaptive Nuclear Energy Evolutions: Towards a World of Neutron Abundance", at the International Institute for Applied Systems Analysis, Laxenburg, Austria, May 26, 1981.

In the summer of that year the company brought together a group of consultants, and paid us to sit and think for three months. The company was then brand new; it had no laboratories, no production facilities, and no products. The consultants could do nothing except think and talk and scribble on blackboards. The company promised to pay one dollar to the inventor for the patent rights to any reactor which we might invent. I collected my dollar, and so did several other people in the group. One of our designs was chosen for immediate development and went into production with the name TRIGA, standing for Training, Research, and Isotope production, General Atomic. The first TRIGA was built, tested, licensed, and sold within less than three years from the day the consultants assembled in 1956. The company is still producing it and still selling it at a profit. The TRIGA is of course not a power reactor; it is mostly used to produce isotopes for medical research and diagnosis, not to produce electricity.

As a follow-on to the TRIGA, General Atomic decided to develop and market a big power reactor called the High-Temperature Gas-cooled Reactor (HTGR). The HTGR is a great reactor. Its high temperature gives it an advantage in thermodynamic efficiency over water-cooled reactors, and its big heat capacity gives it an advantage in safety. Unfortunately the HTGR never captured a substantial share of the market. General Atomic sold one each of two versions of the HTGR. The first was a 40-MWe version, which produced electricity for a utility company at Peach Bottom, Pennsylvania. The second HTGR sold was eight times more powerful, a 300-MW version which is now running at Fort St. Vrain in Colorado.

Now I come to the spring of last year, 1980. General Atomic is still in business and still has dreams of selling HTGRs. A year ago Harold Agnew, newly appointed president of the company, decided to hold a Class Reunion for the Class of 1956. He invited the surviving members of the group of consultants who had started the company in 1956 to come back and have another look at it. Of course we had all in the meantime grown old and dignified, and we were all much too important and too busy to come back for three months and work out some new inventions. The most we could do was to come back for two days and remember our lost youth. And the General Atomic staff told us about their recent activities and about their plans for the future.

The main thing which the General Atomic people had to tell us was the result of two safety analyses of their full-scale HTGR power reactor. By full-scale they mean 850 MWe, two and a half times the designed power output of Fort St. Vrain. Two independent safety analyses of the full-scale HTGR have been done, one by a group of experts in the United States and the other by a group in the Federal Republic of Germany. Neither group of experts was connected with General Atomic; neither group had any commercial incentive to make the HTGR look good. And both groups came out with similar conclusions: in a certain well-defined sense, the HTGR is roughly a thousand times as safe as a light-water reactor of equal power. The meaning of this statement is the following. The experts analyzed billion-year accidents, caused by combinations of stupidity and bad luck more extreme than anything we saw at Three Mile Island. A billion-year accident requires so much bad luck that it is supposed to happen only once in a billion years of reactor running time. A billion-year accident is a hell of a lot worse than Three Mile Island. The reactor core vaporizes, the concrete containment building splits open, the atmosphere happens to have an inversion layer at the worst height and the wind is blowing in the worst direction over a region of high population density. You do not need to believe in the accuracy of the calculation which says that this disaster happens once in a billion years. All that you need to believe is that it is possible to apply the rules of the accident-analysis game fairly, so that a billion-year accident for a light-water reactor and a billion-year accident for the HTGR are in some real sense equally unlikely. The results of the analyses are then startlingly favorable to the HTGR. The billion-year accident of a light-water reactor kills 3300 people immediately and 45,000 people by delayed effects of radiation. The billion-year accident of the HTGR kills zero people immediately and 70 people by delayed effects. The numbers make no claim to accuracy, but the conclusion is qualitatively clear. It is conceivable that a mishandled HTGR may kill people, but it cannot kill them wholesale.

The next question that arises is then, if the HTGR is a thousand times as safe as a light-water reactor, and if public worries about accidents are threatening the very existence of the nuclear power industry, why is there not a crowd of utility executives standing at the door of the General Atomic sales office, waiting to trade in their light-water reactors for a shiny new HTGR? The answer to this question is simple. If a customer should now come to General Atomic wanting to order a full-scale HTGR, the best that General Atomic could do would be to say, "Well, wait a moment. If you can help us raise some government money to finish the engineering development, and if we don't run into any unexpected snags, with luck we could be ready to begin construction in a few years, and if the licensing goes smoothly you should have your reactor on line within ten years after that." This is not the kind of answer which brings utility executives running to place orders.

I told you this story of the two reactors, the TRIGA which was finished and ready to go in three years and the HTGR which cannot be ready in less than 12 or 15 years, because I happen to have been personally involved with them. Similar stories could be told about many other industrial products. The nuclear industry is not the only one which has suffered from a hardening of the arteries and lost the ability to react quickly to changing conditions and changing needs. I believe the difference between a three-year and a 12-year reaction time is of crucial importance. The rules of the game by which public life is governed are liable to drastic and unpredictable change within less than ten years. By rules of the game I mean prices, interest rates, demographic shifts, and technological innovations, as well as public moods and government regulations. We have recently seen some spectacular changes in the rules of the game which the US automobile industry has to play. We can expect such sudden changes to occur from time to time, but nobody is wise enough to predict when or how. Judging by the experience of the last 50 years, it seems that major changes come roughly once in a decade. In this situation it makes an enormous difference whether we are able to react to change in three years or in twelve. An industry which is able to react in three years will find the game stimulating and enjoyable, and the people who do the work will experience the pleasant sensation of being able to cope. An industry which takes twelve years to react will be perpetually too late, and the people running the industry will experience sensations of paralysis and demoralization. It seems that the critical time for reaction is about five years. If you can react within five years, with a bit of luck you are in good shape. If you take longer than five years, with a bit of bad luck you are in bad trouble. That is why I chose for the title of this talk "Quick is Beautiful".

The light-water reactor industry probably made a mistake in going to 1000-MW units. The expected economy of scale seems to have been illusory. Unfortunately, General

Atomic felt compelled to make the same mistake with the HTGR. Just to keep up with the competition, General Atomic concentrated its efforts on the 1000-MW monster which cannot be ready when it is needed.

The market for nuclear power reactors in the United States is at the moment non-existent. Nobody knows whether the market will revive in the future. The hopes of the industry rest on the likelihood that there will be some new oil crisis or some unpredictable change of political mood which will create a massive new demand for nuclear power. When this happens, the demand will be for reactors which are safe, and flexible, and quick to build. The 1000-MW HTGR is safe but not quick. Peter Fortescue, the chief ideas man at General Atomic, has just come out with a wonderfully simple new design for a supersafe 400-MWe power reactor cooled by natural convection, using sodium coolant and thorium hydride moderator (Fortescue, 1980). I am sorry I cannot talk about the details of Fortescue's reactor tonight since I only learned about it this week. Perhaps General Atomic might finally achieve its rightful share of the market, if it could be ready when the time comes with a reactor of modest size, thoroughly tested and debugged, and capable of being mass-produced in a hurry.

When we turn from nuclear power to biology, we see the same historical processes at work. So long as no sudden changes in the rules of the game occurred, all through the soft swampy sluggish hundred-million-year summer of the Mesozoic era, the dinosaurs pursued their economies of scale, growing big and fat and prosperous, specializing their bodily structures more and more precisely to their chosen ecological niches. Then one day, as we recently learned from the brilliant observations of Luis Alvarez and his colleagues at Berkeley, an asteroid fell from the sky and covered the earth with its debris. The rules of the ecological game were changed overnight, and our ancestors, the small, the quick, the unspecialized, inherited the earth.

Let me tell you one more story. In Princeton there are two projects in progress, each of them in its own way trying to contribute to a solution of the energy problem. The two efforts stand side by side on the Forrestal Campus of Princeton University. One of them is the Tokamak Fusion Test Reactor (TFTR), the great white hope of the magnetic-confinement fusion program, a magnificent piece of engineering, lavishly funded by the US Department of Energy. If all goes well, it will cost only $300 million and will be ready to go into operation in a year or two. It will then explore the technology for commercial fusion reactors which will possibly begin producing electricity ten or 15 years later.

The other project, the one with which I have the honor to be associated, is the Princeton Ice Pond. The ice pond is a square hole in the ground with a dirt berm around it and a sheet of Griffolyn plastic lining its bottom. Two men with a mechanical digger dug the hole in January 1980. We rented a commercial snow machine and squirted snow over the hole during the cold days and nights of February, until we had something that looked like a miniature Matterhorn. Halfway through the snow making, we found out that we didn't need that fancy ski-resort snow-machine. We didn't need skiing-quality snow for our pond. We found out that for our purposes a fireman's fog-nozzle which you can buy for $300 will do the job well enough. Our Matterhorn stood high and proud for a few weeks. Then the March sun shrank it down a bit, and the April rains reduced it to a pool of slush, filled up to the top of the berm. We covered it over with an insulating layer of plastic and straw, and on top of that we put an air-supported mylar dome to keep the straw dry. In June a hefty hail-storm wrecked the mylar and so we made do with wet straw for the insulation. I say "we", but you must understand that I am not claiming credit for any of this. The project is run by Rob Socolow and Don Kirkpatrick and Ted Taylor and their students at the Center for Environmental Studies of Princeton University. I am only an unskilled laborer who goes out to help them occasionally. In June we measured the contents of the pond and found that we had about 450 tons of ice with some water underneath it.

All through an exceptionally hot Princeton summer we successfully air-conditioned a building by circulating fresh water from the bottom of the pond. We were melting ice at a peak rate of about 7 tons a day – beautiful cool white ice with crevasses into which we could descend and enjoy Alpine scenery under the blazing Princeton sun. When the hot weather came to an end at the beginning of October there were still about 150 tons of ice left. I am sorry that I don't have any pictures of the ice pond to show you.

The key to cheap and reliable solar energy is to have a cheap and massive storage of heat and cold, massive enough so that it can ride over the annual weather cycle, heat being collected in summer and used in winter, cold being collected in winter and used in summer. The system which we have in mind for practical use would have two ponds for storage, a hot pond containing 100,000 tons of hot water (roughly 1 ha 10 m deep) and a cold pond containing 10,000 tons of ice (roughly 0.1 ha 10 m deep). This would provide heating and cooling for 100 families. We started first with the ice-pond experiment because the money came through in January 1980 just in time for the snow making. It is much easier to make snow in a hurry in winter than to make hot water in a hurry in summer. We hope later on to have an experimental hot pond connected to a large area of cheap plastic air-mattress collecting solar heat. Then in the following winter we shall find out whether the hot pond stays hot. Unfortunately our students are so busy with the ice pond that we missed the chance to get started on a hot-water pond in time for this year's summer.

That is the story of the Princeton Ice Pond. I told you the story because it illustrates what I have in mind when I ask for a technology with a quick response. I am not claiming that solar ponds by themselves will solve the energy problem. Still less am I claiming that the little game we are playing in Princeton has demonstrated the existence of an economically viable solar-pond technology. What I do claim is that solar ponds are an example of a technology free from the rigidities and the decade-long delays which have made both fission and fusion power unable to respond to urgent need. Solar ponds may or may not turn out to be cheap and effective. If they fail, they will fail quickly and we shall not have spent 25 years proving them useless. If they succeed, there is a chance that they could be deployed rapidly on a very large scale. Sites could be surveyed, holes in the ground dug, and plumbing fixtures installed by thousands of local contractors responding to local demand. Plastic liners and pipes and solar collectors could be mass-produced in factories. All this is only a dream, or at best a remote possibility. But there is no reason why a new technology has to develop, as fission and fusion have developed, on a 30-year time scale. All it needs in order to go fast is small size of units, simple design, mass production, and a big market. When I go out to the Forrestal Campus and see those two machines, the $300-million TFTR and our little ice pond, what I see in my mind's eye is a dinosaur and an early primate. I wonder how long it will be before the next asteroid falls.

Let me end my remarks by emphasizing again the moral of these stories. Nuclear energy will flourish only if it is ready to deal with unexpected misfortunes. If we only prepare for the expected future, we are likely to fail badly when the real world does something unexpected. I have spoken tonight about two types of misfortune. One is a sudden and drastic shift in the demand for energy. The other is a sudden and drastic shift in the supply, for example the unexpected appearance of a cheap and practical solar-energy technology. We should never forget that there are about 10 TW of solar energy incident on Austria alone. If nuclear energy is prepared to react quickly to such unexpected shifts, our chances of entering the promised land of neutron abundance will be greatly improved. After all, when King Cavolfiore was slain in battle and the rebel Duke Padella usurped the Kingdom of Crim Tartary, the Princess Rosalba found refuge as a serving maid in the household of King Valoroso the Twenty-fourth of Paflagonia, and there, with the help of the Fairy Blackstick, she married Prince Giglio the rightful heir to the throne and lived happily ever after (Thackeray, 1855). But now I must sit down quickly, before the King's servants throw me out!

REFERENCES

Fortescue, P. (1980). Motivation and prospects for a hydride moderated power reactor system cooled by naturally convected liquid metal. Report GA-A15980. General Atomic Company, San Diego, California.

Thackeray, W.M. (under the pseudonym M.A. Titmarsh) (1855). The Rose and the Ring. (MacMillan, London, 1908.)

PANEL DISCUSSION

Including selected contributions from H. Bethe, F. Culler, W. Häfele, A.A. Harms, J.W. Hilborn, G.L. Kulcinski, and R. Schulten

Häfele. I have asked a few people to join this panel, particularly because some sort of winding up is necessary in order to put things into perspective. Maybe it would be appropriate to approach the panel members now and ask for their comments. As the discussion evolves it will be possible to have the audience participate. But if there is a very salient point don't hesitate to interrupt. This is such a small group that the whole group can be considered to be a panel! I will start by asking Professor Bethe to help us with his comments.

Bethe. I would like to make a plea for the fusion breeder and I don't believe it has been sufficiently strongly emphasized in this meeting. I do admit the necessity of having also the fast fission breeder, but this was very ably demonstrated by the group of Kulcinski, Kessler, and Abdel-Khalik, in their triple-header on the first day, so I won't argue about that. By the year 2000 the world will need a breeder and the only one which will be ready is the standard fast breeder. Whether that applies to the United States I am not sure – that's a different question – but outside the United States I think it is true that the fast breeder will be needed. But one fast breeder can support only about two-thirds of a light water reactor in steady state; this has been figured out very carefully by the Argonne National Laboratory and this means that even in steady state we would have to build almost all new power stations as fast breeders. Furthermore, we won't have a steady state; we will have an expanding nuclear establishment, and fast breeders with the present doubling time would not meet the implied necessities. If you take the Super Phenix, which has a doubling time of 25 years, there is just no chance of the breeders even making their own fuel for the next 30 years in sufficient quantity. Most people agree that fast breeders will be more expensive than light water reactors. The figure of 33% has been mentioned and there is, in the minds of many people, still the safety problem that a fast breeder contains more than one critical mass of plutonium. Now I believe that this is totally irrelevant and that fast breeders are adequately safe, but I would have a hard time persuading the public at large that this is so. A purely fast breeder economy, I submit, would be very inflexible. By contrast, the fusion breeder – one fusion breeder – can support 24 light water reactors of the same power. 24 compared to two-thirds – that's really a qualitative difference, not just a difference in number. And it is a difference especially in the sense of Freeman Dyson's view that quick is beautiful. I realize very well that we can't have the fusion breeder quickly and I believe the analysis prepared by the Wisconsin/Karlsruhe cooperative venture showing we will not have a commercial fusion breeder before the year 2017; maybe a few years earlier, but not very much earlier. So I come back to saying that the fast breeder will be necessary. However, once you have a commercial model of a fusion breeder then you can deploy them as fast as is required; you can surely deploy 20 fusion breeders in 10 years, which corresponds to 500 light water reactors. It will be just a very small addition to the effort which you have to make, and this is the point. It will be ready when it is needed, it will be there, and you can have as many as the nuclear establishment requires. You will also have flexibility: you can feed the produced fissile material into light water reactors, heavy water reactors, and high temperature reactors. It has been amply discussed in this meeting that high temperature reactors could be very useful in themselves because they can substitute for fossil fuel in making heat in addition to making electricity. You can't have any fuel for them if you have only fast breeders and if they can barely sustain their own growth. Pure fusion I believe is not a competitor to the fusion breeder; pure fusion does not make any material for other power producers. Pure fusion is a new aim, but it is the same kind of aim as the fast breeder reactor; you have to build thousands of them in order to supply the world with energy, in contrast to a few tens of reactors in the case of the fusion breeder.

The competition to the fusion breeder is not pure fusion, nor is it the fast breeder. The only competition is the accelerator breeder and, if I stand back from my personal preference, I don't care very much whether we have fusion breeders or accelerator breeders. They both fulfil the same purpose, namely, supplying lots of fission power reactors with fuel and doing it easily and essentially permanently. At the moment probably the accelerator breeder has fewer technical problems than the fusion breeder because fusion has not yet been established. But it will be very important – if the accelerator breeder is to be a serious competitor – to incorporate in that accelerator breeder the same advanced target design which has been developed for the fusion breeder, namely the fission-suppressed blanket. I am not sure whether this is easily adaptable to the accelerator breeder; maybe it is, and if so it should be considered. The idea is that the high energy particles, the protons in the accelerator breeder or the neutrons in the fusion breeder, hit material which does not fission. When the neutrons are degraded into many neutrons per incident particle and are slowed down, then they hit fertile material and are made into fissile material. That is the trick that keeps down the contamination of the blanket; it keeps down the energy in the blanket and, therefore, the amount of energy you have to take away. The fusion breeder in the tandem mirror device is particularly suitable for such an arrangement because it is cylindrical and so, automatically, whatever neutrons are produced go back into the system. So I want to repeat what I said at the beginning, but more strongly: I believe that the fusion or accelerator breeders are not only a good way but the only way to neutron abundance.

Häfele. Thank you very much for that clearly stated view; perhaps it would be appropriate if I ask you, Floyd, to take over?

Culler. I have little disagreement, I think, with the position outlined. There is a rather obvious period of time in the transition period, and perhaps even beyond 2050, when energy will be more dependent on neutron abundance, when the best reactors alone may not be able to keep up with the necessary stocking requirements. Of the systems proposed for the production of abundant neutrons I personally favor the fusion breeder.

I am less certain about my reactions to accelerator breeders, I think primarily because of their lack of success and the lack of good ideas about targets over the last 20 years. If there is an urgency to have plutonium, however, in the near term, the very clear demonstration that accelerators in the proper energy range and producing proper beams are reasonably near-term technical accomplishments would argue for opting for the accelerator breeder. Our opinion, however, is that we possibly should avoid taking on a third line of development in this sequence of actions toward neutron abundance. Until this meeting I was more certain about my conviction that pure fusion, ignition and self-sustained fusion, was the main line to follow for the fusion reactions. The discussions here on the interdependence of fission/fusion systems, however, have changed my views. I think perhaps the fusion breeder deserves more attention.

Going beyond the neutronics to the second line of discussion that was peripherally carried on in this meeting, namely the possibility of developing systems that would make high temperatures available for the necessary transitions in transportable fuels, I subscribe most wholeheartedly to the idea that high temperature heat can be used to modify the nature of natural gas; possibly it can be useful in the production of liquids that will be necessary and useful to displace oil in the international market. I most certainly endorse supporting the research and development necessary to explore the chemistry of the systems involved. I think that it is equally important to pursue the very promising, relatively new routes for efficient electrolysis, using electricity as the second line, and I suppose I am immediately attracted to that as well, primarily because of background. The transition into an abundant-neutron era requires a little more thought. I firmly believe about the implications in the broad area of proliferation. I have also a growing concern about the size of systems that are proposed for the production of abundant neutrons. They are beginning to be beyond the scale that is appropriate for individual companies and, at times, individual nations. This is not necessarily true of the accelerator-breeder systems and I suppose I should favor them for these reasons. I am concerned about the very high development costs. Ed Kintner has remarked quite cogently that fusion and fusion breeders are unique in development and that you are never sure you are going to succeed until you try a very large machine. The experience with fission has indicated to me that the problems of investment and the huge risk that is involved in bringing on any demonstration of large new energy systems may now be straining the economies, certainly of individual companies, and even of nations. Totally in the vein of wishful thinking, I hope that there is some way to scale down the necessary economic sizes of the devices that we must build in the future to provide neutron abundance.

Häfele. Thank you Floyd.

Schulten. It is certainly true that we people from "normal" reactor engineering will have a lot of competitors in the next century who will try to sell neutrons and fissile material to us. It is very good to learn that for fusion breeders or for accelerator-breeders one plant can deliver fissile material for a rather large number of normal reactors. That means that for the next century, perhaps, the question of the costs of these big plants – which will be rather high, I believe – will not play the most important role. I think the feasibility alone will be a very big gain. If one plant of this kind can deliver fissile material for 20 or 30 reactors then such a central plant could cost 5 or 10 billion dollars without being too expensive. In other words, we have a similar situation as with the reprocessing plants.

The other very important point for me is that nuclear energy should also be used for the production of heat. In principle there are three factors influencing the use of fissile material. The first factor is the reactor design. We have learnt in these two days of discussion that fuel inventory plays a very important role. High fuel inventory in a growing nuclear technology means that in the next 50 or 100 years a very large amount of neutrons must be produced to build up all the big inventories necessary. In turn, that means that small inventories for the individual reactors would be a very big advantage; for example, if the inventories could be lower by a factor of two or three. If I compare, for instance, high temperature pebble-bed reactors, which can be loaded and unloaded during operation, with a reactor which can load only once every year then the difference in fuel inventory is about a factor of two. Such effects play a very big role.

The second factor is the thermal efficiency. Electricity production or production of hydrogen by electrolysis gives an efficiency of the order of 30 or 40%. Other heat applications like heat for coal refining or oil refining, or for the Jülich Adam and Eva system, lead to efficiencies of the order of 70 or 80%; for this reason alone, for these applications we need only half of the fuel inventory that would be required for electrical purposes.

And the last factor, which is also very important, is the load factor. For the production of electricity in most countries, load factors of the order of only 50 or 60% are possible. For heat applications and also for the production of hydrogen, it is possible to almost double the load factor because of the possibility of storing the product. Physically, this can lead to load factors of 90%. I have made a quick calculation here: if one compares maximum to minimum inventories in the sense considered here, the resulting factor is 3 to 4. That means we must not only make an optimization with respect to costs but we must also optimize the choice of systems as a whole: the right fuel, fuel producers, reactors, transport systems, and storage systems. All these factors are mutually interdependent, but overall I think we have to work for the best way for saving fuel inventory.

Kulcinski. Well, let me try to keep my remarks fairly short: I will try to contrast them on a good news/bad news basis. On the good news side, I think that we are all in agreement that with respect to fusion we are very close to break-even, or close to making neutrons that we can use for other purposes, whether it is breeding fissile fuel or making synthetic fuels. I think it is fairly clear, perhaps even clearer now than it was a few years ago, that the magnetic fusion side is closer when compared to the inertial side. Perhaps that will change but right now I think it's pretty clear it's magnetic. The second point is that we all agree that DT neutrons will make prolific fissile-fuel-producing reactors and that in fact they can have high support ratios. But I would caution that the support ratio is high only if we consider the thorium cycle. If we go back into the uranium/plutonium cycle those numbers come down considerably. And I think the third area of good news is that from what we have seen in the area of hybrids, fission–fusion hybrids, that the safety issue with respect to the fission part of the reactor seems to be fairly favorable; we have not found any major problems on that side. So there are some very good things to be said and progress is coming along very fast. On the other side of the coin, we are still very concerned about the timing of getting fission–fusion hybrids on line. If there is 20 million metric tons of uranium in the world as envisaged by IIASA then perhaps we don't even need fusion. I know that may not be a very popular opinion but it is, perhaps, a valid statement. If there is 20 million tons of uranium in the world we probably do not need the fusion hybrid and in that case we will probably be looking at pure fusion, if at all.

The second point is that the economics of the fission–fusion system are in a very primitive state. That's not too surprising because the designs of such reactors have not been very detailed. They are certainly not as detailed as the designs for pure fusion reactors and I think that everybody in this room would agree that we are still at a relatively primitive stage in pure fusion in terms of economics. So I think the economic potential of fission–fusion hybrids is not at all clear yet. The third point on that side is that I'm concerned that, if we are going to get the benefit of the high support ratio in the fission–fusion hybrids, we need to have a thorium-based fission-reactor economy and it needs to be moved along very rapidly. Personally I don't see that happening very rapidly at the present time and so we need to keep it in mind.

I don't think that there will be much difference in the date at which we will see pure fusion or fission–fusion hybrids on line from what we know now, especially on the magnetic-fusion side and especially if the tandem mirror turns out to be the most favorable reactor; on that point I agree with Professor Bethe – I think that the tandem mirror looks most promising now. The scaling is very simple there: we just make it longer. So we don't relax the technology that much on a tandem-mirror hybrid by going hybrid compared to pure fusion. I personally think that we will see both systems much sooner, perhaps, than has been indicated here.

The last thing I would say has to do with synthetic fuels; I just have two short comments. It's very clear on the positive side that fusion can, in principle, produce very high temperatures; however, on the negative side, very few, if any, countries have programs on developing materials that will withstand these high temperatures in environments with high neutron fluxes. So it's one thing to talk about the principles of high temperature and it's another thing, in practice, to have a blanket that will operate at such high temperatures in a neutron environment.

Harms. I would like to preface my comments by returning to the title of this conference, "Perspectives on Adaptive Nuclear Energy Evolutions: Towards a World of Neutron Abundance". I think that we would all recognize that the key words in the first part of the title are "adaptive" and "evolutions", and I think there we get back to the tension that I referred to earlier between the dreamer and the realist, since neither a dreamer nor a realist can avoid being adaptive and evolutionary if he wants to remain relevant. In the subtitle of the program we have the key word "abundance"; there's no way of getting around that without technical knowledge and I think in that context we are probably focussing on the essential ingredients, making for a very successful conference. I think it's clear to those of you who have listened to my presentations that my own personal inclinations tend towards options and choices and ideas and concepts, rather than hardware and numbers of reactors by specific years, and I certainly make no bones about this. However, I should put things into perspective. I did spend two years with a consulting engineering firm designing hydroelectric power plants and I can certainly confirm the dictum that any design is out-of-date the minute it is completed. I think this does say something about our aspirations to design systems which hopefully might aid neutron abundance in line with adaptivity and evolution.

Now as a second comment I tend to think that in terms of our nuclear enterprise we may have a 15- to 20-year breathing spell during which we can subject many of these ideas to a lot closer scrutiny than has been possible until now. In a historical perspective I am impressed by what I hear from the "pioneers" – there are many here in this audience who took part in the nuclear program in the 1950s when there must have been 20 or 30 different types of nuclear reactors which were looked at. And many of those types were whittled away until we were left with essentially only two or three. If I relate that to the energy wheel described in the Harms/Häfele paper based on a continuum of choices, I see a personal conflict. At a certain time, we have to make definite selections. And my apprehension is that our traditional methods of making selections seem inadequate. I have heard the comment made by others that our traditional criteria of least cost, for example, are not the determinants we should be using. I have heard the comment made, and we have all heard it, that there is something to be said about the time element involved and the responsiveness of certain choices. We also have heard a very nice presentation on quick being beautiful, also possibly saying that small is practical. Relating this to the element of scale that we have referred to, involving not only geometric size, the temporal effects, the complexity of a system, and the financial and manpower commitment which may need to be made to some one choice, is simply an overwhelming kind of a proposition. Now to some this may be evident; to others it is not and I would like to refer you to a Häfelian phrase in the 850-page book that has just come out of IIASA which in effect says something is either evident or it is not – the implication being that it varies from one person to the next. And therein, I think, perhaps lies a hangup with respect to the issue of choices.

I am very apprehensive if I extrapolate and have to conceive of a broad-based trade in neutrons which potentially could lead to a sort of cottage industry in which the neutron is the medium of exchange, unless something of an essentially supernatural nature could be invoked which would guide the trade, somewhat like Adam Smith's notion of the "invisible hand". This supernatural factor could perhaps provide the required "degree of acceptability" – as a generic phrase for a large collection of terms that we can associate with that expression. I have been very much impressed with the many comments made. But I am somewhat disappointed to hear a note of pessimism which I detect, not only here, but elsewhere in my travels. To put things into perspective, I travel in Germany and Austria quite frequently and the Germans tend to say the situation may be serious but it's not hopeless. The Austrians, on the other hand, have a tradition of saying the situation may be hopeless but it's not serious! This is attributed to a distinguished Austrian dramatist during the early part of the century. But I think there is a degree of truth in it, and in a real sense it seems to me that the curve which Professor Häfele showed, referring to the inflection point of the world population, in a sense characterizes or epitomizes the reality of a very serious transition. We go through such a transition not only in terms of the first derivative changing quite suddenly but also in terms of being translatable into a singularity which, if you approach it from one direction, takes you into a static domain – into the seventh heaven. But if you approach the same singularity from the other direction it takes you into the deepest of doom – an expression that I periodically like to use because it seems to be a reflection of what seems to be happening in the nuclear industry and not as a repeated and persistent reminder of the necessity to reexamine. John Hilborn yesterday, for example, referred to a stack of safety reports which have been prepared in connection with making 200-megawatt heat-producing reactors available. And we have all seen kilometers of shelf space taken up by reports of a safety forum – an acceptability forum – in order to put one reactor on line. A theory that I have is that one reason we have an energy crisis is because we are using so much energy to produce these reports! But the more serious issue, I think, is that the continuing analysis that goes into issuing these reports can potentially lead to a form of paralysis. It's paralysis by analysis. And this can have, I think, a devastating effect: if I rephrase the expression that Professor Dyson used, the fun may have gone out of it. But it would be disastrous if we would not take an optimistic point of view, simply because we have to think of our own generation and subsequent generations. Well-founded optimism can only be recovered, in a sense, by a reexamination of our own intellectual traditions and the degree of necessity for striving towards a goal which we will never achieve; but without striving we will be very poor indeed.

Häfele. Thank you very much, Archie. I, too, have a number of observations: they are of a different nature and I start with something special. I do think that the greatest problem of the fast breeder right now is the relatively high inventory. The importance of inventory has already been stressed by Schulten and by many others; and it is not so much the doubling time but the inventory that is the problem as long as the plutonium comes from the light water reactors and not from the fast reactors themselves. But for the next few decades that will be the situation. If I now hear in the public arena talk of cooling-off periods of ten years and throughputs through the fuel cycle of more than ten years then we are losing perspective and the whole thing starts to look very ill. We are doing this because at present the back end of the fuel cycle is definitely not in hand and I was tremendously impressed by the presentation of Floyd Culler where the present state of reprocessing was assessed against the background of the early experiences and pioneering. It is not so much that this or that piece of chemistry or engineering is missing. What we really have to do is to build the back end of the fuel cycle five times and learn from the experience. Just by doing it we would master the problem and then the next step would become clear automatically. We cannot master the problem of reprocessing facilities or waste-disposal plants by just preparing blueprints. If we do that, and so far we have, we are entering the domain of hypotheticality and leaving the domain of reality. This is already true for this generation of back-end fuel cycle reprocessing and waste disposal and it will be even more true if we can go on to the large-scale uses of nuclear energy envisaged here. So I feel a big vacuum there which must be kept in mind, in parallel with the development, say, of fusion or accelerator breeders. I think this cannot be stressed enough.

During this meeting we have not elaborated much on the overall energy problem. In that respect I have given speech upon speech based on our 850-page book *Energy in a Finite World* during the last two years. Therefore I will not repeat all the caveats here. The point is that the overall energy problems remain. If we continue with the present crippling of nuclear power we will be driven very quickly into the domain of low-grade fossil fuels. I think, none of us has really fully understood what it will mean for whole regions to be destroyed or moved. We have had a little taste of it at Jülich where artificial mountains are being built by overburdens. The largest pit in the world is being created for the sake of producing lignite – brown coal. What is at stake there is a total of less than 1 TWyr over 20 years. And that isn't very much. We are talking of 500 and 1000 TWyr amounts when considering harnessing low-grade fossil fuels. For example, in the Orinoco and Athabasca regions, environmental problems raised to the second or third power of what we normally envisage will be involved. There will be also be emergencies in terms of the response from society, from industry, or from the economy as a whole. It is in this context that I would like to introduce a key notion now, namely that of productivity. We have to make sure that energy productivity is maintained and that the social disruptions surrounding energy do not disrupt labor productivity. Labor productivity could relatively easily compensate for increasing energy prices, in that we could get more GNP out of the same amount of energy, and we could even have growth rates of the size needed for providing the high capital costs. There is nothing worse than increasing capital costs and decreasing economic growth rates: that means running into the sheer ponderousness of the system and coming to a standstill. If, by contrast, the productivity surpasses the increase in energy prices, we will not only have the old conditions but we will also be able to afford the high capital costs of modern energy systems. But what truly bothers me is the fact that even if nuclear power were to come in right now, it would be somehow too late for an orderly approach. I quite frankly admit, much in line with Floyd Culler, that the thing that has impressed me most during this meeting has been the analysis of fission-suppressed fusion breeders made by the Livermore people. This might be a device for overcoming this last problem.

There is one point more. We have listened to marvelous single contributions, but what we haven't done yet is to put it together into coherent patterns. The whole IIASA exercise was meant to design energy scenarios as examples for strategies – not because we know the future. We do not know the future. But the necessary conditions of internal consistency and coherence are very stringent. If we were to apply the scenario-writing scheme it would be a step forward for a group like this if we could identify two or three alternative scenarios where target dates, target values for breeding gains, and target values for capital costs would be spelled out in order to get an understanding of system feasibilities. And especially so when the interconnection of world regions is considered. Western Europe will probably always have to depend on imports and so there will continue to be all sorts of interdependencies as presently exist for oil. Therefore a next step could be to say yes, let's assume for a moment that these nuclear dreams come true: do they fit together?

[*Editors' Note*: After these extended remarks from each panel member, there was a general discussion. Unfortunately, not all of the discussion was recorded, but the following pages contain a number of the interesting comments made, together with occasional notes from the Editors.]

Bethe. I would like to make an attempt, a very daring one, at introducing some priorities. I think our first priority should be to close the fuel cycle. And our second priority should be to develop a thorium-based fission reactor. I don't want it as a breeder: I want it as a modification of the CANDU – a very good converter. Everything else could have a longer time scale and would become much easier if these two priorities were accomplished.

Schulten. In practice it is possible to realize a breeding factor of one for the heavy water reactor at high temperatures. This is very clear: it has been calculated and we know all about it. But the fuel inventory for a thorium breeder reactor on the basis of heavy water or using graphite is in both cases about three tons per gigawatt of electric power, or even a little more. And if the conversion factor causes a reduction, for example in the case of the graphite reactor from the breeding factor of 1.0 down to 0.85, then you have in the core only about one ton of U-233. This means that all the practical calculations that have been made in recent years come down to a practical and most economical conversion factor of perhaps 0.85 to 0.9.

Häfele. So the first priority is closing the back end of the fuel cycle, and then bringing in thorium as number two?

Bethe. Those are my priorities.

[*Editors' Note*: The discussion then moved to the need for providing initial core inventories of fissile material, particularly for a dynamically evolving reactor population. It was observed that the high support ratio of fission–fusion hybrids could play an important role here whereas fast breeders suffer from their relatively tight neutron economy. It was suggested that a linear programming approach might lead to a quantitative and therefore more precise understanding of these interactions. After a plea from Nürnberg for less emphasis on electricity and more attention to the substitution of fossil fuels in the nonelectrical domain, Kulcinski made some remarks on fusion.]

Kulcinski. I think I agree that there is no question that today, in 1981, the tokamak is much further advanced in terms of physics than the tandem mirror. But the tandem mirror has an additional advantage, apart from its simplicity and ease of maintenance, which I think I should mention here, especially in view of our comments about small is beautiful and quick is beautiful. That is, there are only two major fusion concepts that I think we can presently imagine being built in small sizes – small in the sense of being of the order of 100 megawatts. Those two are the tandem mirror and the light ion beam inertial confinement system. Tokamak systems, such as the smallest we could build from INTOR, were of 600 to 700 megawatts: I expect that probably about 1000 megawatts is where one would end up. For the laser system, unless we can get a laser of more than 10% efficiency we're probably talking about 1000 megawatts or so. Heavy ion beam fusion devices also look as if they are going to be large – thousands of megawatts – and stellarators will also be in the thousand-megawatt range. So there is a major contrast between systems which can be or have to be very large and therefore capital intensive, with correspondingly long development times, and those that can be built small and therefore are more easily funded. The tandem mirror is one of the latter systems.

Culler. My primary concern comes from the dollars and the commitments involved, both during the development period and right through to the risk of building two or three machines in a row before they become commercial – and that is essential to get them into the electrical grids. The commitment to 1000 megawatts of steam-producing machine, either nuclear or coal, today costs more than the net worth of most of the utility companies; 2 billion dollars is greater than the net worth of all except Pacific Gas and Electric and maybe five other companies in the United States. As a consequence, they cannot in good conscience put their money in a system that has any risk at all because it is likely to bankrupt them. So my first concern in a very real sense is money, but there are other practical things. The uncertainty of the future market, for example. Moreover, the very large liability of losing power from a 1000-megawatt machine, whatever it's driven by, amounts at the moment to a million dollars a day of lost power. Take the breeder: the breeder will be a three-billion-dollar investment on the line. Now what I've said is that the utilities are not going to buy the first machine or the second machine but they might consider buying the third. Now somewhere in the system the financing for all of that development has to be taken care of and it's unlikely in the present circumstances that the private sector is going to have the necessary funds, derived from power rates, to pay for it. As a consequence I am concerned about size, perhaps primarily because of money; but I am also concerned about all the liability to loss and service and the limitations and applicability – geographically or countrywise – for very large machines. Only the very largest companies and the very largest countries can put 1000- or 1500-megawatt machines on the line, so the market is limited. So when I said small is practical I implied the following.

It may be we're moving into a period when finances and a lot of other factors are combining to force us towards smaller concepts and that even though the economy of scale is still there theoretically it may not be there in practice.

Hilborn. I'm not going to be talking about small reactors here. I wanted to comment on Professor Bethe's second priority since not everyone is maybe that familiar with the thorium/uranium-233 fuel cycle. I could identify the two major development problems, but I'll say first what the problem is not – it's not the reactor. The on-power fueled CANDU reactor is virtually unchanged for the Th/U-233 cycle. But what is different in terms of the fast breeder that you're more familiar with is the low burn-up of the self-sufficient fuel cycle. Yes, we can get a breeding of 0.99 or close to 1 if we assume very low losses per cycle, less than 1% losses per cycle (that's including both ends of the fuel cycle), and a burn-up of 8,000 megawatt-days per metric ton. Now this means a throughput of five or six times the annual throughput of fuel of the fast breeder with its 40 or 50 thousand megawatt-days per ton. So the development problem is as follows: what will the cost of the processing and fabrication be for this enormous throughput? On the fabrication side there is a basic difference between the fabrication of U-233 fuel and of the plutonium-bearing fuels. We are talking about a remote fabrication plant, because we cannot get near the U-233 fuel. If plutonium fabrication goes the route of complete remote fabrication, then that technology will carry over to U-233 fuels and I would be interested in hearing from anyone on the possibility or the state of that technology. But those are two main problems of the self-sufficient cycle: the high annual throughput, and associated costs of the other ends of the fuel cycle, and the remote fabrication of uranium-bearing fuels.

Häfele. I must remark that this sounds like returning to square one. These were precisely the considerations in 1957 and 1958 when the development of the fast breeder with 100 thousand megawatt-days per ton was started, but with exactly opposite goals. The idea then was to avoid frequent recycling.

[*Editors' Note*: After these remarks, Häfele closed the meeting, thanking all the participants for their valuable contributions and looking ahead to the next planned meeting in Helsinki.]

AUTHOR INDEX

MIX
Papier aus verantwortungsvollen Quellen
Paper from responsible sources
FSC® C105338

If you have any concerns about our products,
you can contact us on
ProductSafety@springernature.com

In case Publisher is established outside the EU,
the EU authorized representative is:
Springer Nature Customer Service Center GmbH
Europaplatz 3, 69115 Heidelberg, Germany

Printed by Libri Plureos GmbH
in Hamburg, Germany